THE GRASSLAND OF NORTH AMERICA

PROLEGOMENA TO ITS HISTORY

WITH

ADDENDA

AND POSTSCRIPT

By

James C. Malin

GLOUCESTER, MASS.

PETER SMITH

1967

Copyright 1947, 1956, 1961, 1967
James C. Malin
Fifth printing, with postscript

PREFACE

The area of the earth's surface chosen for this study is that part of the United States designated usually as the Trans-Mississippi West. In its natural state, the feature that gave character to most of the landscape, in contrast with the area east of the Mississippi river, was the vegetational cover of grass rather than forest. Incidentally, the term natural state, as used here, means the condition in which it was found by the European at the opening of the sixteenth century. This rules out of direct consideration the condition of the continent at different periods of geological time, or of anthropological time.

The method employed in the study of this chosen area recognizes the ecological, agronomical, pedological, and geographical factors that provide the areal setting for its history. The sciences bring to the aid of the historian new tools and new methods whose possibilities have been little explored. One purpose of this book is to bring together summaries of the literature in the several borderland fields that seem significant to history, and to give them some application to this specific problem. There is no attempt here to present a formal history of the grassland of North America. There are many works in which aspects of that history are well treated, and, for present purposes, there is no point in mere restatement of such material. The things put into this book are those that seem most pertinent to the main purpose; new methodology, different points of view or emphasis, syntheses of materials not hitherto brought to bear upon this field of history, and some illustrative products of original research. From one point of view, the book may be considered as a series of essays on historiography, materials, and methods, together with sample case studies. A book on fuel and housing in the grassland, as well as other studies, will continue the theme. At some future time, a formal history of the area, based upon these broader considerations, may be in order.

Other aspects of general historiography and of methodology have been discussed in Essays on Historiography (Privately printed, Lawrence, 1946). These two books were originally a single project completed in 1945. The advisability of the separation into two books need not be debated here, but for perspective on some of the terminology used in the present book, attention is called to the Essays. Among other things, the third and fourth of the essays introduce the problem of the adaptation to social theory by physical scientists of certain terminology and concepts of quantum physics. In physics, the behavior of the individual particle is treated as uncertain, but the behavior of large numbers of particles is treated as predictable according to the principle of probability. Historians have held that the individual fact may be established, if not absolutely, often with a relative degree of certainty, but that generalizations from a large body of facts in space and time are always uncertain. Are physicists and historians talking about the same thing, and if so, are not their points of view contradictory? If they are not talking about the same thing, then it is necessary to define the questions more accurately and to reconcile the apparent contradictions. The issues involved are more than the differences between actuality and the record of actuality, because the physicist's data on the behavior of a particle are available to him only as record. Both the physicist and the historian are dependent upon records. There are some differences in the nature of the records, however, because the physicist has prepared in advance his measuring and recording devices, and can repeat experiments and draw conclusions from large numbers of cases. Such a distinction should not divert attention from the fact that it does not get at the whole of the issue of differences or of comparability in facts and methods. Are individuals or events in history comparable in behavior with the particles of physics? If not, then some physicists have blundered in formulating social theories on that assumption. In any case, the emphasis upon the physical sciences in this midperiod of the twentieth century calls attention to the importance of an exchange of ideas on the social applications of thinking derived from, or inspired by, the physical sciences.

The division of the material into two books resulted also in removing altogether from the present book the essay on F. J. Turner, and Halford J. Mackinder, which appears as the first chapter in the volume of Essays on Historiography, and the discussion of the relation of the Turner hypothesis to the four Delphic freedoms, which is included in the fourth chapter. The brief treatment of Shaler in this book is printed in expanded form as the second chapter of the Essays. The discussions of science and of social theory were divided, part appearing here, while other aspects were treated in the third chapter of the Essays.

The importance of the sciences to the work of the historian is a feature of both books, but it is imperative to challenge the misuses to which the sciences have been put in making social applications. At most, science can only describe how, it cannot explain why. Even the how descriptions are of limited validity. The only justification for continued scientific research is to discover more adequate descriptions. In their enthusiasm over planned society, some advocates of statism, implemented by scientism, disregard facts. Between World Wars I and II, and since, too many people in academic life have acquired a vested interest in partisan public policies, many of which are based upon social uses of the sciences that go beyond the facts. The challenge to these abuses is based upon general principles, equally valid in any time or place, and consequently any political implications involved in such a challenge are only incidental to what is necessary to defense of general principles and of intellectual freedom. There can be no compromise in this psychological warfare. To criticise merely the details of particular measures is to indulge only in a futile sniping around the fringes of conflict. To pursue a policy of appeasement is to surrender piecemeal. The course of events in Europe has demonstrated that, once indoctrinated with the ideologies of the totalitarian forms of society, people lose their will to resist. In America, totalitarian ideologies have already acquired a substantial following. Freedom cannot be safeguarded merely by exposing the ideologies hostile to freedom. That is negative. Freedom can be defended only by positive action; by presenting constructively the principles of history to serve as guides along the path of freedom. There can be no concessions on principles; to compromise is to betray freedom.

Three aspects of science, or levels of thought associated with science, may be recognized as of social significance: The subject matter of science, together with its technological applications, is the most obvious and direct in its social impact; the social theories and applications derived from, or allegedly derived from, science, in which science is invoked as authority for social policies; and metaphysical speculation about science which appears in various guises, particularly under the name of philosophy of science. Boundary lines do not separate these necessarily into definite compartments, but the distinctions serve as a convenience for purposes of analysis. Every academic discipline possesses a body of social thought, whether or not its practitioners are aware of it. Apparently, the rank and file of those classed as scientists, have not, until recently, interested themselves directly and extensively in the social implications of science as such. Their awakening to social consciousness found them unprepared, and too often they became victims of thought inimical to the freedom which the sciences profess. It is important that the social significance of all disciplines be recognized explicitly, and that more generally, scientists become acquainted with the body of social thought, not only in their own fields, but especially in its historical setting. Unfortunately, the rank and file of historians seem quite unaware of the body of social thought in the literature of science. That, also, should be remedied. Scientists might profitably address themselves to the problem of how scientific knowledge can be presented in order to make it more effectively available to the uses of the historian. Irrespective of the academic discipline with which the individual is identified, social thought should arrive at a common ground of fundamental principles. The converging approaches should contribute to an enlargement of understanding. It is not a matter of one encroaching upon the field of another; all who are concerned with social thinking are engaged upon a common objective.

Since these books were first written, much interest has developed in the United States in atheistic existentialism, primarily a post-war French cult. The argument of this point of view, as distinguished from Sören Kierkegaard's Christian existentialism, is expounded most conveniently in a handbook by Jean-Paul Sartre, L'existentialisme est un humanisme, (Paris, 1946). He asserts that without God, man is condemned to freedom, and that in that state there is no certainty, or standard of values. Man is only what he makes of himself through action. Irrespective of what that action may be, only action possesses meaning and value. Although there is no direct connection between existentialism and the American uncertainty — frame of reference theory of history discussed in the Essays, the general kinship of thought is evident. Both are based upon a denial of certainty, and find values only in making choices as an act of faith. In so doing, both contribute to an ideological background for totalitarian dictatorship. A critique of between-wars thought in the United States, as presented in the Essays, is convincing that the response of Americans of 1946 and 1947 to existentialism should occasion no surprise. (See also John Chamberlain, American writers, Life, September 1, 1947).

Some readers of this book may prefer to read Chapter 15 first, and for some purposes that may be desirable, but the author decided that perspective is better maintained by the present arrangement. The review of the several sciences in Part One is the first attempt anywhere at such a task. To put it differently, scientists have not made such a survey, even for their own use. The bibliography is extensive, but includes only the more pertinent literature consulted. This is the first time such a bibliography has been compiled anywhere, and this in itself is a justification for printing it. The one compiled by C. C. Adams would be most nearly comparable. Both the survey and the bibliography thereby become available to those who are interested in pioneering in the borderlands of history, ecology, agronomy, and geography. There is no pretense of perfectionism, but if a project of this kind is to serve any useful purpose, the accumulation of materials and thinking over a period of years must be brought to some conclusion and publication. At best, any historical work is only a progress report on the enlargement of knowledge.

James C. Malin.

CONTENTS

THE GRASSLAND OF NORTH AMERICA
PROLEGOMENA TO ITS HISTORY

CHAPTERS   PAGES

PART ONE: THE SCIENCES AND REGIONALISM

1	The grass problem	1
2	Plant ecology	5
3	Ecology	20
	Animal ecology	
	Insect ecology	
	Plant-animal relations and biotic equilibrium	
4	Climatology	34
5	Geology and geography	40
6	Soil	48
	Soil science	
	Soil microbiology	
7	The grass formations: The ecologists' view	62
8	The grassland and early explorations: Southern	82
9	The grassland and early explorations: Central	102
10	Factors in grassland equilibrium	120

PART TWO: HISTORIOGRAPHY

11	Science and social theory	156
12	The communications revolution	169
13	Pioneering toward grassland regionalism	173

        William Gilpin   177   E. W. Hilgard   212
        A. D. Richardson 192   F. H. King   221
        Cyrus Thomas   195   W. E. Smythe   225
        Josiah Strong   196   H. W. Campbell   227
        N. S. Shaler   197   E. C. Chilcott   242
        J. W. Powell   202   H. L. Shantz   247
        W. D. Johnson 207   Isaiah Bowman   249

14	Soil physics and tillage	252
15	Webb and regionalism	259
16	Methodology for history of social change: Population studies	278

THE GRASSLAND OF NORTH AMERICA

CHAPTERS		PAGES
17	Agricultural studies	292
18	Land tenure, operator turnover, and farm organization	312
19	The community and the city	316
20	Harmonization of culture	323
21	An open system	331
	Bibliography of literature	336
	Bibliography of maps	395
	Appendix: Grassland map studies	398

Addenda

22	Introduction	399
23	Ecology and History	408
24	Man, State of Nature, and Climax	417
25	Soil, Animal, and Plant Relations	434
26	History of Geographical Area	470
	Postscript	487

Chapter One

THE GRASS PROBLEM

      The heart of North America is a Grassland. This region is being given a new significance by reason of the Air Age and its reorientation of the world outlook in terms of circumpolar land-masses, — just a century after the Anglo-American forest man met for the first time the problems of living in a grassland. The eastern extremity of the grass country is a triangular prairie peninsula between the Great Lakes and the Ohio river, the point extending east as far as Indiana, with detached outlying openings farther east in Ohio and Pennsylvania, south in Kentucky and Tennessee, and north in Michigan. From this eastern forest boundary, the grassland extends westward through passes in the Rocky Mountains across the continental divide to the southwestern deserts, and farther north, to the Palouse prairie of the inland empire of the Pacific Northwest. To the northward, the grassland extends through the prairie provinces of Canada to the northern forest boundary line, and to the southward into the Mexican desert plains.

      The region is occupied by a wide variety of grasses with their associated plants, and of animals; forms of life mostly characteristic of, and peculiar to, the environment. There is present also, a minority representation of forms of life typical of other regions, especially along the boundary transitions. The grassland is complete in itself, a relatively stabilized product of nature, the outgrowth of climate, soil, vegetation, animals, and microorganisms, all interacting together. This general kind of region is not unique to North America, but is a characteristic formation, of greater or lesser extent, occupying parts of all continents, unique only in respect to the particular species of biological forms that enter into the combinations.

      Why are the prairies and plains a grassland? As early as 1870, Charles A. White, of the Iowa Geological Survey, commented that "the question of the origin of the prairies has become more hackneyed perhaps, than any other of the speculative questions which North American geology affords," and in addition he challenged the assumption of the forest men: "There seems to be no good reason why we should regard the forest as any more a natural or normal condition of the surface than the prairies are" (White, Report 1:132, 133). "There is comparative uniformity on the nature of [the]... floral covering [of the prairies]", observed Shimek (1911), "they are marked ... definitely by the presence of a flora which is wholly distinct from the smaller (chiefly herbaceous) flora of the forest." The essence of these arguments is that this uniformity of vegetation is a positive characteristic, an evidence of completeness in nature, and not a sign of something deficient.

The use of the term prairie is not uniform, but there is good precedent for distinguishing between prairie and plains, the word prairie being limited to that portion of the grassland with forest along the streams lying east of an irregular north-south line falling mostly between the 96 and the 98 meridians, the tall-grass, or true prairie. In their westward movement across the continent, the Anglo-Americans met the prairie first in the small occasional openings in the forest, as in Ohio and Kentucky— "oak openings" and "barrens"—, "but it was not until the white man crossed the Wabash river ... that he beheld the prairies in all their splendor, and all their monotonous magnitude" (Shimek, 1911).

There were fairly heavy forests along the streams and in some areas some distance back from the streams. The uplands, varying in topography, were covered with tall grasses, the most conspicuous being the bluestems. The Big Bluestem occupied the lower prairie and the Little Bluestem occupied the high prairie. Indian grass was present and in some habitats locally dominant. Gradually to the westward the proportion of forest to grassland diminished, until forests in any proper sense of the word gave way to the almost uninterrupted grass covering. West of the prairie boundary, not far from the 98 meridian, is a mixed grass low plains, tall and short grasses. Among the latter through the central region the most important are the grama and the buffalo grasses. Farther west, roughly, west of the 100 meridian is the short grass country in which the grama and the buffalo grasses were dominant. Still farther to the west and southwest, the grasses gave way to the typical desert forms of life.

Why are the prairies and plains a grassland? As yet there is no fully satisfactory explanation. A quantity of literature on the question accumulated during the nineteenth century, before scientific skills were applied to the search for a more comprehensive answer. This was reviewed carefully in 1911 by Shimek, both chronologically and in terms of the alleged factors responsible for the "treelessness of the prairies." The word "treelessness" was characteristic of the literature in question, which, with few exceptions, assumed as axiomatic that the presence of forests was natural and the absence of them was an unmistakable sign of deficiency, an abnormality of nature. In popular and scientific opinion, prairie fires had been held responsible for "the treelessness of the prairies." Shimek rejected the prairie fire argument: "Prairie fires were an effect rather than a cause, and where acting as a cause were local." The accidents of seed distribution, the grazing and trampling of the bison, and other alleged causes were declared "of remote interest and not to be taken into account on any attempt at the explanation of the prairie as a whole." He pointed out that the Iowa prairie afforded a conspicuous example of grass growing on diverse geological formations, and on different soils, using the term soils in the sense then current. Furthermore, the total annual rainfall was not in itself the determinant, because Iowa prairie and forest were found on opposite sides of the same ridge, and oak openings had existed farther east in the heart of the forest.

The growth of forests or the growth of grass, in Shimek's opinion, was the result of a combination of climatic factors. The artificial growth of trees could not change climate, and furthermore, the artificial growth of trees was possible only where the relative balance of factors adverse to trees was slight and man's efforts might be sufficient locally to redress such unfavorable influences. By the same token, when such human influences were withdrawn, in due course, nature would eliminate the trees and restore the grass (1913).

Shimek's was the most comprehensive historical review of the available literature to that date, and in its negative conclusions eliminated from the discussion most of the irrelevant factors alleged as causes, or at least challenged them explicitly. On the positive side, his conclusions were less satisfactory. The relatively new science of ecology was devising new methods of attack, and was developing an essentially new outlook on vegetation and animal life in relation to environment. As Americans had met the problem of the prairie early in the nineteenth century, they had harmonized their folkways to the realities of the changing situation through their own native resourcefulness, and prior to the time when science had much to offer. In the twentieth century, science was approaching the same issues, devising more exact techniques of analysis, going over much the same ground, and arriving at conclusions which enlarged the sum total of understanding of the grassland.

Part one of this book is devoted to a survey of the sciences that seem to have a bearing on the relations of man with his geographical setting — considering them in both time and space. The development of each of the sciences is approached historically. To the historian, this seems the best means of establishing perspective, both on the science itself and on the interrelations of the sciences with the main course of social change. The reader is informed by this procedure on what scientific knowledge was available for social utilization at any particular time.

These chapters are probably not what scientists in each of the respective fields would write, and what the scientists would write might not be of the most direct value to the historian. In any case, the important fact is that the historians and the geographers should possess a working knowledge of science if there is to be any understanding of cultural adaptation to the grassland. These chapters may serve also as a guide to further reading of the scientific literature, and if the summaries or interpretations are inaccurate in fact or emphasis, the underlying studies may be examined for verification and correction.

Chapter Two

PLANT ECOLOGY

The forms of life of each natural geographical region (vegetation, animals, and microorganisms) are primarily characteristic of the particular region or type of region. They are closely interrelated in an unstable equilibrium with each other and with climate, soil materials, physiography and time. Such a description of the natural environment of man is the product of twentieth century science crystallized out of scattered nineteenth century beginnings.

Livingston (1921) designated three phases in the evolution of this twentieth century point of view on vegetation, and in some degree a similar framework may be applied to the other forms of life. Early nineteenth century biological leaders, from Alexander von Humboldt to Adolph Engler, were interested primarily in the distribution of the forms of life over the earth's surface and in the composition of the different populations, a process of description and classification. The second phase began with Darwin's Origin of Species (1859). In biology Darwin's emphasis on the idea of dynamics meant reinterpretations of accumulated knowledge, among which was the relation of structure to the conditions of environment, but with the emphasis on structure to the neglect of environment. The third phase was the study of "natural assemblages of plants" — vegetation — and of animals. This was a combination of pre-Darwinian study of distribution and post-Darwinian study of relations with environment, and to this is given the name ecology. Asa Gray (1859, 1872, 1878), in America, was credited with inspiring Adolph Engler to lead the study of vegetational history and development, migrations and survivals (Gleason, 1922). Grisebach (1838) is credited with formulating the concept of plant communities; Holts (1885-1887) with first recognizing fully the importance of the development of vegetation by succession of stages to a climax formation. The decade of the 1890s produced an outstanding triumvirate of ecologists, Warming, Drude, and Schimper. J. E. B. Warming (1891) was credited with giving the first consistent account of succession on sand dunes and in his major work in 1895, Plantesanfund, rewritten and translated from the Danish into English (1909) under the title of Ecology of Plants, an introduction to the study of plant communities. Warming's book presented the new point of view in so comprehensive a manner as to elicit the comment (Cowles, 1909) that it was "the most important ecological work in any language." Because of influence on Roscoe Pound and F. E. Clements, their estimate (1897) of Oscar Drude's work Deutschlands Pflanzengeogrophie (1896) is significant to this narrative— his "concept of plant formation is undoubtedly the correct one," his book a classic, and his conclusions final for the present. The third landmark in the new approach was A. F. W. Schimper's Pflanzengeogrophie auf physiologischer grundlage (1898) translated into English in 1903 as Plant geography on a physiological basis.

While Europe was engaged along these lines of scientific innovation, Americans were devoting their attention primarily to the older type of activities of describing and cataloguing the modern distribution of forms of life. In defense of them it is said (Gleason, 1922) that this was a natural and proper objective, because in a biological sense the American continent was largely unexplored. They were making an inventory of a new country. Only upon such foundations could the other types of scientific investigation be built. In 1889, C. H. Merriam began his experimental biological survey work on the San Francisco mountain, later applied to other regions. In his own account (1894) of this new departure he claimed,

One result of this first survey was the complete overthrow of the principal faunal areas previously recognized in the United States, and a radical change in our conception of the principles involved. In ascending the mountain a succession of climatic belts was traversed, similar to those encountered in journeying northward from the Southern States to the polar sea, and each belt was found to be inhabited by a distinctive set of animals and plants.

The next step was to generalize from the zones found on the mountain and to apply this zone system to the geographical distribution of life on the North American continent. Merriam's own summary reported that,

The principles of geographic distribution of terrestrial animals and plants in the Northern Hemisphere were clearly recognized in 1889; the correlation of the life zones was completed in 1892; the laws of temperature control were formulated in 1894. The work remaining undone relates to details....

It appears, therefore, that in its broader aspects the study of the geographic distribution of life in North America is completed. The primary regions and their principal subdivisions have been defined and mapped, the problems involved in the control of distribution have been solved, and the laws themselves have been formulated.

There is no question that the contribution made was revolutionary so far as it went, but it was limited in its outlook to modern distribution, it was a single factor hypothesis, and it was static. The new point of view being developed in Europe was concerned not only with modern distribution, but also with a question, of more significance scientifically, of how and why modern distribution came about and its relation to contemporary problems. According to Merriam's static point of view he had finished his task, except for some details; but according to the new dynamic concept, he was just ready to begin the investigation which should explain how and why, and test the validity of his life-zone hypothesis.

At the turn of the century the time had arrived in America for a new orientation and a synthesis on the basis of the accumulation of scientific materials and ideas. There was little that could be truly new in the process, and many minds were engaged. It is difficult to know how to apportion credit for specific developments. All were building on the work of the late nineteenth century Europeans, particularly of Engler, Warming, Drude, and Schimper. C. C. Adams (b. 1873, Ill.,) was most directly in the line of development from the Asa Gray influence on historical distribution and development. He was applying (1902) the dynamic approach of the European school and utilized the findings of geology and paleontology. He found two centers of refuge for fauna and flora of North America during the glacial invasions, the southeast and the southwest, and from them came the redistribution northward over the continent during post-glacial times.

In the investigations of vegetational processes of recent times two centers of study were conspicuous, the Universities of Chicago and Nebraska. At the former institution under John Merle Coulter a number of men received all or part of their training: H. C. Cowles (b. 1869, Conn.), B. E. Livingston (b. 1875, Mich.), and in part E. N. Transeau (b. 1875, Pa.), who finished his work at the University of Michigan. In Nebraska, of the students of C. E. Bessey, E. F. Clements became the outstanding leader, and H. L. Shantz received his training there. These two great teachers, Coulter and Bessey, have been placed for the first time in their proper historical perspective in Rodger's biography, John Merle Coulter (1944). A considerable part of the work of this group of Middle West Americans dealt with problems relating to the forest and grassland along the zone of transition or with the grassland itself. They were young men who were contributing to the creation of a new field of science. When publishing their first major contributions Adams was 29, Cowles was 30, Clements was 24, Livingston was 29, and Transeau was 28 years of age.

H. C. Cowles had begun his training as a geologist under Chamberlin and Salisbury at the University of Chicago and later turned to botany under Coulter. The result was a combination for the first time, in his study of Lake Michigan sand dunes (1899), of physiography with botany, according to the dynamic view of vegetation. Two years later in "The physiographic ecology of Chicago and vicinity" Cowles is said to have given the concepts of succession and climax their "first adequate expression." To him the climax was not permanent:

The condition of the equilibrium is never reached, and when we say that there is an approach to the mesophytic forest, we speak only roughly and approximately. As a matter of fact we have a variable approaching a variable rather than a constant.

This cautious avoidance of anything like a rigid closed system was said to be an important characteristic of his influence (Cooper, 1935).
  Clement's first major work (1898, 1900) was done jointly with Roscoe Pound as senior author, but his ideas were not fully formed. The quadrat method of quantitative study of plant communities was first devised for this study and became a standard procedure for certain types of quantitative investigation. In 1904 Clements "made the first attempt to organize the whole field of present-day succession, and to connect the structure of vegetation with its development...." Here appeared for the first time his organistic idea of plant communities: "The concept was advanced that vegetation is an entity, whose changes and structures are in accord with certain basic principles in much the same fashion that the functions and structures of plants follow definite laws." The next year he stated it again: "The formation was regarded as a complex organism, possessing functions and structures, and passing through a cycle of development similar to that of a plant" (Clements, 1916). In 1916 Clements published what was said to be "the first systematic monograph of the phenomena of succession...." Three important criticisms were directed at this notable work, the exclusive climate theory of vegetation, a new terminology that was "too complete:" and his use of an extreme form of the concept of organism in dealing with plant communities (Tansley, 1916). "The first quantitative study of the reactions and successions of a great grassland vegetation" was given by Shantz in 1911 (Clements, 1916).
  The study by Transeau, of 1905, correlating climate centers with vegetation centers, illustrates the process of synthesis of the contributions of others as the basis for a new and significant step in the enlargement of scientific knowledge. Referring to Adams's studies of "centers of distribution", as well as his own, he pointed out four such forest centers in eastern North America; the Northern conifer forests of the Laurentian plain; the deciduous forest centers of the Middle

Atlantic, the Southern conifer forest center of the Atlantic Gulf States, and the Insular Tropical climax formation of Southern Florida centering in the West Indies. All but the last possessed a common grassland border where the plant societies of the great plains met these forest societies.

Referring next to the work of Cowles, Transeau emphasized the rôle of physiography in contributing to the plant successions. Part of the problems of analyzing vegetation of eastern North America could be explained by physiography as Cowles had done in his studies, but part of it could be explained only on the basis of the meeting and intermingling of plant societies of two or more geographical centers. Thus the first three stages in plant succession of the southern Great Lakes region were attributed to the northern conifer forest associations influenced by physiographic factors, but the fourth, or climax stage, was determined by the invasion of the beech-maple forests from the central deciduous forest center. Without competition from other shade-making and shade-resistant deciduous trees that did not thrive that far north from the deciduous center, the beech-maple society became dominant over the northeastern conifers. The prairie peninsula of Illinois and Indiana was a projection of the grassland between the northeastern conifer and the deciduous forest centers. Or more accurately, it had been the transition zone on the periphery of the three vegetation centers, and still was, except that the beech-maple forest had crossed over or around the grassland peninsula in its northward expansion and crowded out the earlier conifer succession to the northward of the prairie peninsula. Around the northern edge and to the south of the prairie peninsula only the oak-hickory combination of the deciduous forest center had occupied the western central periphery and around the south edge reached west as far as eastern Kansas. There the plains grasses met in the transition zone the shrub and still farther east the oak-hickory trees of the eastern deciduous forest center. On the Minnesota border and on the Texas border the plains grasses met different forest societies, the northeastern conifers and the southern conifers.

Transeau's major contribution was to formulate the next step in analysis; the correlation of climate centers with these vegetation centers. No single factor of climate was sufficient, so he hit upon the idea of the precipitation-evaporation ratio as a single index for mapping the climate and vegetation regions. To determine this ratio, he divided the annual rainfall by the depth of evaporation in inches: "The depth of evaporation depends upon the temperature of the evaporating surface, the relative humidity of the air, and the velocity of the wind. These are the same climatic factors which most powerfully affect transpiration, and which must be of great importance in determining the geographic range of plants." Rainfall data were available for substantial periods of time in the eastern United States, but the only evaporation data on record that could serve the purpose were those of T. Russell (1888) collected for a single year, July 1, 1887 to June 30, 1888. Transeau realized that not all the water that falls is available to plants, and that evaporation data from a water surface even if adequate for a period of years were not an accurate measure of water given off by plants, but nevertheless "the figures have a comparative value in both cases and when combined probably give a fairly correct idea of the distribution of these climatic factors in the eastern United States." Transeau does not appear to have appreciated how abnormal a climatic year of heat and drouth was represented in the Russell data for 1887-1888, and it is probably just as well, or he might have been discouraged in using it at all. As will come out later in this discussion, possibly this abnormality was after all by coincidence a factor of safety, according to the hypothesis of R. J. Russell relative to the ecological significance of climatic extremes. Transeau plotted his ratio values on a map of the eastern United States which he explained as follows:

It will be noted that the Great Plains are marked by a rainfall equal to from 20 to 60 percent of the evaporation called for. The prairie region where the forests are confined to low grounds is identical by a ratio of from 60 to 80 percent. Its limits as indicated show a remarkable agreement with the actual distribution of the prairie. The region indicated by ratios between 80 and 100 percent is more or less coincident with the occurrence of 'oak openings', 'open forests', and 'groves' on the uplands, and dense forests on the low lands.

The southeastern area where the rainfall is from 100 to 110 percent of the evaporation, corresponds to the region of the deciduous forest center. The distribution of the ratios above 110 percent in the region of the coastal plain is remarkably similar to the position of the Southeastern Conifer forest center. In the southern Appalachians the ratio also rises above 110 percent and coincides with the occurrence of the southern extension of the northeastern forests. No data are available for the mountainous parts of Pennsylvania, so that this apparently isolated area may be climatically connected northward along the higher mountain crests. The Northeastern Conifer forest center is marked by ratios above 100 percent and centering in the St. Lawrence basin. It is probable that the northern limits of this formation will not be indicated by rainfall-evaporation ratios, for the factors commonly accepted (Schimper, 1903, p. 168) as determining the northern limits of forests are very different from those causing the boundaries of other formations. It should also be stated that since climates are constantly changing and effects may be behind their causes, no map of present climatic conditions can hope to do more than approximate the present distribution of plants. Geographic and historical relations must be constantly borne in mind.

The preceding studies deal with the larger phases of the relations between vegetation and environmental factors such as historical geology, physiography, migration, and succession of plant communities, and climate. The next phase in the development of ecology was derived from the twentieth century emphasis on physiology and the correlation of this with the others. The goal of this approach was to reduce ecology definitely to a quantitative and objective scientific basis with a view to making quantitative determinations of the rôle of the several factors involved. In 1905 the first compilation of relative humidity data was prepared, by W. B. Stockman, covering the period 1888-1901 for 130 stations. It made possible the study of relative humidity in relation to plants. Livingston became a leader in America in the work on evaporation in relation to plants, particularly because of his invention (1904) of the porous cup atmometer for measuring evaporation as a climate factor. This instrument and its later modifications was widely used in America and extensively in Europe where it displaced for many purposes the older Picke atmometer (1872). Livingston's work (1906, 1908) on desert plants inspired many others to study evaporation in relation to several aspects of plant communities and succession (Transeau, 1908; Fuller, 1911, 1912; Gleason and Gates, 1912). Livingston (1906) was a leader also in the study of transpiration, root systems, root physiology, and soil moisture, but the most important early study as related to the grassland was that of Briggs and Shantz (1911, 1912, 1913, 1916) on the water requirements of plants in the great plains. In a region of limited rainfall the root systems of plants were very large in proportion to the surface parts — in fact most of the plant was in the soil and as the significance of this relationship became more fully appreciated, the roots and the soil processes became more important as subjects of investigation.

The most comprehensive study of The distribution of vegetation in the United States, as related to climatic conditions, and the first of its kind for the United States, which employed on so extensive a scale the new physiological and quantitative techniques, was that of Livingston and Shreve in 1921. They emphasized that it was "to be regarded only as a beginning along a line that holds forth very great promise." In respect to the climatic factor they warned that methods and interpretation were "woefully inadequate." In this connection, they insisted that new methods were necessary for obtaining and interpreting climatic records. The ecologist was interested in physiological effects, not in meteorological causes.

A further limitation of the work was pointed out by the authors. Data were available primarily only for environmental conditions above the surface of the soil and consequently the book was limited to that scope. Little that was satisfactory had been done with roots and soil in relation to plant distribution. On the root question, Livingston had pointed out, in 1909, that the neglect was such "that we are at present more densely ignorant of these organs than of any other equivalent portion of the plant," and concluded that from such investigations as had been made recently there was evidently a need for a new hypothesis of soil fertility. Livingston had said in 1917

During the last quarter-century we seem, indeed, to have been mainly engaged in discovering new problems, rather than in solving the problems dealt with by earlier workers. In many cases the problem has been analyzed into several component problems, each one of the latter appearing as difficult now as did the less thoroughly analyzed problem to the writers of a few decades ago.

As the years passed many advances were made in answering some of the questions and in further redefining the unanswered problems. As the characteristics of xerophytes, or so-called drouth-adapted plants, are a matter of special interest to the low-rainfall areas of the grasslands, a few of the leading contributions to plant physiology dealing with the relations of plants to water will be pointed out in this discussion. It is important to notice, among other things, that drouth-resistance as commonly used in agricultural discussions has a meaning quite different from its use in plant physiology. Originally proposed by Kearney and Shantz in 1911, the latter had settled, by 1927, upon a classification, which placed drouth-adapted plants in four groups: drouth-resisting, drouth-escaping, drouth-evading, and drouth-enduring. As used in such a classification, drouth-resisting plants were those that stored their water-supply within the plant itself either above or below ground, and during drouth were enabled to continue growth for long periods of time. Most conspicuous were the cacti, but some non-succulent plants belonged to this group such as African grassland trees cited by Shantz. The drouth-escaping plants include such examples as the desert ephemerals that grew during short favorable seasons, matured, and seeded quickly. The plant itself possessed no drouth-enduring characteristics, but depended for perpetuation upon seed which survived the drouth. The drouth-evading plants conserved moisture in one of several ways, by small areal parts, or by wide spacing, or by efficiency of transpiration, or by restriction of growth. Many of the cereals of dry-land agriculture belonged to this group. The divergences in the physiological characteristics of plants in this category were illustrated further by variations in leaf temperature. On a hot, bright day the leaf temperature of alfalfa might be several degrees below the atmospheric temperature, while the leaf temperature of sorghum might be above atmospheric temperature. Alfalfa possessed a high transpiration rate and sorghum a low transpiration rate. The drouth-enduring plants made additional adjustments including thickened, leathery, or woody leaves. During drouth they made no growth, and with prolonged drouth might drop their leaves or part of them and even twigs might die, but when water was again available they revived quickly. Desert shrubs were most typical of this group. Maximov (1929) objected to the inclusion of cacti among drouth-resisting plants because they lacked the characteristic xerophytic feature of high osmotic pressure. It is evident from this classification that there was no single answer to the problem of the relation of plants to water. Each of the several groups met their water requirements in different ways and there was the widest difference among the members of each class.

Maximov (1929) summarized the results of approximately a generation of research workers on the water-relations of plants, along with his own conclusions. His book marked a major revolution in the field. In respect to the anatomical characteristics of xerophytes, he concluded that all cells were smaller, the networks of veins were denser because of the smaller cells and about the same number in the spaces between the veins, the cell walls were thickened, the number of stomata were greater, and special protective devices were present, such as a thickened cuticle to reduce intercellular water loss. In respect to physiological characteristics, Maximov (1929, 1931) rejected the traditional views of xerophytic adaptation which were based on the assumption of morphological changes to reduce intensity of transpiration, and emphasized instead the greater intensity of transpiration and assimilation, a higher osmotic pressure, and an increased capacity to endure wilting. He was convinced that the secret of drouth endurance lay in the protoplasm, which experienced a change analogous possibly to the hardening process associated with frost resistance. In order to correct some misunderstanding of his emphasis on greater intensity of transpiration (1929), he restated his position (1931), to emphasize that because of the several types of xerophytes the same characteristics were not common to all.

Experiments on osmotic values conducted at the Desert Laboratory of the Carnegie Institution represented an important field of investigation of the properties of protoplasm in relation to water. The general scope of the work can be followed through the annual reports of Shreve and Mallery, in the Yearbooks, (Particularly Yearbook No. 33, 1934). These studies emphasized that in creosote bush (Larrea) the osmotic values varied widely within the same day; that in general they reached their minimum at the end of the rainy season, their maximum at the end of the dry season; that the variation in values between plants was widest during drouth, individual plants responding differently to the progressive impact of extreme conditions; and that the greater the range of osmotic values, the more successful and uniform the growth of the creosote bush (Mallery, 1935).

In rejecting the assumption that the primary functions of the stomata were related to the control of water loss, attention was turned anew to the positive aspects of these structures. Among the criticisms directed at research in plant physiology, one was the neglect of the most important of all plant processes, photosynthesis. In this field the hypothesis of Scarth (1927, 1932) became generally accepted (Seifriz, 1938) that the stomatal movements in plants were controlled by the effect of light on bio-chemical processes tending to produce alkalinity of the cell protoplasm which increased osmotic pressure and opened the stomata, while conversely darkness reversed the process. These assumptions related stomatal movements, therefore, to the basic physiological processes of photosynthesis.

The older theory of the functions of the stomata as regulating transpiration, particularly, the assumption that the stomata closed to reduce or prevent water loss, was in direct conflict with the requirements of photosynthesis which required free exchange of gases for the conversion and assimilation of nutrients in the process of plant growth. The view held by Maximov and others resolved substantially this difficulty as applied to those xerophytes that possessed a high rate of transpiration. Maximov (1931) explained that the closing of the stomata as a defense against water loss did not occur until the wilting point was reached. The real test of drouth resistance was not the transpiration criterion, but the test that came with wilting; "The capacity to survive long periods of drought and dehydration of their tissues without injury, or with only slight injury." At this stage structural features such as thick cuticle did promote economy in the use of the water that remained in the plant, but the principal defense was in the protoplasm itself. Needless to say the views of the Maximov group aroused much adverse criticism, but they constituted a major landmark from which new advances were to be measured. One of the several significant criticisms was that of Wood (1934) based on Australian material, that xerophytism must be viewed on a broader basis than water relations alone. In a sense, however, this was not so much a disagreement as an advance, because he directed attention also to the problem of protoplasm.

In its transit around the globe, modern civilization was concerned mostly with the temperate zones where problems of light and temperature attracted less attention than water requirements of plants and animals. The extension of man's activities into the higher latitudes and into the tropics emphasized those neglected factors. Plant physiology gave to them a new significance in all environments. With respect to light, plants may be classified as long-day, short-day, intermediate, and day-neutral. Some plants require long and some require short seasons of a particular day-length and temperature to grow, to develop toward reproduction, and to mature seed. In the high latitudes the days are long and the growing season short. In the tropics, plants live under conditions of approximately equal periods of light and dark— short days—, and nearly continuous high temperatures throughout the year. Plants of the same genetic value may react quite differently to dissimilar environments, the factors of one region releasing characteristics inhibited in another. Geographical races or strains within the same species of native vegetation may show different responses to light and temperature, when moved from one latitude to another. A plant promising in one environment may be a disappointment in another, and the reverse (Whyte, 1946; Yarnall, 1942). Problems of distribution, migration, or adaptation, as applied to plants living in natural conditions or in domestication, require much rethinking when described in this framework of developmental physiology.

Plant physiologists distinguish between two aspects of the life of a plant, "growth" and "development." By growth is meant increase in vegetative size, and by development is meant the progress of the plant toward reproduction; flowering and seed formation. The German botanists, G. Klebs, and G. Gassner, were pioneers in the developmental approach, their more significant contributions being published first in 1918. Among Americans, the significant publications of H. A. Allard and W. W. Garner on light relations start in 1920. Those of the Russians, led by T. D. Lysenko and beginning about 1928, dealt with both light and temperature. Particularly significant was the broadening of range and volume of contributions in all research centers after about 1935. The most recent survey of the accumulated literature, which affords some perspective on its practical significance, is that of R. O. Whyte, Crop production and environment, issued late in 1946. Most of the following discussion is derived from Whyte's book as a basing point, being supplemented by other studies particularly pertinent to the North American Grassland.

Sharp differences are conspicuous in the theoretical aspects of developmental plant physiology. This is more or less characteristic of most new fields of investigation. The Russians, led by Lysenko, formulated the theory of phasic development; that growth and development follow definite phases which must proceed in rigid sequence and that one phase must be completed before the next can continue. Outside Russia, this view is largely questioned, or rejected outright. In nature, the factors that would initiate these phases can hardly influence a plant in the exact succession required by the Russian theory.

Vernalization is the name given to the method of modifying plant development so that the production of the seed may be advanced in time. A plant of winter habit may be made to produce a crop if planted in the spring, or a plant of spring habit may be made to mature earlier than otherwise. Klebs and Gassner applied low temperature treatment to seeds that would be the equivalent of the influence of wintering. Lysenko carried the procedure farther, applying his phasic development theory. Under the rigid sequence of phases, the Russians held that the process was irreversible. English experiments came to the opposite conclusion (Whyte, 1946). The system of developmental physiology raised the question of when the influence of temperature began, and answered that it might begin during the formation of the seed on the mother plant. This introduced the concept of physiological age as distinguished from actual age of the plant dated from the time the dormancy of the seed was broken. In practical application, the study of the growth and development of a plant must take into account the environment under which the seed was produced. This helps to explain the perculiarities of behavior of seeds introduced from different environments, and the contrasts in the behavior of introduced seeds with the seeds produced in the new environment.

The disagreements among scientists over vernalization are a reminder of the discussions among the farmers of central Kansas (1866, 1867) of the effect of wintering upon wheat varieties. It was generally agreed that winter wheat must undergo a wintering process if it was to produce a crop. Some argued that the seed should be planted even if conditions were not favorable for its growth, because the seed would thus be exposed to cold and the result would be equivalent to the winter effect upon the growing plant. They discussed also the latest date at which wheat could be planted, if it were to complete the desired cold-conditioning. One variety, Odessa, imported from Russia, was popular temporarily during the late 1870s, partly because of the claim that it would make a crop either as a winter or as a spring wheat.

The subject of genetics, as related to developmental physiology, is particularly confused. The formal genetics derived from Mendel and Morgan, based inheritance upon characteristics associated with genes. The developmental genetics, as interpreted by the Russians in terms of the phasic theory, insisted that the individual phases of growth and development were inherited separately. On this assumption breeding should be directed to phasic analysis and combination of phasic period lengths which would fit particular environmental conditions. The criticism directed at the formal genetics maintained that the latter based its choice of gene characteristics on morphological or growth factors which were valid only in the environment of the experiment. Although Whyte (1946) expressed doubt of the validity of the rigid phasic sequences insisted upon by the Russians, he pointed out that breeding on the basis of physiological characteristics opened up new possibilities yet to be explored and tested.

    According to developmental physiology, when the term adaptation to environment is used, it may involve one or more of several things such as breeding, selection, and agronomic management. The meaning of the term in this sense differs from the popular usage, and does not include the Lamarckian meaning. It is doubtful whether the individual plant can adapt itself to a change in environment. During the several phases in the growth and development of a plant, there are varying capacities to resist cold, drouth, insects, or fungus diseases. The objective would not be necessarily to breed for greater resistance to unfavorable conditions in the popular sense, because the possibilities in that direction seem to be limited. The purpose in both breeding and adaptation would be to discover and combine the growth and development factors that would afford the optimum timing to utilize favorable and to avoid unfavorable conditions. This includes timing for greater resistance against insects and fungus diseases. The desired result would be achieved partly by breeding, partly by selecting individual variants (mutations), which demonstrated most effectively their ability to survive, and partly by agronomic management such as choice of planting dates. Timing of the growth and developmental phases is the point to be emphasized, not an actual change in the plant in the Lamarckian sense. This does not exclude, however, consideration of the difficult problem of hormones, and the so-called hardening process in relation to cold and drouth, which require further investigation and interpretation.

    The discussion of the influence of length of day (photoperiodism) on the growth and development of plants was the contribution of Garner and Allard in 1920, and it inspired many others to study the effects of light and darkness on plants and animals. The most extensive studies of native grasses of the North American grassland in relation to drouth and light, with some reference to the temperature, are five papers by Olmstead dealing with grama grasses (1941, 1942, 1943, 1944, 1945). The last three deal with photoperiodic responses. The third paper (1943) reported on investigation of six species of grama grass; slender, side-oats, Rothrock's, hairy, black, and blue. The experimental material for five of these came from southern Arizona, and that for the blue grama from Montana. The plants of southern origin were either short-day plants, or showed tendencies predominantly in the direction of the short-day or intermediate classes, while the blue grama from Montana seemed to be day-neutral (indeterminate), or to some extent exhibited long-day tendencies.

The fourth of the Olmstead papers (1944) reported on twelve geographic strains of side-oats grama grass. This species is found in the temperate zones of South America and as far north as $50°$ in North America, yet its center of distribution seems to be southwestern United States and northern Mexico. Fundamentally, therefore, it would appear to have been originally a short-day species, which had achieved a differentiation into strains with an adaptation to a wide range of light and temperature. The geographic strains used in the experiment came from different latitudes, southern Texas and Arizona to North Dakota, were morphologically and physiologically distinct, and yet were members of the same species.

The fifth of the Olmstead papers (1945) reported on the photoperiodic responses of clonal divisions of three latitudinal strains of side-oats grama grass. By making clonal divisions rather than growing plants from seeds, the material of the experiment represented genetically identical populations for each of the three strains: San Antonio, Texas; El Reno, Oklahoma; and Cannonball, North Dakota. The strains from the three latitudes were adapted to the light and temperature differences of their geographical origin, exhibiting the capacity for seasons of long vegetative growth in the southern material and of short vegetative growth in the northern material, combined in each case with the capacity for development toward reproduction and maturing of seeds. The widest range of variability in photoperiodic responses was found in the strain from the middle latitude of Oklahoma. This led to the suggestion that on this basis acclimatization of this strain might be possible both north and south of the native latitude of the strain. Investigations of blue grama grass have been made, but have not been published, and investigation of other species when undertaken should make important contributions to the understanding of the distribution of native grasses.

As applied to agricultural crops, developmental physiology has already contributed much information for the guidance of the agronomist. Although corn is a short-day plant its cultivation has been pushed northward, especially for fodder. There the day length permits vegetative growth, but the short season does not permit development for reproduction and ripening of seed.

As a matter of agronomic management of domestic crops, the seed can be produced further south. Sugar beets afford a striking example of the possibilities of making choices. Sugar beet seed is grown in the Pacific Northwest latitudes favorable to reproduction and the root crop is grown in southern California and Arizona. The bio-chemical composition of sugar beets may be controlled in some degree to bring the sugar content to its highest peak through limitation of irrigation water at the proper stage of growth. As a matter of practical pasture management, the Bluestem Pasture owners of Kansas recognized what plant physiologists later demonstrated as a common occurrence, that the nutritive peak of vegetative growth occurs during the transition to seed production, and diminishes rapidly as the seed forms and matures. Commercial pasture contracts run accordingly from April 1, to October 1, and out-shipments of grass-fat cattle begin in late July leaving the pastures vacant by the end of September. Similarly, bluestem grass is cut for hay in late July, and prior to seed formation. The short grasses of the great plains behave differently, however, the leaves curing on the plant and retaining high nutritive value for winter pasture, a fact that was early recognized in relation to the beginnings of the livestock industry of that region.

  The small grains, including wheat, are mostly long-day plants. Hard spring wheats occupy the most northerly position on the western grassland, and the hard winter wheats occupy the central area. Wheats do not find the southern grassland suitable, and there upland cotton, a short-day, warm-temperature plant, provides the staple crop. During the early years of the twentieth century, M. A. Carleton, under the auspices of the USDA imported durum (macaroni) wheats from southern Asia, recommending them especially for the southern portion of the grassland. In practice, however, the durum wheats were successful only on the northern parts of the grassland, where, among other things, they were subject to long-day conditions.

Chapter Three

ECOLOGY

Animal ecology

Thus far in this survey plant life has been the center of interest. This is partly a convenience in dealing with complex materials and partly because there is so large a body of pertinent literature. For animals, both large and small, there is relatively little literature in a form suitable for this survey. In 1937 the American edition of Richard Hesse, Ecological Animal Geography was offered in English. It was translated from the German and in part rewritten by W. C. Allee and Karl P. Schmidt, with a preface stating that its original publication in 1924 marked a new era as it was the first attempt to apply ecological principles to the study of animal distribution on a world-wide scale. In reviewing a group of biology textbooks in 1942, Dexter pointed out again that in ecology, zoology lagged in both research and teaching.

Hesse devoted one chapter to the characteristics of animals of the "dry, open lands" of the world. Animals of such areas were described as tolerant of dry air, relatively independent of water, or regular watering, and provided with means of protection against or resistance to wind and to sharp fluctuations of temperature, illustrated by the burrowing habits of rodents, snakes, and ants. To both birds and animals he attributed conspicuously the cursorial habit, quiet rather than noisy behavior, dependence upon sight rather than sound for protection, the flocking habit, or life in communities. The grass and grain eaters were the key industry animals of the native grasslands, converting vegetation into food for the carnivorous animals. Insect species were numerous and their numbers enormous, with three types dominating; the grasshoppers, ants, and termites. In the economy of nature, the grasshoppers were stressed as eaters of the dry, hard vegetation, turning it into live form. Rodents were said to exceed all other mammals in both species and numbers, consuming 75 per cent of all the food available.

One of the earliest attempts at a scientific description of the American prairie fauna was that of Allen (1871). His (1870, 1871) was the typical reaction of the forest man to the grassland:

With all the beauty and the novelty of the primal flora of the prairies, the traveller, after a few weeks of constant wandering amid their wilds, is apt soon to experience a monotony that becomes wearisome, the full degree of which he scarcely realizes till the soft green sward and the varied vegetation of cultivated districts again meet his eye.

He held to the theory, reiterated by Chapman (1931), that "the diversity of the animal and vegetable life of a given region ... [is] dependent upon the diversity of its physical features," using the plains, the prairies and the forests to illustrate the three degrees of variation in topography, flora and fauna. Upon entering the prairie from the eastward Allen said that the mammals of the forest did not disappear but were restricted to wooded areas and therefore were relatively less prominent. The prairie mammals met there were markedly different. In transition the animals of the two regions mingled. The forest lynx, panther, and bear became rare, while rodent, skunk, mink, fox, and wolf were relatively abundant and together with the prairie species the kit-fox, badger, prairie wolf, and striped skunk, the ratio of these types of animals to the whole population was conspicuous. Among the rodents met for the first time were the pouched gopher and two species of ground squirrels of the prairie. The same principle applied to the birds, new ones named were the Cerulean Warbler, Swallow-tailed Kite, Prairie Hen, Turkey Buzzard, and Sand-Hill Crane. Reptiles and mollusks were said to have been rare, on account of annual prairie fires. Data on fish were not available. In respect to insects their characteristics were referred to as similar to the flora upon which they depended so largely; "No country, however, it is hoped, is richer in Orthoptera (grasshoppers), either in species or individuals." He concluded the discussion with the verdict that in prairie fauna, as in flora, "there is a simplicity and uniformity that gives to both a comparatively low and uniform character" in contrast with forest life.

Among the first of the twentieth century ecologists to study the fauna of the prairie, Ruthven (1908) insisted that "The prairie is the most interesting biotic region of North America," but emphasized that "the vertebrate life has never, as a whole, received examination." His expedition into Iowa in 1907 led to the conclusion drawn from comparative lists of plains, prairie and forest birds, gophers, and snakes, that

Most of the forms which inhabit the prairie region either extend also into the eastern forest region or into the plains region, or rarely both, few being confined to the prairie region.... The intermediate character of the new environmental conditions makes the prairie region an extensive area of transition between the plains and eastern forest regions, but that the environmental conditions are not either intensive or extreme enough to mold the forms into a peculiar fauna.

Shackleford (1929) made studies of the small animals of the high and low prairies of Illinois. Those of the former migrated to some extent to the latter during dry seasons, but those of the low prairie did not migrate to the high prairie. The animal population exhibited more uniformity on the high prairie, while on the low, it varied from place to place from year to year. Similar determinations have not been made systematically for other portions of the grassland, but topographical, soil, and vegetational variations were probably accompanied by analogous but widely varied fluctuations in animal populations.

A study of the eastern house wren, by Kendeigh (1934), indicated a wintering range extending as far as eastern Texas and a breeding range extending northward and westward to Indiana and eastern Wisconsin. The range of the western house wren lay west of that line, belonging primarily to the grassland, wintering from Texas and California south into Mexico, and breeding as far north as southern Canada.

In his study of rodents in California, Grinnell (1923) found in that state seven genera with burrowing habits, or about one-fourth of all California mammal species. In tracing the extent of distribution eastward he indicated the first break at a line or zone running north and south at approximately the 100 meridian. This marked the principal area occupied by the same forms. From the 100 meridian eastward the representation in genera and species diminished until only two species of one genus (Citellus, ground squirrels) were recorded in Indiana, which, according to vegetation distribution, lies at the eastern tip of the prairie peninsula. The area between is the tall grass and the mixed grass prairie. The forest regions had their own peculiar population. Grinnell attributed these limitations on burrowing rodent distribution to atmospheric humidity, rainfall, sharp alternations of dry and wet seasons and probably "a relatively greater abundance of plants with nutritious roots or thickened underground stems (corms, rootstocks)."

The prairie dog possessed a particular significance to the grassland as a regional indicator. The Rocky Mountains marked its western boundary of distribution, except in the southwest, where it extended as far west as Arizona. Its eastern limit was near the 97 meridian in Kansas and Nebraska bending somewhat west of that line in Texas, but in the North following approximately the Missouri river through the Dakotas and Montana (Merriam, 1901). These animals had multiplied and spread as a result of increased food supply from agricultural crops, and from the killing of their enemies by man, the coyotes, kit foxes, badgers, ferrets, weasels, minks, owls, and snakes (Merriam, 1901; Shelford, 1940). The coyotes, represented by twelve species, had a wide range of distribution from the central Mississippi valley to the Pacific coast and from Costa Rica to Athabasca (Lantz, 1905). Jack rabbits were present with two species having as an eastern limit of distribution the western part of the tall-grass prairie in Minnesota, Iowa, and eastern Kansas and extending westward to the Pacific coast in California and to the forest regions of the Pacific Northwest. The white-tailed species occupied the northern part extending into the Canadian plains, and the black-tailed species occupied the southern part. In the middle was an area of over-lapping in southern Nebraska and in northern Kansas, extending westward to Nevada (Palmer, 1897).

Different species of the Tetraonidae (ptarmigan, prairie chicken and grouse) occupied tundra, forest, and grassland separately or in combination in varying regional distribution (Pitelka, 1941). The greater prairie chicken (Tympanuchus cupido) occupied the area from the forest edge of the prairie peninsula westward into but not through the high plains, with a southern boundary of abundance at about the south line of Kansas, and scattering occurrence farther south. The lesser prairie chicken occupied southern Kansas, northwestern Oklahoma, the Panhandle of Texas, and the eastern edge of New Mexico. The sage grouse (Centrocercus urophasianus) ranged eastward into the short-grass plains and westward into the desert of the Great Basin and northwest into the Palouse prairie. In much reduced numbers they still populate much of the designated area of the grassland. Some other related species were not restricted to a single form of vegetation, particularly the sharp-tailed grouse, which ranged as far south as $37°$ in the grassland and desert and occupied northern forest as far as the tundra regions.

For subdivisions of the grassland, the animal distribution can be analyzed in more detail. For the northern grassland Visher (1916) made a study in what he called biogeography in which he described all major forms of life together; vegetation, large animals, birds, smaller animals, and insects. To emphasize the variety, rather than the uniformity of the region, he broke it down into its vegetational and soil types, showing the wide range of difference in each, animal life as well as in grasses and forbes: The buffalo-grama grass association; the needle grass, or sandy loam steppe, association; the wheat grass, or clay steppe, association; the bunch-grass, or dry soil steppe, association, of which there were two phases, the rugged areas, slopes, not sandy, river bluffs, and moraines; and secondly, the sand dune areas; and the low shrub group association. This kind of treatment is particularly important to counteract the prevailing misconceptions of the uniformity and monotony of the grassland. The only thing that was uniform about the grassland was that the vegetational covering was grass as distinguished from trees, but within limits of the grasses as plant life, there was a great variety of species and combinations of grasses and forbs. The grassland had fewer layers of vegetation in its structure than the forest and possibly it was this characteristic that gave the impression to the forest man of monotonous uniformity.

In describing the animals of the northern grassland, Visher said that the mammals possessed two or more of a list of characteristics; ability to run swiftly, the burrowing habit, acute long-range vision, coloration adaptation of gray or tawny, ability to do without much water, daily activity confined to early morning, evening, or sometimes nighttime, ability to hibernate, lack of gregariousness except for protection or warmth in some cases as antelope and bison, and largely herbivorous habits. A similar method was used in describing birds which possessed two or more characteristics; ground nesters, singing on the wing, louder calls than forest species, social flocking less prominent, ability to withstand a strong wind, females and nestlings mostly protectively colored, ability to withstand intense heat and scarcity of drinking water, and acute long-range vision.

In North Dakota, as in most of the low-rainfall grassland area, the burrowing forms of animals are conspicuous and in this instance gave to the state the nickname "Flickertail" state. The Franklin ground squirrel occupied only the northeastern-most portion of the state with a western boundary to its range near a line from the Mouse river running southeast down the Dakota river (Bailey, 1926). The Mississippi pocket gopher area lay east of a line from Devil's Lake down the Dakota river while the Dakota pocket gopher occupied the country west of that line. The Richardson ground squirrel, "the Flickertail", was limited to the country east and north of the Missouri river. The major race of the thirteen-lined ground squirrel was similarly limited but a pale western form occupied the area south and west of the Missouri river. The same river marked the eastern boundary of the plains form of the prairie dog in North Dakota. In respect to these species of animals the state was divided by the Dakota and the Missouri rivers into three ecological areas. The Dakota river lay between the 98 and the 99 meridian, but the Missouri river did not approximate any meridian.

At the southern end of the grassland, the 98 meridian marked a fairly close approximation to the western range of some eastern animals or the eastern range of others (Bailey, 1905, 1928). The Texas woodpecker and the Texas rattlesnake territory lay just west of the 98 meridian with a western boundary between the 100 and 101 meridians, and a northern range extending just north of the Red river. Such plains animals as the bison, the plains race of the white-tailed deer, the plains jackrabbit, and the plains prairie dog reached their western limits just east of the Pecos river watershed in New Mexico or just west of it (Bailey, 1931).

In the central grassland, it has been customary to designate the 97, the meridian of Fort Riley, Kansas, or some line or zone between that and 99° as the approximate division line between eastern and western forms. Blair and Hubbell (1938) listed 73 species, or 91 species and races of mammals in Oklahoma. Of the 73 species, 22 (30 per cent) were eastern forms that reach western limits within the state, and 22 species were western forms that reach eastern limits within the state. Six species were northern forms that reach southern limits, and two species were southern forms that reach northern limits in the state. No survey of similar completeness exists for Kansas and Nebraska, but Hibbard's (1944) revised viewpoint abandoned the 98°-99° boundary in Kansas and proposed a pattern in which the eastern forms extended up the valleys of the principal streams somewhat west of such a line, and the western forms extended down the upland divides between the streams to points east of such a line. This is the same pattern of distribution for animals which Schaffner (1926) had designated for central plains vegetation and tended to bring ideas on animal distribution into harmony with Schaffner's suggestion of the coincidence of boundary between the tall-grass and the mixed-grass prairie at the eastern extent of the range of the prairie dog and the harvester ant. Clements and Shelford (1938) took over the idea from Schaffner. It fitted also into the findings of Shackleford (1929) relative to the influence of topography on faunal distribution between high and low prairie of Illinois.

Insects

In 1927 Hayes remarked that "the insect fauna of the prairie has been given so little consideration, ecologically, that it is with some hesitation that [this] ... discussion is attempted." When Osborn's Meadow and Pasture Insects was published in 1939, the point was stressed both by the author and the reviewers that little had been done with grassland insects. Osborn emphasized as a retarding factor, the "relative obscurity and constancy of the damage," and the conviction that nothing could be done about it. In reviewing Osborn's book, King (1940) emphasized that it was the first genuine treatise in the field and that the grassland comprised about 40 per cent of the land surface of the world as a natural or climax community, besides the area planted to grass as an agricultural crop. The understanding of insects was important not only to the grasses, but to the crops subject to injury by insects from the grasses. Allen's general surveys of flora and fauna of the prairie, 1870 and 1871, included a brief mention of insects, especially commenting upon the number of grasshoppers. Allee (1927) pointed out that in some land communities, the most abundant and important animals were insects, although in land communities generally, Shelford (1931) held that mammals and birds were most important.

The grassland and the shrub transition zones, according to Shelford (1915), were the original habitat of the principal insect pests of the forage and garden crops and of small fruits and orchards. Native to the original vegetation of these areas, with the clearing of the adjoining forest and undergrowth and the planting of cultivated crops and fruits, the insects changed hosts and many multiplied as a result. In addition to the native grassland insects the central area was the crossroads of insect migration from northwest, southwest, southeast, and importations (Webster, 1903; Van Dyke, 1919; Hayes, 1924, 1927; Smith, 1925; Blair and Hubbell,1938). Of Oklahoma's 254 species and races, 90 (35.4 per cent) were eastern forms which reached western limits to their range within the state; 89 (35 per cent) were western forms with eastern limits within the state; 32 (12.6 per cent) were plains species extending into the state from north and south, but having both their eastern and western range limits within the state; 62 (24.4 per cent) reached northern or southern limits within the state (Blair and Hubbell, 1938). Riley county, Kansas, was designated by Hayes (1924, 1927) as the ant center of the United States and Smith (1925) made a similar claim in respect to the Neuroptera and Mecoptera, the forms from the four directions meeting and overlapping in that vicinity. Hayes (1927) suggested that this might be true of other faunal groups when adequate information was available. Such a generalization may be too sweeping as applied to Riley county, or to Kansas because the Oklahoma data show a substantial meeting of north-south range limits within that state. A more substantial accuracy would be achieved probably by designating the central grassland (Oklahoma-Kansas-Nebraska) as the area of transition and such a picture would contribute much to the understanding of the striking variety and abundance of insect life in that area.

Climatic factors affected the several kinds of insects differently; wet years were favorable to the Hessian fly, but unfavorable to the chinch bug (Shelford and Flint, 1943), and the reverse. The westward diminution of rainfall probably acted as a limitation on the distribution of the Hessian fly (Hayes, 1927). The idea that temperature controlled the development of insects was challenged by Shelford (1926). Especially in unusual seasons, when control was most needed, this hypothesis was most likely to fail. The relation of insects to weed control was illustrated by the white beard tongue, a weed with an increase potential of 4000 fold. Insects destroyed 55 per cent of the seeds within two months of the flowering of the plant, constituting the most important single factor in holding that weed in check to a point of approximately a single plant annually replacing its predecessor (Brandhorst, 1943). Chapman (1931) cited what must be an unusual example of an insect, normally unimportant, attacking and destroying an invading plant, thereby contributing to the maintenance of the integrity and stability of the biotic equilibrium. The cacti of the plains region multiplied rapidly during all prolonged drouth periods, and gained unusual publicity during the depression-drouth hysteria of the 1930s under the ministrations of those who were ignorant of the biological principles in operation in the grassland. The situation did focus attention, however, upon making the controlling factors the subject of investigation and scientific record, an opportunity which was practical for the first time as this was the first prolonged drouth since the biological sciences were adequately developed and staffed to make such studies (Cook, 1942; Turner and Costello, 1942).

The cacti afford an unusually clear case of the coordination of a number of multiple factors in effecting the unstable equilibrium in nature. In an investigation centering around Hays, Kansas, of the relation of insects to the cacti (Opuntia humifusa), Cook (1942) found three kinds; one a bug similar in appearance to the box-elder bug (Chelinidea vittiger Uhl) which had little apparent importance, several species of mealy bugs (Dactylopius sp-) which were of greater importance, and a moth (Militara dentato) which seemed to be a decisive factor. In its larval stage, this moth occupied the central stems of the cacti, eating the plant from the inside. Moist warm weather was favorable to the rapid build-up of the moth population, and dry hot weather was unfavorable to its growth. Rodents, and especially jack rabbits, ate the cactus fruits and scattered the seeds. Thus grass, cactus, animals, and insects interacted upon each other under fluctuating weather conditions, especially successive years of similar weather where the effects became cumulative.

Cacti and grass were not so much competitors or enemies as complementary to each other. In prolonged moist periods grass provided the soil cover, with a minimum of cacti; during prolonged dry periods unfavorable to grass, the cacti took over and provided the soil cover.

This occurred irrespective of the extent of grazing. It
was associated with the weather fluctuation and was not
the result of overgrazing. Grazed and improved tracts
showed the same trend (Cook, 1942). The multiplying
clumps of cacti held the soil from extensive blowing,
catching in the clumps the dust and grass seed. As soon
as favorable grass-growing weather returned the seed
sprouted, the cactus clumps serving as a protective nurse
crop to the new growth of grass (Malin, 1942, p. 25 note).
The grass repaid the cactus by harboring insect life which
attacked and killed the cactus, leaving the grass in full
possession during the moist period. With the return of
dry years the grass was weakened or died out and insects
died also, permitting the cacti to multiply, and animals,
particularly jack rabbits, being limited in other food
and water supply, ate more cacti and this scattered the
seed. Thus alternately a series of dry years built up
the cactus phase and a series of moist years built up the
grass phase of vegetation, both coordinated with animal
and insect reactions. This emphasized how foolish it was
during the 1930s to become excited over the supposed danger of the cacti permanently dominating the plains.

Grasshoppers received more publicity than any
other insect of the grasslands, but only occasionally
were they an outstanding menace. The chinch bugs probably did more consistent and serious damage (Malin, 1944).
The dramatic aspects of the swarming phase of the so-called Rocky Mountain locust in 1874 so impressed the
public mind, ignorant of the characteristics of this form
of insect, that tradition alleged frequent and periodic
swarming and devastation (Hafen and Rister, 1941, p. 429).
A survey (Ball, 1937) of Colorado and Arizona grasshoppers established a record of 130 species in each state.
Some 40 species were beneficial in attacking the worst
weeds of the range, some 70 were not important, 5 or 6
were of major importance in attacking cultivated crops,
and 10-12 in damaging range grasses. Of 40 species studied by Isely (1938), 37 were non-migratory and were usually rated as of no economic importance, from the standpoint of crop damage, but of this second group several
species were positively beneficial as weed eaters, and
the grasseaters had adverse significance only when there
was a shortage of grass. To the mind of the average city
dweller, conditioned by grasshopper scourge stories, the
idea is novel indeed that a grasshopper could be beneficial. The use of grasshopper machines and poison made
spectacular headlines, but dealt with results not causes
and such benefits as they brought came after the damage
was done in part.

The grasshopper was subject to many predators;
fungi and parasites, land animals and birds, and to some
degree biological controls were possible (Ball, 1937;
Sweetman, 1936). In irrigated districts, insects were

more generally under control (Ball, 1937). In non-irrigated areas, insect damage was more serious, where control measures proved ineffective because of waste, eroded, or inefficiently managed land. Also, the non-irrigated regions involved much greater land areas, wider differences in environment and more variety in species (Ball, 1937). Some species of grasshoppers died if not provided with grass for food, while others, Melanoplus differentialis, possessed a wide range of adaptability to other plants (Isely, 1938). Climate, particularly rainfall, also seemed to be a determining factor in the distribution of some species according to studies in northeastern Texas (Isely, 1937). Different species propagated at different rates, the population being built-up during favorable years, became a menace during unfavorable years. A specimen of Melanoplus Mexicanus might multiply theoretically in three years to 15,000 while Melanoplus differentialis might produce 500,000 (Ball, 1937).

The migratory locust presented a problem different from the solitary grasshopper. Uvarov (1911, 1921, 1928) elaborated the theory of phase which was based on the assumption that some species were biologically unstable and under certain conditions assumed the solitary form (grasshopper) and under other conditions changed character so completely as to appear to be another species, the migratory, swarming phase (locusts). This theory of phase was originated in connection with Asiatic grasshoppers, but has been applied to the American species Melanoplus Mexicanus atlantis (synonym Mexicanus Mexicanus), the solitary non-migratory grasshopper, and to the Melanoplus spretus, the swarming Rocky Mountain locust so spectacular in 1874. At any rate M. spretus is practically extinct, never having appeared again in numbers after the seventies, but entomologists warn that the swarming phase might reappear (Parker, 1925; Hebard, 1925; Uvarov, 1928).

On first contact with the grassland, the forest man was impressed by the number of grasshoppers. They were numerous both as to species and numbers. What appeared to him as abnormal, however, was only natural. Hebard (1931) had listed 197 species, 12 subspecies, and no migratory phase found in Kansas. The grasshopper was a typical insect of the grassland and a product of the normal working of biological forces. Every natural region possessed characteristics favorable and unfavorable to the several forms of life, but with an overall long-time tendency to establish and maintain a balance. Grasshopper outbreaks of serious character were of different kinds and were only an exhibition of temporary upsetting of the prevailing unstable equilibrium. Other natural regions with different vegetation and animal life each had their share of pest problems resulting from temporary disturbances of equilibrium, and some pests, especially imported types, sometimes become a permanent threat to forest as well as to grassland regions.

Plants, animals, and environment

American zoologists discussed less fully the factors which determine the distribution of animals than the botanists had done for vegetation. Merriam's (1890, 1892, 1894, 1898) temperature zone hypothesis received much support among zoologists. The early challengers were Adams (1902) whose work on distribution centers of flora and fauna ran contrary to the life zone maps, and Transeau's (1905) work on climate centers in relation to vegetation was even more explicit. The direct critical analysis and fuller demonstration came later (Shelford, 1911, 1932; Ruthven, 1920; Livingston and Shreve, 1921; Dice, 1923, 1943; Kendeigh, 1932; Smith, 1934; Pitelka, 1941; Hibbard, 1944; summary of the literature, Daubenmire, 1938) and emphasized that there was little correlation when the Merriam transitional zone was applied in detail to particular areas outside those on which Merriam's conclusions were based. The influence of the governmental bureau, The Biological Survey, of which Merriam was head, tended, however, to perpetuate his point of view even after it was rather generally rejected. Revision and restatement of the hypothesis was made by Hall and Grinnell (1919), and Grinnell (1935), in which a rather wide allowance was made for variations under the influence of factors operating locally. It was on the Pacific coast and in the more rugged mountain areas where temperature changes were relatively abrupt, in respect to space and time, that Merriam's hypothesis had its greatest following. Under such special conditions the single factor of temperature might prove to be the principal limiting factor under the law of minimum, especially with the benefit of the allowances recognized by Hall and Grinnell. This would not prove, however, the general validity of the system, or its applicability to other geographical surroundings. On the non-mountainous, or relatively level grassland sound physiological conclusions required a broader base which recognized water, light and soil relations as independent variables. The plants or animals that survived were those that escaped the fatal effects of the extremes but not necessarily the same one, or combinations of two or more of the life factors, depending upon the range of circumstances.

Botanists tended to follow more generally the theory that food controlled animal life, an assumption that subordinated it to vegetation. Walker (1903) emphasized atmospheric moisture, and others sought the solution in hydrogen ion concentration (pH) of the soil. Little attention was given to light (Chapman, 1931), much less among zoologists than among botanists. Allee (1927) rendered a ringing verdict on single-factor procedures in commenting that "There has been a constant chase for a possible single factor index which would be a short cut to environmental analysis. In the very nature of the case, this hunt has met with the failure it deserves." Shelford (1913) had urged as early as 1912 that plant and animal life must be studied together as a unity, and Vestal (1914) developed and demonstrated the idea in his study of Illinois animals. Later ecological work tended rather generally to the broad view that life must be studied as an interacting whole, differences being more in details and theoretical interpretations than in the general hypothesis (Visher, 1916; Allee, 1927; Shelford, 1931; Phillips, 1931; Chapman, 1931; Kendeigh, 1934; Clements and Shelford, 1938; Weese, 1941).

The botanists used the terms plant associations, communities, or assemblages, and the zoologists used similar terms, but in dealing with plants and animals together in the aggregate, new terminology emerged, much of it too complicated for common use, and some of it carrying implications of a subjective rather than strictly objective scientific nature. Shelford (1931) said that "all the life (plant and animal) is a biota"; Clements (1916) offered the term biome, excluding habitat which he viewed as cause with the biome as effect (Clements and Shelford, 1939). Phillips (1931) used the term biotic community. Stemming from the Clements-Shelford terminology, Weese (1941) divided the North American continent into six biomes, with some allowances for areas that did not fit into the system: The Tundra Formation (Sedge-Musk Ox Biome); The Coniferous (Evergreen) Forest Formation (Spruce-Moose Biome); The Deciduous Forest Formation (Oak-Deer Biome); The Grassland Formation (Grass-Bison Biome); The Sagebrush Formation (Sagebrush-Jack rabbit Biome); and the Desert Formation (Creosote Bush-Kangaroo Rat Biome). Dice (1943) offered the term biotic provinces and mapped them for North America. The geographers offered physiographic provinces along with maps (Fenneman, 1931; and Atwood, 1940). For the historian none of these terms seems altogether satisfactory for the study of a natural geographical region, but acquaintance with them is essential to the reading of the literature, especially that of the Clements-Shelford school. Among the several difficulties encountered are the implications of uniformity and rigidity which are so far from reality that some zoologists and botanists tend to emphasize variations and the unusual rather than normals and averages, multiple rather than single factors, restating for the purpose Liebig's law of minimum (Chapman, 1931; Taylor, 1934; Rubel, 1935).

The search for physiological explanations of the relations of animals to water, drouth resistance, have been conducted by zoologists, but with less success than that culminating in Maximov's (1929, 1939) treatment of plants. Sumner (1925) doubted whether there were "xerophytic" animals in the sense that there were xerophytic plants. Although desert animals tended to lighter coloration than animals of moist regions, Sumner questioned the adaptational value of such a factor to burrowing, nocturnal animals, and furthermore, why should concealed parts, such as the bottom of the feet show a change in coloration. Dice and Blossom (1927) concluded that there was correlation between animal and soil color. A large number of animals, both mammals and insects escape, rather than resist climatic extremes, by burrowing in the ground or hiding during hot weather, and by burrowing and hibernation during the winter (Sumner, 1925). In Riley county, Kansas, Hayes (1927) found that insects had burrowed as deep as thirty inches in the soil for the winter, coming out again in the spring, a process by which they escaped cold. On the main issue of drouth resistance, it is evident that generalization awaits fuller knowledge.

The influence of light upon animal behavior, in the form of length of day, received its first experimental treatment from Marcovitch in 1923. Marcovitch, Eifrig (1924), and Rowan (1926) and others received their inspiration from the pioneer work of Garner and Allard on photoperiodism in plants. Marcovitch demonstrated that the reproductive development of plant lice was controlled by the shortening of the length of day in Tennessee. Rowan attributed the migration of birds to physiological changes in reproductive organs induced by the factor of change in day length. He pointed out also that chickens increased egg production with the spring change in length of day. The work of Brissonnette (1930—) opened the period of comprehensive studies of light in this relationship.

Park (1940) made the initial attempt to coordinate the accumulation of literature on nocturnalism in animals and to trace the development of the problem for the benefit of the ecologist. It is important to know what animal activities took place in the dark and their relations to the biological equilibrium. About 1919 a new era was opened in the study of the relations of ultraviolet light to rickets in children. Vitamin D deficiency was more than a dietary problem, light being the more fundamental controlling factor. This introduced the problem of color, pigmentation of skin or hair of animals in relation to light. The effect of sunlight in killing bacteria and other microorganisms had long been known, but exact analysis and application were slow. The physical basis of solar radiation, the interruption by ozone, the differences in the effects of the extreme ends of the spectrum, infra-red and ultra-violet, daily variation, seasonal variation, differences in latitude, all became biologically important to the distribution of life on the earth (O'Brien, 1943). In 1929, the Smithsonian Institution created a new Division of Radiation and Organisms to deal with the relations of light to living organisms (<u>Annual</u> <u>Reports</u> 1929—).

Chapter Four
CLIMATOLOGY

This seems an appropriate point at which to interrupt the story of ecology in order to present some aspects of the development of climatology in relation to the problems of biology. The daily weather record was kept traditionally in terms of rainfall, temperature, and wind movement. Meteorology attempted to explain in terms of science the laws governing weather behavior, with a view to immediate usefulness as in navigation and agriculture. The climatologist dealt with the series of daily weather records and he assumed that the mean or average of these daily variations determined the general prevailing condition called climate. The stress upon usefulness emphasized the idea of forecasting the weather changes, either for short periods or over long periods of time. This latter aspect stimulated attempts to discover some definite rule of periodicity in recurrence. Extremists yearned to reduce the whole subject to a fixed law and order that would govern man and civilization. In this pursuit of a goal of wishful thinking, theories outdistanced facts.

The weather data in the traditional form of averages, deviations from the mean, maxima and minima, etc., did not provide adequate material for the ecologist. In the study of physiological reactions to climate, humidity and evaporation were important and little data of that nature were available. Questions arose whether the methods of recording such data were adequate. Doubt arose whether the averages or so-called normal conditions were determining factors in the survival and competition of living organisms. It was experience with such data and methods that caused Livingston (1921) to call for new approaches and Shelford and Flint (1943) to point out that for practical chinch bug control studies weather data were still inadequate.

After the beginning made by Transeau, there were many attempts to formulate a single index which would express by a strictly quantitative procedure the several factors which enter into the mapping of climatic regions. In Germany, W. Koppen presented one, first in 1918, but more fully in 1923, and revised successively through 1936, that recognized rainfall, temperature, humidity, and used the coldest period of the year as a basis of division between hot and cold climates. R. J. Russell (1926, 1931) devised some modification and applied it to the western United States. W. Van Royen (1927) made different adjustment and applied it to the United States and Canada east of the Rockies. Thornthwaite (1931), devised a four-factor procedure: a precipitation-evaporation (P-E) index; a temperature-efficiency (T-E) index, and seasonal distribution of the P-E index, and summer concentration of the T-E index. In his map of the North American climates and of the world climates he applied only the first three. Trewartha (1937) made other modifications of Köppen, applying them to the mapping of world climates. There were valid objections to all of these procedures because the exact relations of the factors to plant growth were not fully understood. Purely quantitative methods broke down and resort was made to practical adjustments which necessarily were subjective rather than quantitative. One of Thornthwaite's seasonal distribution types, was rainfall deficiency for all seasons, which was applied to the great plains. A critic, S. B. Jones (1932), queried, deficient for what?; certainly not for native vegetation. However, too much emphasis should not be directed at the deficiencies of these systems because, as P. R. Crowe (1936) pointed out, they were new departures in climate analysis and as such opened a new era.

As a further step in avoiding subjective concepts such as the ideas of adequacy and deficiency, the present author suggests a new set of terms, quantitative in their implications, to designate climates; wet, high rainfall, mid rainfall, low rainfall, and dry. These words substitute for super-humid, humid, sub-humid, semi-arid, and arid, and at the same time, express the purpose of classification in simple common names (Malin, 1947).

Another kind of challenge to old procedures is found in the concept of climatic years and the significance of extremes and variability. At various times the idea had been advanced that the effect of extremes of climate might be more decisive in relation to vegetation than the normals or mean temperatures and rainfall. Shimek (1911) had given it some stress in his analysis of why the prairies were treeless. Jenny (1941) summarized the results of a fourteen-year record of rainfall and erosion in Missouri during which time there had been 420 rains of sufficient volume to cause run-off. In 28 of these two inches or more of water fell during 24 hours, and were responsible for over half of the soil erosion during the whole fourteen-year period. As a system for dealing with climate classification, however, R. J. Russell (1932, 1934) led in this new departure, pointing out that especially in erosion, although occupying a short time and infrequent in occurrence, extremes might have a greater significance than normal conditions. He then proposed the idea of climatic year, using the concept of desert year, steppe year, etc., in mapping climates, emphasizing the frequency of recurrence of each in contrast with the system of normal climates which smoothed out the exceptional years in the averages. In respect to the climatic year and vegetation, he said:

Mature plants are less likely to succumb to the vicissitudes of climate than are young ones. Though an occasional desert or tundra year may upset vegetational balances temporarily, recurrence at such a rate as once a century very likely has little bearing on the distribution of forest or grassland. Yet, if the recurrence is so frequent that individual plants are prevented time after time from reaching maturity to withstand such extremes, it becomes highly significant. The westward extent of the forest toward the Great Plains of the United States is more probably related to the recurrence of desert years than it is to the distribution of normal precipitation.

The second concept proposed by Russell (1934) was that of nuclear and transitional climates. The former was one that experienced no exceptional years denoting a different climate class. Between two nuclear climates was an area of transition within which occurred exceptional years of extremes. Applied to the eastern United States the mesothermal nuclear climate occupied the southern Atlantic and Gulf states and the microthermal climate occupied the Great Lakes states and those lying westward. The line dividing the intervening transitional zone on a 50 per cent frequence of climatic years ran from New York City southwestward down the Ohio river and across the plains to central New Mexico. This system abandoned the traditional assumption of fixed climatic boundaries based upon statistical averages. It revealed the more important fact of variability in a wide transitional zone between major climatic regions and the general conformities of both natural and domestic vegetation with such climatic designations.

The idea of variability was applied in a different manner by Kendall (1935) employing a modified Köppen formula to the mapping of annual climate for each year, 1914-1931, and then making a composite map by superimposing all the boundary lines of the year-climates on one map. This procedure presented vividly the fact that climatic regions were not bounded by a fixed line, but by a maze of lines occupying a wide transitional zone. As applied to the North American Grassland and the Prairie Peninsula the maze of north-south lines intertwined up and down the Prairie Plains, and the east-west lines between the Great Lakes and the Ohio river and eastward across southern New England. The Prairie Peninsula lay in the area of right-angle intersection of these two sets of lines.

The Scotch climatologist, P. R. Crowe (1933) introduced another new concept which avoided the ideas of adequacy and critical, both of them subjective terms, and dealt with variability, frequency, and "indices of probability." After applying it first to European material, he applied it to the great plains of the United States: "The essential feature is the degree of change experienced between one period and another" — contrasts in distribution and frequency — "rainfall relativity."

Still another innovation was that of E. E. Lackey applied to rainfall in Nebraska in 1935 and to the great plains in 1937. He based his system on the median rather than the mean in relation to maxima and minima and on the frequency of variability in percentage with the median at 50 per cent. Five maps in the annual variability series were made, one for each frequency interval of 20, 40, 50, 60, 80 per cent, and lines were drawn from the Canadian boundary to southern Texas connecting specified rainfall amplitudes. Thus the line showed nineteen inches of rainfall 80 per cent of the time fell in central Kansas; 60 per cent near Dodge City; 50 per cent between Dodge City and Garden City; 40 per cent near Garden City, and 20 per cent in eastern Colorado.

A project somewhat similar to Kendall's was executed by Thornthwaite (1941) using his own system as applied to annual and crop-season climates for each of the years, 1900-1939, but he did not make the composite map. Instead he applied the idea of frequency of occurrence over a forty-year period (1900-1939) of six types of climate years, one map for each type; arid, semi-arid and drier, sub-humid and drier, humid, and superhumid. Each map showed in colors the areas included, in six degrees of frequency. To all residents of the grasslands, variability of climate was axiomatic, but the climatologists had at long last given it both mathematical and cartographical demonstration.

Many climatologists concluded that the weather ran in cycles, but they did not agree on the length or even the definition. Brunt (1937) insisted upon the definition of cycle given in the Oxford English Dictionary: "A period in which a certain round of events or phenomena is completed, recurring in the same order in succeeding periods of the same length." Others thought of cycles as recurring events over periods of varying length. The important point is that the first definition would eliminate definitely practically all argument over weather cycles, while the second definition rendered the whole argument virtually meaningless because of the introduction of the unpredictable factors of approximation and coincidence. Furthermore the weather trends of one region did not necessarily coincide with some other region, even an adjoining one. A climate that was favorable to one form of plant or animal life was not necessarily so to others. The concept of universal optimum efficiency climate collapsed with the simple question, for what plant or animal?

The idea of cycles came to be most generally associated with sun-spot frequency of approximately eleven year (11.13 years) periods. Here the popular idea was one of absolute cycles of exactly eleven years, without realization that no such periodicity existed. The sunspot frequency varied from seven to seventeen years, and the eleven-year figure was a mathematical average — a fiction — which would have about as much practical value to agricultural planning as the tradition of planting in the light or the dark of the moon. The whole question was subject to controversy but the fact remained that the burden of proof rested upon the advocates of periodicity and cause-and-effect relationship. Among other things it should be conclusive that when climatologists cannot predict from one year to the next even approximately the weather behavior, its relation to crops, or its relation to insect pests, there cannot be any scientific basis upon which to predicate a climatic theory of civilization. As applied to particular problems, MacLulich (1936) challenged the sunspot periodicity as applied to the fluctuation of numbers of lynx and varying hares in Canada, and Cross (1940) went further in demonstrating not only that there was no such periodicity in the red fox in Ontario, but that the numbers of foxes fluctuated on a basis of natural regions within the province.

A revision of the viewpoint expressed in some of the agricultural publications of the federal government is found in R. J. Russell's chapter "Climatic change through the ages" in the departmental Yearbook for 1941, Climate and Man. In this, Russell rejected explicitly the cycle hypothesis, substituting the term fluctuation:

> Man had observed that climatic conditions fluctuate rather widely from time to time at a given place, and in seeking to understand such natural phenomena he has been tempted to explain such fluctuations on the basis of recurring cycles. As yet, however, no definite proof has been advanced to contradict the opinion that all such relatively short-term climatic changes are nothing more than matters of chance. The world pattern of climates today is the product of climatic variations, not the expressions of recurring mean, or normal, conditions. The extent of desert climate will not be the same next year as this. The humid margin of the desert is the product of an ever-changing distribution of extreme aridity. The time may come when such changes will be well enough understood to be of definite forecast and economic value, but it is likely that such information will be the fruit of long-continued and patient research.
>
> Interest in changes of geologic proportions will remain intellectual....

Many years must pass, however, before the careful presentations such as R. J. Russell's, will rectify the influence of the misrepresentations of the drouth of the 1930s.

The most recent attempt at a theoretical explanation of the climate of North America is that of I. R. Tannehill, Drought, its causes and effects (1947). The thesis is that on a broad scale the Pacific ocean controls the weather of the continent. The Pacific high pressure area off the California coast in winter is associated with rainfall; while the shift northwestward of the enlarged Pacific high pressure area in summer is related to a dry California coast. Not actual temperatures but relative or contrasting temperatures over the oceans and the continents control air mass movements. This basic principle, applied consistently, imposes a substantial rethinking about the uses of weather data. The task is not easy. Only an over-simplified version can be indicated here. In winter, the northern North American continent is colder relatively than the oceans west and east. The eastward moving Pacific air mass passes rapidly over the mountains into the Canadian sink, while its continued movement eastward is retarded by the resistance of the Atlantic-continental temperature contrast. The outcome is a movement of the Canadian air mass southward, where it meets normally the Gulf and the Atlantic air masses moving northward and into the continental area, producing rain. In summer, the temperature contrasts are opposite, the Canadian air mass moves eastward, reinforced by hot air from the southwest, tending to deflect the moist Gulf air mass eastward and to leave the great plains dry. Unusual temperature contrasts react to create extreme drouth conditions.

What controls the northern Pacific high pressure area? Tannehill relates this problem to solar radiation. As more is known about sunspots than about other aspects of solar radiation, the influence of the sun is discussed primarily, but tentatively, in those terms. But Tannehill is not naïve about sunspots, and points out that there is as much confusion about them as about drouths; rainfall variations do not follow exactly the sunspot cycles; local changes in temperature and rainfall appear quite irregular and do not coincide with sunspot numbers. The author repudiates, however, any form of single factor explanations. As the rainfall of each local area is controlled by its own peculiar conditions taken as a whole explanations of drouth must vary from place to place. He admits the inadequacy of the data and the weaknesses in the explanations, and calls for data collected on a different basis suitable to the purpose in hand. Tannehill's general hypothesis is the first that appears to be at all tenable, and pending further research and testing, it merits serious consideration, but expectations must be kept within the limits indicated.

Chapter Five

GEOLOGY AND GEOGRAPHY

Geology

The beginnings of American geological work were largely in the hands of self-trained, or largely self-trained, amateurs. Such formal training as many of them had as background for their invasion of this field was often medicine or chemistry. Among the early great names were William Maclure, and Amos Eaton, but more important was Benjamin Silliman, professor of chemistry and natural science at Yale College, and founder (1818) of the American Journal of Science.

As the continent was new to white men, the first work was exploration, collection of specimens, description, classification, and determination of stratigraphy. Little of such work had been done west of the Mississippi river prior to the middle of the nineteenth century and most of that had been incidental to military, geographical, and railroad reconnaissance. Among the leading men associated with the first geological surveys in the area were G. W. Featherstonhaugh, D. D. Owen, James Hall, G. C. Swallow, the Shumard brothers (G. G. and B. F.), Jules Marcou, and L. Lesquereux. After the Civil War, although each of the western states set up its own geological survey, the major projects in the west, upon which public interest centered, were those sponsored by the federal government; F. V. Hayden, "Geographical surveys of the territories," Clarence King, "Geographical survey of the fortieth parallel," J. W. Powell, "United States geological and geographical survey of the Rocky Mountain region," and G. N. Wheeler, "Geographical surveys west of the One Hundredth Meridian." All of such work was finally consolidated in 1879 into one organization, the Geological Survey, Department of the Interior, with Clarence King as the first director, and J.W.Powell succeeding to the headship the second year.

Geology was the first of the sciences to establish itself in public confidence sufficiently to secure government aid in a large way. Geological surveys were authorized first by the states and later by the federal government. Such support was motivated, of course, by practical applications of the science. The first objectives apparently were the discovery of mineral resources and the manner of their occurrence; iron, lead, copper, and other metals, and coal, salt, gypsum, and other nonmetallic minerals. Some of these subjects were considered systematically in such major works as J. D. Whiting, Metallic wealth of the United States (1854), and J. P. Lesley, Manual of coal (1856), and Iron manufacturers guide (1859). In Mississippi, the geological survey (1858 -) was turned to agricultural objectives by E. W. Hilgard who quickly realized that the state did not possess great mineral wealth.

The Pacific railroad surveys authorized in 1853, as well as others both before and after, which were associated with railroad projects, included a combination of objectives; geographical, geological, and agricultural. In the western mountain states, the discoveries of gold and silver influenced largely the direction of most of their state surveys, iron and coal and other minerals being given a minor consideration in all early survey work. The problem of water supplies in the low rainfall and desert areas, which intervened as a barrier to be crossed in maintaining communications between the east and the west, directed attention during the 1850s to water resources and their relation to geological formations. After the successful completion of the first drilled oil well in 1859 in Pennsylvania, oil and gas became major objectives, and near the close of the nineteenth century came to dominate geological thinking in those mid-continent states not largely favored with other mineral resources.

The glacial problem was of particular interest to the upper Mississippi river and Great Lakes drainage basins. The first definite suggestion of the glacial hypothesis came in 1825 from Peter Dobson, a New England cotton manufacturer, but did not receive recognition of scientific men until the French and English geologists, Louis Agassiz (1840), and Sir Roderick Murchison (1842) championed the idea, and Agassiz, after establishing himself in the United States (1846) led in giving it a substantial confirmation based on field work (Merrill, 1924). Maps showing continental evolution in relation to glacial action are conveniently arranged in the Schuchert-Dunbar, Historical Geology (1942). The oldest and lowest in position of the gracial deposits was the Nebraskan, then the Kansan, the Illinoian, and lastly, the Wisconsin.

As already pointed out (Ch. 2) these periods of glaciation exerted important influences on the distribution of plant and animal life, as well as laying down materials for soil formation, and contributing largely to the physiography of the area concerned. The glacial hypothesis was important also in explanation of the great plains. In this area the rock formations were covered to various depths with the outwash of debris from the Rocky Mountains, the streams being fed with enormous quantities of water from melting glaciers. The loess was attributed to the blowing of these outwash deposits by the winds

(Elias and Bryan 1945). The high points of the underlying rocks either were not covered fully or the subsequent erosion exposed them. In this manner may be explained the relative scarcity of rock outcrops in the plains country, most of the area remaining covered, often to great depths (Frye, 1946). These general facts are essential to an understanding of the grassland which is so largely an area characterized by porous transported soil materials, the product of glaciers, water and wind.

During the early nineteenth century American geologists depended for theory largely upon the two competing European schools of thought. The German geologist A. G. Werner (1749-1817) explained the rocks of the earth's crust as formed by sedimentation from a great ocean that covered the earth, the Neptunian theory. The Scotch geologist, James Hutton (1726-1797) found his explanation in heat as well as water, the Plutonian theory. The latter body of theory came to predominate during the mid-nineteenth century.

Although Americans made relatively little contribution to theory, there were several examples of independent thinking. G. W. Featherstonhaugh's proposal (to measure the silt load of the Mississippi river), growing out of his Ozark expedition of 1834, was important as suggesting a basis for estimates of geological time and as presenting a significant approach to physiography. Edward Hitchcock presented a beginning of the systematic study of physiography in America in his book Surface Features (1856). He had been studying the problem of rivers and was uncertain as between theories; whether to explain river gorges and canyons as arising from fractures in the earth's crust through which rivers flowed as offering least resistance, or to assume that the rivers cut their own channels by the erosive power of water. Hayden is probably entitled to the credit for advancing, as late as 1872, an adequate theory of canyons by showing the capacity of a river to cut through its rock bed during a period when movement of the earth's crust was elevating the plateau through which the river flowed (Merrill, 1924). Applied originally to the Montana area by Hayden, this idea was elaborated by Powell (1875), G. K. Gilbert (1877), C. E. Dutton (1880), and W. M. Davis (1889).

It is clear from this brief sketch that when the grassland was first being occupied there was no body of scientific knowledge which might serve as a guide to the occupation of the area. The tall grass prairie was largely occupied, at least by the first wave of pioneers, prior to the Civil War. More permanent establishments as far west as the 100 meridian, or mixed grass region, occurred in the 1870s. The high plains boom and collapse was the experience of the 1880s and 1890s. The geologists, botanists, zoologists, or other branches of science were little, if any, in advance of the settler in the search for understanding of this strange region.

At the close of the nineteenth century, geological thought was dominated by the concept of the earth as a cooling, shrinking, dying world, a point of view derived from the tentative suggestions of Laplace, in the eighteenth century, in respect to the origin of the solar system. In the minds of nineteenth century geologists, the nebular hypothesis was transformed into a rigid system of scientific orthodoxy, as something substantially proved. This was illustrated clearly by the standard American textbook on geology (six editions, 1866-1897) of James D. Dana. The opening of the twentieth century brought a challenge to the pessimism of the dying earth theory in the work of T. C. Chamberlin and F. R. Moulton (1902—), and T. J. J. See (1896-1910). These new theories, the planetesimal and the capture theories, instead of assuming that the planets were formed by separation from a central molten body, cooling rapidly to their present form, proposed that they had formed in spiral nebulae by the consolidation or capture of smaller bodies by larger bodies acting as nuclei (Willis, 1942; Chamberlin, 1929; See, 1896-1910). In England, Sedgwick, and Jeans independently offered similar theories (Jeans, 1944). According to these hypotheses it was not necessary to assume that the earth was a molten mass or relatively very hot, and in consequence there was necessity of revising assumptions in respect to a rapidly cooling, dying world. In 1916, Jeans (1944) proposed the tidal theory which assumed that a passing star drew off a filament of gaseous matter which consolidated. To the various theories offered in the field of cosmogony, radium, radio-activity, and the atomic theory, unfolding in the course of the twentieth century, continued on new lines the procession of revisions of cosmic hypotheses and revealed sources of energy undreamed of by nineteenth century scientists. Whatever the ultimate destiny of the universe, temperature changes that would influence in any substantial way the course of modern civilization would require possibly ten thousand million years, and that prospect should not cause alarm to the twentieth or any nearby century.

Geography

In discussing the nature of geography as a discipline, Barrows (1923) presented it as the mother of sciences; of astronomy, botany, zoology, geology, meteorology, and anthropology. As these children became separate branches, geography found itself obligated to redefine its jurisdiction. Irrespective of whether this view was altogether correct historically, the development of scientific specialization during the later nineteenth century imposed upon the geographers the problem of a restatement of scope and objectives. The so-called modern geography was launched in the universities of Europe during the last quarter of the nineteenth century, but for the most part did not gain general recognition until the twentieth century. In the United States, the modern geography was fostered by geology, growing out of physiography or physical geography and the work of Hayden, Powell, Dutton, and Shaler, and more particularly that of W. M. Davis, who in 1889 formulated his theory of the erosion cycle.

From this point there was a two-fold differentiation in the United States; the separation of geography from geology and physiography, and the search for a definition of geography in terms of the relation of man and the earth, human geography. At first geographical determinism was conspicuous in human geography (F. Ratzel in Germany, and Ellen Semple in America), was gradually modified, and eventually, by the early 1920s, was largely repudiated but not very successfully eliminated in practice. Peculiarly, however, American geographers were slow in responding favorably to the concept of possibilism — Vidal de la Blache and Lucien Febvre — that always there is more than one solution in any environmental situation, and therefore choices are always possible. The position of the geographers was weak to the degree that they failed to provide an adequate positive alternative to determinism. People are more important than the physical environment. People can make choices. Even submission to determinism is a matter of choice. Physical environment provides certain limiting factors which people must recognize, but Liebig's law of the minimum should not be confused with determinism — the element of choice remains within any particular framework of limiting conditions.

In spite of the uncertainty concerning the scope of geography, a number of fields were cultivated which extended into or across borderlands of other disciplines: economics, sociology, anthropology, soil science, botany, zoology, political science, and history. Economic geography was developed most fully as it seemed to possess the most immediate utility, and therefore, justification of geography in its own right, as a discipline separate from geology. When Barrows defined geography as human ecology, this approach found itself in competition with sociology defined as human ecology which was also invading the borderlands. Barrows (1923) proposed to limit geography by defining it as human ecology, taking over into geography much of the thought of the plant and animal ecologists.

Sauer (1925, 1927, 1941) led in challenging the trends of the day to limit the scope of geography, characterizing the period 1923-1939 as the Great Retreat. In restating his views of geography, he urged a return to the study of land forms as an essential factor, and a study of historical geography (1941) as well as the present, which would introduce an adequate recognition of the factor of time and the genetic development of areas of the earth. To accomplish this he would invade the borderlands of both anthropology and history. The extensive literature of the period between World Wars I and II on the scope and methodology of geography not only challenged environmentalism, but emphasized the weakness of the concept of natural regions, especially where associated with the idea of organism. The region was not a fixed entity, but depended upon the factor chosen as a standard of measurement, and the boundaries stubbornly refused to conform to theories even for single factors, and attempts to combine several factors proved even more conflicting. It was the same problem met by the plant and animal ecologists, the climatologists, the soil scientists, and the agronomists.

Some of the outstanding characteristics of American geographical thought were summarized in a review article by the English geographer E. G. R. Taylor (1937) dealing with an accumulation of American college textbooks. In contrast with English books, he emphasized that these were general rather than specialized, and conspicuously subjective rather than objective; "they seek to establish particular theses or to present particular points of view rather than provide a detached record of observed facts." Taylor was sound and severe in his analysis of the misuses of social theory — drawing conclusions "beyond the facts," which were not geography, social science, or history. Hartshorne (1939) made the most comprehensive analysis of the trends of geographical thought, and one of his points was in line with Taylor's in emphasizing that geographers had been much better trained in geology than in the social sciences and that they felt that they would be held to account from the geological point of view but were free to speculate in the social field. Hartshorne's study made clear also that regionalism and environmentalism as they had been practiced, both in Europe and America, were substantially discredited and that the sound aspects of those points of view could be protected only by restatement and the application of a more scientific methodology. The disturbing aspect of Hartshorne's analysis of methodology was in the weakness of the case in defense of the existing procedures, and in the confusion of views, although he made a definite attempt to bring his discussion to a conclusion that would avoid mere frustration and defeatism.

The decade 1935-1945 was particularly fruitful in productive emphasis on fields heretofore little cultivated; conservation, political, and historical geography.

Sauer's (1941) plea for historical geography has already been mentioned in combination with his advocacy of a new emphasis on physiography, with a view to tracing the evolution of cultures in relation to geography.

In much the same spirit Whitaker (1940) proposed the study of the conservation problem "viewed as worldwide in scope and prehistoric in origin." To this end he surveyed the history of conservation thought stemming from the work of the American, George P. Marsh (1801-1882) and his monumental book, Man and Nature; or physical geography as modified by human action (1864), revised and translated into Italian 1870, and revised again for a second American edition in 1874. In spite of the fact of Marsh's influence on European thought, he was largely forgotten in the land of his birth and the credit for formulating conservation thought was assigned to others.

Political geography received a new emphasis in Europe about the opening of the twentieth century which centered around the idea of closed space. As one outcome of the communications revolution of the last half of the nineteenth century (mechanical power applied to communications) a new impetus was given to the neo-mercantilist imperialism during the last quarter of the century. All the new land opened to occupation by the discovery period of the fifteenth century had been appropriated and occupied and man must order his living within the limits of a known world. This kind of geographical determinism was expressed in its most challenging form by Mackinder (1904), and it was utilized by the German geographers in combination with the organistic concept of the state advocated by the Swedish geographer Kjellen in their development of geopolitics. In the United States, Turner's frontier hypothesis (1893) was used, or more accurately stated, misused, during the 1930s in justification of the application to the American scene of policies grounded in the idea of closed space — i. e., the passing of the American frontier meant closed space (Malin, 1943, 1944, 1946).

Among the principles or corallaries derived from the concept of closed space in the two-dimensional sense were the doctrines of the necessity of control of the most valuable world space; of a certain land mass area as the geographical pivot of history in the sense that in it was embodied the monopoly of world power; the shifting of the idea of the measure of national and imperial strength from accumulated wealth to the capacity to produce; and the reduction of human population to a status of mere units of manpower in the competition among totalitarian states as Going Concerns (Mackinder, 1904, 1905, 1919). The validity of these ideas came to be challenged in various quarters, but for the time being they dominated the thought of the World War II generation. From one point of view it was the historic sea-power interpretation of history (Mahan, 1890, etc.,) challenged by a new landmass interpretation, with a still newer third-dimentional factor of air-power introduced as the final determinant. In another sense it was the antithesis of two ideas, the concentration of power in one place against the strategic dispersion of power, but again with air-power as the determining factor. The use made of his ideas by geopolitics under German leadership certainly was not anticipated by Mackinder, who through these years was thinking of the geographer, not only as a town and regional planner, but also as a participant in world planning (1931, 1943) for peace as the means of checkmating the consequences of struggle for power and world domination by the occupant of the heartland, the geographical pivot of history.

In summarizing the status of political geography in 1935 Hartshorne emphasized the lack of interest by the geographers of English speaking countries. In Great Britain only two or three major works had appeared, and in the United States only one, Bowman, The New World (1921), a product of the World War I peace making period. Consequently, there was little discussion by Americans of the place of political geography in the general field. During the next decade, 1935-1945, American geographers produced substantial works on political geography, but they were influenced largely by German geopolitical thought. In other words, it was not yet clear that political geography was firmly and independently rooted in American soil.

The idea of two-dimensional closed space in the sense of occupation of the earth's surface, together with all the unfortunate consequences of this idea for the world during the first half of the twentieth century, did not mean necessarily a closed world and the end of opportunity for man. There was a less pessimistic interpretation of the same body of facts. Science tended to introduce the idea of three-dimensional space and then air communications seemed to climax the process — in a sense reopening space. Instead of earth-bound man fronting on the oceans, the Atlantic and the Pacific, air communications focused attention on a circumpolar world, all nations fronting to the north. This meant a major redistribution of power in the course of the twentieth century (Malin, 1944). Whittlesey (1945) traced brilliantly the history of man's sense of space and emphasized the term fourth dimension to describe the time factor as a function of space. To the three-fold aspect of this factor of time he gave the names, velocity, pace, and timing, all of which take a new significance in consequence of the realization of three-dimensional space through the several kinds of mechanization culminating in air communications.

Chapter Six

SOIL

Soil science

Soil science is largely a product of the twentieth century because it is only in most recent years that a reasonably satisfactory theory of soil and its formation has been proposed. Justus von Liebig (1840) had explained soil fertility in terms of chemical elements as plant nutrients stored in the soil. These could be absorbed by the plants and replaced by man in the form of chemical fertilizer. There was an important truth in Liebig's theory, but it was not an explanation of soil. In the United States, E. W. Hilgard (1833-1916) anticipated much of the modern theories which are credited to the Russian school of V. V. Dokuchaev (1846-1903), and his followers, N. M. Sibirtzev (1860-1899), and K. D. Glinka (1867-1927). It is fundamental to the concept of soils to distinguish between the parent soil materials, derived from the disintegration of geological formations, and soil as a complex body derived from the operation of several soil forming factors. In equation form, soil may be defined in principle as the product of parent materials, climate, organisms, topography, and time. This does not disguise the fact, however, of wide differences in emphasis and interpretation of this or a similar formula. The followers of the Russian school placed climate first and minimized the others in varying degrees of extremes. Some virtually ignored parent materials, and also relegated organisms, vegetation, animals, and microorganisms, to a minor rôle of dependent variables.

Until the early twentieth century European and American soil scientists looked primarily to geology and chemistry as the parent sciences, and their studies reflected that emphasis in soil analysis and classification according to geological origins of the materials, the manner of their distribution (residual, or transported by water, wind, or glacier), and their chemical composition. Under the leadership of Milton Whitney (1860-1927) the United States soil survey work, begun in 1899, emphasized texture (the sizes and proportions of different sizes of soil particles and their combinations in the surface layer). He attributed the adaptability of a soil for a given crop primarily, if not exclusively, to texture.

In California, Hilgard followed the general trend in emphasizing geological origins and the chemical composition of parent materials, but also he struck out on new lines. Going to California in 1875 as Professor of Agriculture and Director of the Agricultural Station, after many years in Mississippi, he grasped quickly the fact of the importance of climate and made the study of soils of arid regions the most conspicuous feature of his long, active California career (1875-1905), climaxing it with his book Soils, their formation, properties, composition, and relations to climate and plant growth in the humid and arid regions (1906). As indicated by the title, he treated climate as an independent soil forming factor. He distinguished between soil and subsoil, and although ill-defined, he attributed "the differentiation of soil and subsoil as due partly to the action of organic matter and micro-organisms, partly to physico-chemical causes...," or in more complete form it was the result of vegetation, moisture, access of air (oxygen), temperature, "and the presence of the several organisms which in the course of time take part in the process of soil-formation." One chapter each was devoted to humus, soil organisms, and vegetation. As respects vegetation — soil relations, he missed the most significant point in viewing vegetation as controlled by soil, rather than distinctive types of vegetation, grasses and forests, contributing as soil formers to equally distinctive soil groups — i.e., mutual interaction between soil materials and vegetation.

In describing the physical effects of the percolation of water, he pointed to the fact that the diffused colloidal clay was carried with it to a certain depth into the subsoil and that this clay layer was more retentive of moisture and "plant food substances in solution" than the surface layer. The chemical effects of the percolation of water he recognized as dissolving and carrying lime down through the soil; in arid climates depositing it in a lime accumulation zone of the subsoil, but in humid climates carrying it off in drainage. His illustrations showed diagrammatically the differences in the resulting soil profiles in arid California climate and in humid climate. The superior fertility of the soils of arid regions was explained similarly by the theory that nutrients were not leached out by rainfall. In estimating the significance of his contribution to soil science and agriculture his associate R. H. Loughridge (1917) concluded:

Among his California activities there stand out prominently his studies on humid and arid soils, in which he was the first to point out their differences in depth and in physical and chemical characteristics; he was the first to explain endurance of drouth by culture crops in sand soils and why sandy soils are among the most productive in the arid region and least so in the humid....

While Professor Hilgard was not the first to make a soil survey or a chemical analysis of the soil, he was the first to interpret the results of analysis in their relation to plant life and productiveness. He was also the first to maintain that the physical properties of the soil are equal in importance to the chemical in determining the cultural values.

This review of Hilgard's work has been given at some length because of the distortions by the enthusiasts of the newer Russian-American school of soil scientists of the United States Department of Agriculture. Marbut (1935) said that "up to this time [of the Russian influence] no suggestion that the characteristics of soils were due to other agencies [than geologic] had received recognition in the United States." In the departmental Yearbook Soils and Men (1938) Kellogg made some amends. Outside of the federal governmental publications, Hilgard was given substantial recognition as in the writings of Joffe (1929, 1936), and Jenny (1941), but an adequate evaluation of his whole career is yet to be written — (cf., further discussion in Ch. 14).

The soil work done in Russia, isolated because of the language barrier, was largely unknown outside of that country until the translation into German of Glinka's work in 1914. C. F. Marbut, of the United States department of agriculture, led in popularizing this point of view among soil scientists in the United States. In 1927 he published a translation into English from the German edition of the major portion of Glinka's The Great Soil Groups of the World and their Development. The work of Dokuchaev on the genesis of the Russian chernozem (black earth) was not a study of isolated samples of soil in the laboratory, but a field study of the profile of the soil body for itself. From this came the recognition of three layers or horizons in the typical soil profile representing different stages of physical and chemical weathering of the parent material at different depths; different amounts and conditions of organic elements incorporated; and the degree of development depending upon the time factor operating in soil formation. To the Russian school, climate was the dominant, almost exclusive independent factor in soil formation as vegetation and other biological factors were viewed as a product of climate. Soil was defined in terms of soil-forming processes, therefore, and not in terms of parent materials.

Attempts to apply the Russian system to the soils of the United States met difficulties because of the fundamental differences in environment. In Russia and in most of eastern Europe, climate belts ran in a general east-west direction, dry and hot at the south and wet and cold at the north. In the United States, the regions east of the Mississippi river were of approximately the same humidity both north and south, the temperature providing the major variable. West of the Mississippi river to the Rocky Mountains the surface rose in altitude to the westward and the rainfall diminished; thus there were three important variables, altitude, rainfall, and temperature, the third roughly running at right angles to the first two. The Great Basin interior provided further differences and the Pacific coast the most unusual variety of all. In American hands, the Russian formula, with its exaggerated emphasis on climate, was substantially reconstructed, and soils were defined on the basis of soil characteristics rather than soil forming processes. The first presentation of soil classification on the American plan was by Marbut, in 1922, and his most complete statement was in "Soils of the United States" (1935) for the Atlas of American Agriculture (1936). Later statements of the work of the United States Soil Survey have appeared mostly under the name of C. E. Kellogg, chief of the Division (1936, 1937, 1938, 1941), and have modified Marbut's conclusions.

Preoccupied with the work of soil description and classification, it was not until after 1930 that the new point of view took shape for popular presentation. Wolfanger's book The Major Soil Divisions of the United States (1930) presented the view of regional distribution of soils like regional distribution of natural vegetation and climate. The principal governmental publications that presented the broader outlines were Marbut's, "Soils of the United States" (1935) a section in the monumental Atlas of American Agriculture (1936), Kellogg's Development and Significance of the Great Soil Groups of the United States (1936), and of widest circulation, the departmental Yearbook, Soils and Men (1938). Among the commercial soil textbooks were Lyon and Buckman (Second Edition 1929, Third Edition 1937, and Fourth Edition 1943), Joffe (1936), Millar and Turk (1943), and most original and stimulating of all, Jenny, Factors of Soil Formation (1941). To these should be added the European books of Robinson (1932, 1936), and de Sigmond (1938). Kellogg's popularization, The Soils that support us (1941), had a large measure of political and social propaganda intermixed with science.

Most of the new school of soil scientists treated the soil as "a natural body" or an organic entity, an object of nature in equilibrium with environment. This concept carried with it also the idea of age in relation to immature, mature, and degraded soils. The idea of a mature soil implied somewhat the same thing to the pedologist as climax formation did to the botanist. Jenny (1941) was an exception, taking the ground that soil was "a physical system", not a natural body, and as an open system it might be added to or subtracted from and that there was "an element of organization" in every soil.

Although he defined mature soils in terms of equilibrium with environment he warned that assumptions regarding maturity are all inferences, theories, not demonstrated facts. In Buckman's revision (1943) of the Lyon and Buckman book he carried further than others the idea of the biological nature of soil, "a colloidal-biological concept", in which the genesis of soil began with "a biochemical phenomenon", the operation of microorganisms and higher plants, living and decaying matter, in the soil materials.

A soil profile was a vertical cross-section of soil from the surface to the underlying bed rock. It was divided into layers or horizons and each such horizon was subdivided. The A or top horizon, at its surface, A_1 horizon, accumulated organic matter, the lower part, A_2, showed the results of the leaching out of certain soluble and fine material, the layer of eluviation. The B horizon was the layer of accumulation (illuviation) of deposits from above. The C horizon was parent material, physically disintegrated, but on which soil forming processes had not yet operated. The D horizon was the bed rock. The solum was the name given to the A and B horizons when considered together as soil. In the soil forming process, climate operated as a factor in dissolving compounds and translocating them and fine material, organic colloids, from one part of the soil to another or in removing them altogether. The problem of lime, emphasized by Hilgard, illustrates the process, when in low-rainfall climates the dissolved lime was deposited in the lower horizon, while in high rainfall climates it might be removed altogether. The zone of clay concentration in a high-rainfall climate was the product of the downward movement of colloids. The leaching out of more soluble materials left the insoluble forms, alumina, and iron, as conspicuous residual material in leached soils.

The biological factor entered into soil forming processes in a conspicuous manner as each type of vegetation possessed different kinds of root systems, fibrous or woody, penetrating to different depths, and living or dead, influenced the soil both physically and chemically. The grasses and the forests as soil formers contributed substantially different characteristics. Grass soils contrasted as distinctly with forest soils as the grass and forest vegetation above ground. The whole range of animal life played its rôle, that on the surface, and the rodents and insects that burrowed, and at death contributed to the organic fraction of the soil. Most neglected until late, was the soil population of microorganisms. The emphasis on the surface vegetation was relatively obvious, but it was the belated emphasis on the second and third aspects that gave particular significance to the Lyon and Buckman (1943) phase, that the genesis of soil begins with "a biochemical phenomenon." Topography contributed its part to the soil forming process through its influence on moisture distribution in the soil and consequently on the distribution of the biological factor upon and within the soil mass.

It is inconceivable to think of process and dynamics except in terms of duration. Soil formation required relatively long periods of time, but there was no conclusive data of experimental character which would determine the length of time necessary. Jenny (1941) gave examples of the results of formative processes in particular cases of known duration from a few years to 230 years. The net result of such examples was to throw doubt upon the extreme requirement of a thousand or even thousands of years claimed by some soil conservation enthusiasts. It was at this point that Jenny's (1941) idea of soil as an open system rather than a natural body gained significance. The use of the word age in connection with a natural body carried with it an implication of a life cycle of youth, maturity, and degradation that is not subject to experimental proof — it went beyond the facts. Any change in the components in an open system would only initiate another phase in readjustments of unstable equilibrium with environments, not maturity or decadence.

There are many systems for the classification of soils (Robinson (1936), but for the United States, the Federal Soil Survey set the principal model. Taking the country as a whole, Marbut (1922, 1935) divided the soils into two parts on the basis of characteristics of the solum (A and B horizons taken together) calling them pedocals (ped = soil; cal = calcium) were the soils of the low rainfall regions with a lime accumulation horizon. The pedalfers (ped = soil; al = alumina; fer = iron) were the leached soils, without lime, exposing conspicuously the alumina and iron. The chemical reaction (pH) of the pedocals was neutral to alkaline, that of the pedalfers was acid. The line dividing these great solum groups followed approximately the eastern rim of the watershed of the Red river of the North from the Canadian boundary southward through western Minnesota, turning southwestward at the Iowa line into Nebraska, then southward passing somewhat west of Lincoln, Nebraska, and Wichita, Kansas, then slightly west of south past El Reno, Oklahoma, to south central Texas, slightly west of the 99 meridian, where it made a sweeping eastward curve to the Gulf coast near Corpus Christi. Although it must be shown on a printed map as a line, it was more accurate to think of it as a transitional zone. Except at the northern end, it was not far from the 98 meridian, the line so much discussed in the literature of the plains region as the line dividing the prairie from the low plains.

The Great Soil Groups or Provinces (Marbut, 1935) were subdivisions of the solum groups. In the pedalfer area, except the Prairie Peninsula, from north to south they were tundra soils in Canada; podzols near the borders of the two nations associated largely with the northern evergreen forests; gray-brown podzolic soils of the oak-hickory deciduous forests, and red soils and yellow soils of the southern hardwood and evergreen forests regions. All soils formed under forests were of limited or even inferior fertility, a fact which runs counter to popular traditions. The prairie soils constituted a group by themselves in the region of the Prairie Peninsula and southward through the prairie-forest transition country to the Gulf. Intermixed, however, were gray-brown podzolic soils which occupied most of the eroded valleys. In the pedocal area, just west of the pedocal-pedalfer division line, lay the chernozem belt forming a north-south belt from near the Rio Grande into Canada.

Farther west were the dark-brown soils of the plains and the brown soils of the plains-desert transition, and the gray soils of the deserts and the Great Basin. The Pacific coast region has a wide range of soils; from gray-brown podzolic to gray-desert.

The prairie soils were the subject of much discussion among pedologists. It was much like the problem, among botanists, of grass in a high rainfall climate that, by all the known rules, should have produced forests. The prairie soils are chernozem-like soils formed under a grass cover in a mid-rainfall region, but they lie in a high rainfall region, and are surrounded by a network of valley soils of the gray-brown podzolic group which theoretically seem to be proper soils for the entire Prairie Peninsula. Marbut (1934, 1935) insisted that they are normal soils, while Joffe (1936), more directly in the Russian tradition, called them degraded chernozems. At any rate, it is agreed that the prairie soils are unique, that no other substantial area of similar soils exist elsewhere in the world. They possessed the remarkable fertility of a grassland soil, were only slightly acid, and received ample rainfall to make them highly productive. As the prairie belt ran north and south from near the Canadian line to the Gulf, the temperature range, produced differences, the southern line of Kansas marking roughly an irregular division line between southern prairie soils and northern prairie soils.

The American chernozems were similar to the Russian chernozems (black earths) which were the original inspiration to the development of the modern Russian school of pedology. They were formed under a grass cover, moderate temperatures, and mid-rainfall, just enough to produce at medium depth, a substantial lime accumulation zone and maintain a neutral to alkaline reaction. The mass of fibrous grass roots and the annual deposits of grass leaves accumulated the maximum of organic matter in the A horizon, and the soil assumed an almost perfect crumb-like structure. They were the richest of all soils in plant nutrients, the only limiting factor to their productivity being rainfall which was sometimes deficient. As Robinson (1936) pointed out, the highest ambition of the agriculturist who dealt with forest or other types of soils was to create and maintain an artificial chernozem. The wide north-south temperature range resulted in differences which were indicated by a division line south of the Arkansas river in Kansas between the northern and southern chernozems. The dark-brown soils of the high plains were much like the chernozem soils, except that the lime zone lay nearer the surface, there was less organic matter in the top horizon because of less heavy growth of grass. They were also highly fertile. Only where adequate rainfall was uncertain was productivity rendered hazardous. The division line between northern and southern dark-brown soils ran southwest from a vicinity of Dodge City into northern New Mexico. The brown soils of east central New Mexico, Colorado, and northwest were intermediate between the dark-brown soils and the gray soils of the desert.

In the classification of soils of the United States into Great Soil Groups, Marbut assigned such designations only to soils which he held were mature. By his standard of maturity, nearly half of the soils were designated as immature. In his maps (1935) for the Atlas of American Agriculture, one was designated for the Great Soil Groups, and on it the immature soils were not shown. A second map indicated the immature soils. The two must be studied together. Usually only the first of Marbut's maps was reproduced in books using his soil maps. The use of both maps is particularly important to the mixed grass prairie-plains where a large part of the soils were designated as immature. Furthermore, some of these were among the most productive of the central grassland. Most pedologists took the view that only in early stages of soil formation did parent materials largely influence its characteristics. In more developed soils, the mark of the parent materially lessened until in mature soils that influence was practically non-existent, climate and vegetation fully determining the soil properties.

In the Yearbook of the United States Department of Agriculture, 1938, Soils and Men, a substantial revision of the Marbut classification was presented, introducing in part new terminology as well as reclassification. The southern prairie soils became reddish prairie, but the area was broken up into several types of soils. The chernozems were retained, the northern dark browns were renamed chestnut, but the southern chernozems and southern dark browns were substantially rearranged and the name reddish-chestnut given to the major portion.

The brown and desert gray soils underwent the most extensive revision and one that is traceable only by reference to the detailed maps.

In dealing with the low-rainfall regions, Shaw (1937) still held to classifications of soils which emphasized parent materials, and Graham (1941, 1943) concluded that in Missouri the parent materials continued to influence plant growth even in mature soils and that soil classification should recognize this fact. Robinson (1936) noted a swing back to greater recognition of parent materials. The whole subject of soil science was so conspicuously in its formative stage that it was important to recall Robinson's (1936) admonition, that all soil classifications were provisional. The problem of minor elements placed a new emphasis on parent materials of soils as well as upon occurrence of mineral nutritional diseases of plants and animals. The importance of the subject was such that the editor of Soil Science sponsored a symposium of twenty-one papers dealing with sixteen minor elements which were published in the July and August issues, 1945, including a map of minor element distribution in the United States which affected plants and animals.

Soil microbiology

The least known of the aspects of soils in their relations with plants centered around the problems of microbiology and gave special significance to the biochemical concept of soil. The population of the soil included animals, both large and small, and plants, both the roots of higher plants and the plant micro-organisms. Summarizing from Lyon and Buckman (1943), the animal division included macroorganisms living primarily upon the plant materials, rodents (gophers, squirrels, woodchucks, etc.), insects, millipedes, sowbugs (woodlice), mites, slugs, snails, and earthworms; macroorganisms (largely predatory, living on other animals), moles, insects, (many ants, beetles, etc.), mites, centipedes, and spiders; microorganisms, either predatory or parasitic, or living on plant materials, nematodes, protozoa (amoeba, ciliates, and flagellates) and rotifers. The plants included the roots of higher plants, algae, fungi (including mushrooms, yeasts, and molds), actinomyces, and bacteria.

On the physical side rodents and insects, along with many of the macroorganisms, were constantly engaged in the process of translocation and pulverizing of soil and their channels and galleries contributed to aeration and drainage, constituting a natural tillage. Humus was largely synthesized through the feces of insects, millipedes, sowbugs, mites, slugs, snails; such animals as use plant material thereby initiating the process of decomposition which was continued by bacteria and fungi. Their dead bodies in turn contributed further to the organic content of soil. Actinomyces, next to bacteria, were the most numerous of the soil population, and in bulk were greater. Possibly fungi and certain animals were even more important than bacteria.

W. Kette (1865), according to Waksman (1932), was the first to recognize the relation of microorganisms to soil fertility. The first great contribution related to bacteria and the nitrogen problem and belong to the decade beginning with 1886 in association with the names of Hellriegel, Wilfarth, Frank, Beijerinck, and Winogradsky. Almost another generation passed before further fundamental advances were made relative to protozoa (E. J. Russell, 1909, 1913), algae, fungi, actinomyces and nematodes. Soil microbiology was little more than a test tube or laboratory science until the 1920s when H. J. Conn and Winogradsky attacked the problem of direct methods of examination of microorganisms in the soil. Sergi Nikolaevitch Winogradsky, rated by Waksman (1925) as one of the two greatest soil microbiologists of the period, was exiled from Russia by the Bolshevist revolution and in 1922 found refuge in France, with the opportunity to continue his work as head of the soil microbiology section of the Pasteur Institute at Paris. In 1925 he presented his direct method which became the most important project in the new methodology. It was his contention (1925, 1928) that with the laboratory method exclusively, based on isolation of the organism from the soil, pure culture, and bacteriological media, no real soil science could be established; and that as an independent science soil microbiology must be carried out under conditions as near nature as possible, "in the soil itself." He cited the example of the Azotobacter, which was active in isolation, but remained "obstinately at rest in the midst of the soil." Waksman (1925, 1932), Romell (1930, 1935), and Jacot (1936) emphasized the same theme through the next decade.

The reaction of the soil solution favored bacteria when it was on the alkaline side and fungi on the acid side (Joffe, 1936; Vandecaveye and Katznelson, 1940; Jenny, 1941). From this the conclusion was clear that any combination of conditions which tended to change the pH of the soil was reflected in the composition of the soil population which must adjust itself to fluctuating biochemical environment. These principles had a practical bearing, therefore, upon the problems of the grassland and the differences in soils formed under a grass cover and under a forest cover. Wolfanger (1930) pointed out that studies had not been made on an extensive or systematic scale with reference to the relations between the great soil groups and microorganisms. Waksman (1932) had a chapter on the microbiology of peat soils (Ch. 26) and forest soils (Ch. 27), but none on the grassland soils. In neither case, however, could these treatments go much beyond a recognition of the problem. Joffe (1936) lamented how little was known of microorganisms as soil formers, their distribution and behavior in the soil profile, but concluded that each climatic zone had its specific population. Jenny (1941) said practically the same thing, emphasizing that all soils were subject alike to this independent biotic factor, to these microorganisms dispersed through the atmosphere, but for unknown reasons the response of the several soils was different.

Investigations of the relation of microorganisms to the different horizons of the same soil led to a variety of conclusions (Waksman 1932). Vandecaveye and Katznelson (1940) found in a majority of soils tested a larger population in the B than in the A horizon. Other investigators sometimes arrived at different results and Newman and Morgan (1943) not only found the subsurface population smaller, but less adaptable, introduced plant material being less rapidly and less extensively decomposed. There was rather general agreement that the soil population contributed essentially to the structure of the soil (Waksman 1932, 1936; Jacot 1936; Jenny 1941; Lyon and Buckman 1943) and therefore to the success of tillage and control of soil erosion either by wind or by water.

The problem of soil fertility was largely bound up in the problem of microbiology. Russell's (1909, 1913) protozoan theory was based on the assumption that protozoa consumed bacteria, under some circumstances to such an extent as to destroy fertility. Cutler and Crump (1935) assigned a different rôle to protozoa; that their presence kept the bacteria at the level of highest efficiency. Waksman (1937) doubted both theories because direct examination of soils did not show either the number or volume of protozoa great enough to serve such purposes. It was evident that there existed within the soil populations a complex balancing of forces, because theoretically each could soon multiply sufficiently to overpopulate the soil. Some were balanced against each other, some destroyed others, and some were beneficial in relation to higher plants. Because of antagonistic relationships, it was sometimes impossible to introduce organisms into a soil, or if possible, only by changing the soil environment (Waksman, 1941; Jacot, 1940).

An investigation by White (1941) of prairie and forest soils in Wisconsin offered interesting suggestions to the problem of the grasslands. He found that by introducing into the prairie soil a fungus, a mycorrhizae from red pine roots, the growth of trees was promoted. The subject needs more extensive investigation under a wider range of soils and environments, but if the conclusion should prove of general significance, it would only change the form of the question. Instead of, why the prairies did not grow trees, the question would be, why prairie soils did not grow mycorrhizae? Seifritz (1938) discussed the relationship between soil acidity and plants concluding that plant growth was largely determined by the pH of the soil, but after a particular vegetation was established, it in turn influenced the soil acidity. He did not include the problem of microorganisms in his discussion.

The same question of grass and trees was approached by Joffe (1936) from a quite different direction. As a result of lysimeter experiments in the podzol zone he found that the soil reaction was acid during the winter, spring, and early summer, but in late summer and fall it became less acid or nearly neutral. His suggestion to the prairie discussion, therefore, was that should this seasonal cycle hold true for the grass region, the naturally alkaline reaction would be intensified in the fall creating conditions definitely unfavorable to the sprouting and growth of tree seeds, but highly favorable to grass seeds. Stone (1944) cautioned about soil reaction studies in relation to plant growth, pointing to defects in the techniques of most older literature, the complexity of interacting factors, and the fact that vegetation influenced the pH of the soil as well as the fact that soil reaction limited species of plants. In trying to sum up any possible conclusions in the matter, the historian is reminded of Livingston's observation of 1917, that instead of solving the problems of twenty-five years ago, they had only succeeded in analyzing them into several component problems, each of which was as difficult as the original problem.

Soil fertility and productivity

The modern soil scientist makes a distinction between the terms fertility and productivity of soils. Kellogg (1941) said:

soil productivity may be most clearly conceived as being a response to management, yields alone are not adequate guides to soil productivity— the management under which the yields are obtainable must be known as well.... Soil fertility is included in the concept of soil productivity but refers only to the content, and availability of chemical compounds in the soil that influence plant growth.

Jenny (1941) took the ground that productivity was measured in yields per acre and might be stated in a mathematical formula: Productivity equaled the product of climate, plant, man, soil materials, and time; while fertility was associated historically with Liebig's idea of soil and plant nutrition and became therefore a concept of plant nutrients in the soil which might be taken from the soil by plant growth and restored by fertilizers.

There was little difference here between the two statements of soil fertility, but there was substantial difference between the definitions of productivity. The factor of management in Kellogg's definition might be interpreted as equivalent to the factor man in Jenny's and it might be interpreted to include most of the other factors, but in that form it might be too vague to be of much value as a definition.

Tyulin (1938) attacked the problem of fertility from a different direction, emphasizing colloids as being associated with soil fertility, particularly the first humate fraction of group 1 colloids. Atkinson, Turner, and Leahey (1944) found similar results and concluded that the relationship was closer than between yields and percentage of nitrogen in the soil. Robinson (1936), an English soil scientist, argued that fertility had meaning only in terms of particular crops, and therefore the only generalized definition could be the capacity to grow satisfactorily the largest number of the crops suitable to a particular climate. He advocated a new attack on the problems of plant nutrition from the standpoint of plant physiology in relation to the properties of the soil.

The maintenance of the original fertility or the destruction of the original fertility of the soil were themes of popular agitation among conservationists and historians in dealing with agriculture and often were based, obviously, on misconceptions. To maintain original soil fertility meant the development of a type of agriculture which would combine in domesticated and artificial forms all the factors, biological and physical, which originally produced the soil in the state of nature. Among other things, the forest soils must grow an exact equivalent and a grassland soil an exact equivalent of the original vegetation. Of course, such an idea of exact equivalents was impossible to achieve. Also tillage methods would have to be adapted to duplicate exactly the natural tillage effected by wild animals, especially those burrowing in the soil. Even if these things were possible, they would be undesirable because some soils in the natural state were unfit for the successful growth of the crops which man required and from the beginning it was necessary to add factors which would produce new equilibriums for crop production. No one would want to restore their original fertility for that would render them relatively useless.

In the light of experience of man with soils long in cultivation, one is tempted to inquire whether the assumption of the soil scientist was sound or only exaggerated that assigned to surface vegetation so important a rôle as was implied in the use of the terms forest and grassland soils. If vegetation was so largely a determinant in soil formation, should not the removal destroy those assigned characteristics? To what extent did removal of forests or grass destroy or change the microorganisms within the soil peculiar to each of the Great Soil Groups? Was it possible that the biochemical factor associated with microbiology was more of a determinant than the surface vegetation?

Jenny (1941), and Robinson (1936) both recognized explicitly that soils changed with the removal of the original natural vegetation. Robinson's experience was primarily with English soils that had been cultivated for centuries and continued fertile and productive. Jenny emphasized that such a change in vegetation was a modification of one of the major factors in the soil formation equation; so-called soil was reduced to the rank of parent materials, and a new stage in the process of soil formation was initiated. Both writers stated the problem and the principles, but neither had the data to chart the course of the soil changes which had taken place in soils long in cultivation or which would take place in new soils just being brought into domestication. Both phases of the problem were vital for the guidance of the historian.

So long as soil science was the work of a relatively few workers, and those dominated by the USDA, there was little opportunity to deviate from the course marked out as official. The extent to which the situation was changing is emphasized by Kelley's (1946) restatement of "Modern concepts of soil science." From the five concepts discussed, three aspects are indicated here; the emphasis on parent materials, on the ready response of soil to chemical, physical, and biological forces, and on the revised estimate of the clay factor in its bearing on plant nutrition.

Chapter Seven

THE GRASS FORMATIONS; THE ECOLOGISTS' VIEWS

Introduction

No comprehensive study of the whole of the North American grassland has yet been written, but a number of studies of different phases and segments are now available. Some of these are of high quality, but none have yet given a complete answer to the question why the region grows grass rather than trees or why different parts of the region grow different grasses. The following is a summary of those that seem most significant, although it is manifestly impossible to draw a hard and fast line.

The studies of the grassland might be divided into five groups on the basis of point of view and emphasis. The theoretical ecologists, of which Clements and his associates were the most conspicuous examples, wrote of vegetation climax in terms of what they thought was the ancestral condition prior to the invasion by the white man — the original or natural state. The practical ecologist dealt with vegetation as he found it. The range specialist centered his interest on the volume and value of the forage yield to livestock and on the ways and means of maintaining the grass yield, or restoring, or increasing it. The crop indicator studies, such as that of Shantz (1911) which was the pioneer ecological study of the plains grassland, had as their objective the determination of what grasslands should not or could not be plowed up for crops, and if put to agricultural uses, what crops would be best adapted to successful production. The systematists collected specimens, described and classified them with relation to the plant world and to their geographical distribution. They were interested in where the species had been found growing — its range — not its ecological status as a dominant in plant communities, or its relation to succession and climax, or its economic value. Their maps of species distribution differed substantially from the maps of the ecologists which showed where the same species appeared as a dominant. It is from studies representing these diverse points of view that the story of the grassland must be reconstructed for the purposes of historical study. It should be noted that none of these studies gives satisfactory treatment of the animal life, or of soils in their relations to the grassland problem, and soil microbiology is virtually ignored.

Biome or province interpretations

Except Clements and Shelford and associates, most authors who dealt with the grassland as a whole recognized three divisions from east to west between the eastern forest and the Rocky Mountains; the tall grass or the true prairie, the mixed grass prairie-plains, and the short grass plains. The Clements-Shelford (1939) treatment extended the area under consideration to the Pacific coast, dealing with the North American grassland on a continental basis and recognizing six divisions. Also they dealt with it as a biome, the vegetation and animal life taken together, but not including soil microbiology. The two western units were the Palouse prairie of the Pacific Northwest, or Inland Empire, and the California prairie. The other four divisions occupied the area eastward from these and the true deserts as far as the forest regions of the Mississippi valley. The north-south line dividing this interior area of grassland was located near the 100 meridian; and the east-west dividing line was near the 33 parallel or an irregular line from the Red river to the south boundary of California. Northeast of these intersecting lines was the true prairie; southeast, the coastal prairie; northwest, the mixed grass prairie; and southwest, the desert plains. Taken together these designations applied to the area extending westward to within about one hundred miles of the Colorado river in Arizona, the boundary line running northwestward through Utah to the continental divide in Wyoming and thence northward along the mountains to central Alberta. The true desert was quite small, the extreme western part of Arizona and southeastern California and southwestern Utah and southern Nevada as far north as the Palouse prairie in the northern part of those states. If their interpretation was correct, then there was no discontinuity of any of the six divisions, except the California prairie which was cut off at present by the desert and mountains. The Palouse prairie would be connected fully with the interior grassland as a unity.

In treating the North American grassland as a biome Clements and Shelford were approaching the problem from the theoretical point of view of climax formation on the ancestral prairie, prior to disturbance by white men. They concluded that the present grassland, as the scientist found it in the twentieth century, was not the original grass climax. As a biome, they characterized it as a Stipa-Antilocapra biotic formation (needle grass-pronghorn antelope). The binding dominant grasses, meaning those that appeared as dominant in three or more of the six prairies, were designated as follows, in order of decreasing wideness of range:

Sporobolus cryptandrus	— (Sand Drop-seed)
Koeleria cristata	— (Prairie June grass)
Stipa comata	— (Needle-and-thread grass)
Stipa viridula	— (Green needle grass)
Agropyron smithi	— (Bluestem or western wheat grass)
Bouteloua gracilis	— (Blue grama)
Bouteloua curtipendula	— (Side oats grama)
Bouteloua hirsuta	— (Hairy grama)
Elymus sitanion hystrix or	— (Bottlebrush squirrel tail)
Elymus sitanion jubatum	— (Big squirrel tail)
Poa scabrella	— (Pine blue grass)
Festuca ovina	— (Sheep fescue)
Andropogon scoparius	— (Little bluestem)
Buchloe [Bulbilis] dactyloides	— (Buffalo grass)

The second ecological study to apply the biome concept to the North American grassland was that of Carpenter (1940), The Grassland Biome, whose objective was to describe it as it "existed in a natural equilibrium." He differed from Clements and Shelford (1939) by delimiting the area included to the interior grassland between the Rocky Mountains and the eastern forests, excluding also New Mexico, east of the continental divide and southern Texas, in fact most of Texas was ruled out of the grassland vegetational classification. The Pacific areas were excluded provisionally awaiting further investigations of animal relations but the Texas-New Mexico-Arizona area was definitely assigned to the desert complex. Carpenter was in agreement with Clements and Shelford, however, in giving little recognition to soil population and the problem of microbiology. Carpenter emphasized climate and geological history as factors determining grass regions. The biome as a whole was characterized by its plants and animals, and was divided from east to west into the traditional three divisions, but with his own ecological descriptive terms:

Andropogon-Bouteloua-Bison-Canis (Grassland) Biome.

1. **Andropogon-Bison-Canis** (Tall grass prairie) Association.

2. **Andropogon-Bouteloua-Bison-Antilocapra** (Mixed grass prairie-plains) Association.

3. Bouteloua-Bulbilis-Bison-Antilocapra (Short grass plains) Association.

In giving descriptions of the biome and the constituent associations, Carpenter introduced each with a table of terms of other authors the nearest equivalent to his own. This emphasized the wide divergence among ecologists both on what constituted a dominant plant or animal and on the proper geographical boundaries of the unit under consideration. While Carpenter emphasized the bluestems and gramas as the characteristic grasses of the whole region, Clements (1920) insisted on needle grass and grama, and Hendrickson (1930) insisted on needle grass and June grass. Similar disagreement appeared relative to animals, Carpenter selecting buffalo and wolf; Bird (1930), antelope and buffalo; and Clements and Shelford (1939), antelope. In describing the social characteristics of the 26 leading grassland mammals only five were described by Carpenter as living in herds or packs (buffalo, elk, prairie dogs, Richardson's ground squirrel, and wolf), the others being either solitary or living in family groups.

In parts of his discussion of the tall grass prairie Carpenter elaborated constructively upon two fundamental concepts which had been first clearly presented by Schaffner (1926). The problem of the western boundary of the tall grass prairie was determined by most plant ecologists solely on the criterion of grasses. This had led to disagreement, especially the insistence by Clements and associates that the mixed-plains in this sense was non-existent, but was mistaken as such because it was a disclimax true prairie. Schaffner had used animal ecology to coordinate with plant ecology in determining the western limit of the true prairie in central latitudes as the zone where it met the eastern limit of the range of the prairie dog and the harvester ant (Pogonomyrmex occidentalis). The second concept is related to the first and was the double relationship of the dominants to environment; first to topography and second to east-west variation associated with diminishing rainfall and rising altitude. In the eastern prairie, at the three topographical levels, high, sloping, and low prairie, the grasses were: on high prairie, big bluestem (Andropogon furcatus) with little bluestem (Andropogon scoparius) on the dry and sandy areas; on sloping prairie, big bluestem; and on low moist or wet grounds, big bluestem, switch grass (Panicum virgatum), and cord grass (Spartina spp.). In the mixed prairie-plains, some of the grasses of the wettest low grounds of the tall grass prairie had dropped out; some of the grasses of the slope level dropped one level to the low grounds of the mixed prairie-plains along with survivors of the eastern levels; the grasses of the high levels dropped in part to the slope level; and the top level of the mixed prairie-plains acquired a new group of grasses, those most conspicuous in the short grass plains farther west; buffalo grass, grama grasses (Bouteloua spp.), and on sandy or other locations little bluestem persisted. In the short grass plains the buffalo grass was found in the lowest levels along with some bluestems; at the slope levels it appeared along with the gramas (Bouteloua spp.); and at the highest levels, the gramas (Bouteloua spp.) were most conspicuous.

The longer one studies the problem of vegetational boundaries, and particularly this tall-grass mixed-grass boundary problem, the greater becomes the fundamental significance of Schaffner's distinctions. No line on a map can be drawn to represent in any realistic manner the actual conditions found in nature. Isolated hilltops, slopes, and low lands exhibit the transitions on both sides of such a theoretical marginal zone, and only in the broadest sense could insistence be made that, even in the zone itself, a continuous zigzag line could be drawn to outline the dovetailing or interlacing of ridges, and valleys, and slopes growing the respective species of grasses adapted to them.

Carpenter worked out more completely than former studies the subdivisions (faciations) of the tall grass prairie; the central and most typical prairie; the northern and southern faciations, and the eastern (postclimax) faciation. Within each faciation he listed the dominants of the three topographical levels, high, sloping, and low prairies. Carpenter's treatment of the eastern segment of the prairie calls for emphasis; from the deciduous forest biome point of view it would be a proclimax, and from the grassland biome point of view it would be a postclimax. This peculiar double position of the area is explained by the emphasis on long-term geological climatic fluctuations; the theory that grassland once reached much farther eastward, but is now retreating before a slowly advancing forest. Studies that laid the foundations for this point of view included Adams (1902), Sampson (1921), Gleason (1922), Woodward (1924), Sears (1926, 1932, 1935), and Fuller (1935).

Another aspect of Carpenter's study that can be given only a passing reference here is analysis of seasonal sub-dominants for this tall grass prairie. His was the only general study of the tall grass association in which this is done, and Carpenter did not carry it out for the other associations of the North American grassland because of insufficiency of data available. Other aspects of his study were limited similarly for these latter associations, particularly the animal relations.

The mixed grass paririe-plains association received less intensive study than the tall grass prairie, and the disagreements relative to its right to recognition as an independent association confused the literature. Schaffner (1926) and Aikman (1935) took the ground that it was a stabilized association and likewise that the western grass association was a short grass association, disagreeing with Clements, Weaver, and Shelford. Carpenter's treatment of the mixed grass prairie-plains followed rather closely the Aikman analysis. As already pointed out the eastern boundary in the central latitude was determined or confirmed by Schaffner's (1926) coordination of plant and animal evidence at the eastern range limits of the prairie dog and harvester ant. This meant that in the central latitudes it fell near or east of the 98 meridian, in southern latitudes nearer the 99 meridian, and in North Dakota it followed the ridge west of the Dakota river curving northwest in a crescent across central Alberta. The western boundary might be designated roughly as the 100 meridian, but that simplifies the description too much. In southern latitudes it followed an irregular line as far as the 101 meridian, with a southern limit about the 33 parallel. In the north central latitudes the line dipped far westward up the Republican river valley as far as the 102 meridian, and again dipped westward north of the Platte river around the Nebraska sand hills as far west as the 103 meridian; but in the Dakotas the line was the Missouri river curving northwest into Montana where it departed from the river into southwestern Saskatchewan and southeastern Alberta.

The list of dominant grasses assigned by Carpenter were blue grama, hairy grama, little bluestem, and buffalo grass. In the northern faciation additions to this base list include prairie June grass, porcupine grass, needle-and-thread grass, and dropped from the list was buffalo grass. In the southern end of the mixed-grass association the buffalo grass became increasingly important.

The short grass plains association adjoined the mixed grass on the east and its western limits was the western boundary of the grassland biome as already described. The dominant grasses were blue grama and buffalo grass. In the extreme north the buffalo grass dropped out while needle-and-thread grass was added, and in South Dakota bluestem wheat grass was added.

Dice's (1943) concept of biotic provinces was the culmination of the development of a series of concepts beginning explicitly with Ruthven's (1908) use of the term "biotic region." It was defined as "a considerable and continuous geographic area ... characterized ... by one or more ecological associations ... also by peculiarities of vegetational type, ecological climax, flora, fauna, climate, physiography, and soil." This was substantially different from Clements's (1939) biome, particularly in respect to its emphasis on continuity of area and climax, limitations which greatly reduced the size of "biotic provinces" as compared with the grassland biome either Clements or Carpenter (1914). The differences extended also to the important factors emphasized by Dice of physiography and soil. The central grassland of the United States and Canada was divided into five biotic provinces, Illinoian and Texan compared in many respects with the Clements true prairie and coastal prairie respectively; the Illinoian province extending west of the 100 meridian and the Texan extending

west to an irregular line lying mostly between the 98 and
99 meridians. These western boundaries recognized precipitation efficiency in response to northern and southern great plains temperatures. The great plains area was
divided into three provinces from north to south; the
Saskatchewan to southern Wyoming, the Kansan to eastern
New Mexico and northwestern Texas, and the Comanchean including southern Texas. To the westward the desert or
desert grassland was designated as Navahonian, including
most of New Mexico and Arizona; Mohavian as applying to
the Arizona-California desert; and the Artemisian centering on the Utah-Nevada desert. The Palousian province
included a restricted bunch grass area, and the Montanian
and Coloradian names applied to the northern and central
Rocky Mountain regions.

Tall grass prairie studies

The portion of the grassland most fully studied
ecologically was the tall grass prairie. Although the
several authors place different evaluations upon the work
done; there was general agreement on some of the items.
Transeau (1935) listed the "major contributions to the
study of prairie vegetation" as Shimek (1911, 1925),
Weaver (1934), Sampson (1921), Schaffner (1926), and
Gleason (1922). Omitted from the list was a short paper
by Cowles (1928), and to it must be added Transeau's own
synthesis "The Prairie Peninsula" (1935) and additional
titles of significance: Fuller (1935), a group by Weaver
and associates, particularly Fitzpatrick (1934), and Albertson (1937); Clements and Shelford (1939); and Carpenter (1940). Part of these have been reviewed already.
Fuller's (1935) paper was a restatement and synthesis of the problem of the post-gracial geological
period and its bearing on the relations of forest and
prairie as developed from pollen counts of peat bog deposits. He challenged the views of the extremists of the
climate cycle school concluding that there were three
climate phases; a period of increasing warmth accompanying a glacial recession during which coniferous forests
of the northern type were established in the southern
Great Lakes region; a long period of maximum warmth with
deciduous forests of oak, hickory, maple, and elm dominants; lastly a poorly defined period of decreasing
warmth, with only slight changes in forest limits. According to this, the conifer period was less than one
quarter of post-glacial time, with fully three-fourths
of that period occupied by deciduous forests, undergoing
only minor changes in character during the more recent
period. The presentation of animal evidence in favor of
the other view was made by Schmidt (1938), "Herpetological evidence for the postglacial eastward extension of the

steppe in North America." Some of his distribution maps correlated closely with the prairie peninsula boundaries, some extended farther east and others seem to have no relation. In conclusion he suggested the importance of the study of insect evidence for what it might contribute.

Cowles's (1928) paper had a special significance. It was short (only three pages) but was a compact and masterly restatement, by a nature scholar of outstanding distinction, of the problem of prairie persistence and for the first time introduced explicitly into the prairie discussion the issue of soil formation in the Glinka-Marbut sense. The earliest recognition of the relation of the new soil concept to the grassland riddle was Shantz's (1923) application of it to the great plains. The disappointing aspect of this suggestion is in how little advance has been made on this basis.

Transeau (1935) gave the most complete restatement of the grass-tree problem, the origin and the persistence of the "Prairie Peninsula," although his professed purpose was only "to locate more accurately the eastern extension of the prairie at the time of settlement." The study was organized around two summaries; a statement of twenty-one problems which must be taken into account in order to explain the prairie; and a statement of sixteen factors (error in number in printing which indicates 15) that contribute to an explanation. In connection with these problems and factors Transeau made two important and unique contributions in his rainfall distribution patterns and his series of tree distribution maps. Many systems have been devised to illustrate rainfall pattern, but for the particular purpose at issue, these are effective in showing differences between the grassland area and the forest area. In the former the moisture was mostly in the growing season, while in the forest area it was distributed generally through the year. The tree distribution maps, seventeen of them, one for each species, presented the most accurate data available on that subject. The soil problem was recognized fully. The prehistoric influences were assigned at least as important a rôle as present day factors and he accepted the view of a postglacial dry period when the prairie extended farther east than at present, although warning that pollen procedures had been challenged and awaited further investigation. His conclusions had been based, however, on soil profile studies as well as pollen analysis of peat bogs. On most of the issues involved in the prairie discussions, characteristically, Transeau emphasized the limitations on available knowledge. Two types of evidence were omitted from his long lists, animals and microorganisms of the soil, both of which emphasized that he was dealing with the subject strictly as a plant ecologist.

The most intensive studies of the tall grass prairies, from the standpoint of existing structure, were made by Weaver, Albertson, Fitzpatrick, and associates. Weaver (1931) declared that "the bluestems are the aristocrats of the prairie," and with Fitzpatrick (1932) selected the ten dominants of the tall grass prairie of eastern Nebraska, western Iowa, northeastern Kansas, and northwestern Missouri — "the Prairie Center." On the low prairie they were big bluestem, Indian grass, slough grass, switch grass, nodding wild rye; on high prairie, little bluestem, porcupine grass (Stipa spartea), prairie drop-seed; and on both moist and dry locations, side oats grama.

Soils and grass associations

Shantz (1923) and Marbut (1923) proposed a definition of the great plains in terms of soils. Marbut's eastern boundary was the line dividing the pedocals from the pedalfers, the great plains including the soils west of that line which possessed, if mature, a lime accumulation zone and a dark color. At the west the boundary was the line dividing the brown soils from the lighter colored soils, a line which fell in the Rocky Mountain region from New Mexico northward. In the South Pass region of Wyoming, it was near Casper and Sussex, the country west of that being excluded from the great plains. To the southward of Las Vegas the line ran west across the Pecos river and then looped back again above Roswell and turned southeast to a point on the Rio Grande river west of Sanderson, Texas. Shantz correlated the natural vegetation to this great plains soil area, although he pointed out that the grass associations both in the east and west extended beyond these soil boundaries. Within the great plains as defined by Marbut, Shantz located the western boundary of the blackearth belt, or chernozem soils, as the western limit of the tall grass association with the short grass plains lying in the area of the dark brown and brown soils. Roughly the tall grass portion of the great plains as defined here lay between the 98 and 101 meridians, except in the eastern Dakotas, and western Minnesota where it was farther east. Shantz was unable to explain the presence of the tall grass on the chernozem soils because the most typical tall grass area was the prairie soils in the pedalfer region to the east. In 1924 and 1938 he still adhered essentially to the position presented in 1923.

Mixed grass prairie-plains studies

Between the tall grass and the short grass lay the much-disputed transitional mixed grass prairie-plains. Schaffner (1926) used a three fold standard for determining the mixed grass association boundaries. He made an important departure from the Shantz interpretation, although accepting the basic idea that vegetation and soil correlated approximately in the central latitudes. The grass association, approximately 100 miles wide, occupying most of the chernozem soils, however, he designated as mixed grass prairie-plains rather than tall grass prairie, a grass association that possessed an individual identity although transitional. The dominants were Andropogon furcatas and Andropogon scoparius "together with an abundance of deep rooted plants of the bean family and a great array of composites on ordinary levels." The short dominants entering from the west were the two gramas (Bouteloua gracilis and Bouteloua hirsuta) and buffalo grass (Buchloe dactyloides). In this connection it should be emphasized that Schaffner used also the concept of an irregular dovetailing of grass associations of high and low topographical levels, which eliminated the idea of an absolute line. In the middle latitudes, therefore, he indicated that Marbut's pedocal-pedalfer division line and the eastward limits of the range of the prairie dog and harvester ant were not far apart, mostly east of the 98 meridian. The third standard was Transeau's precipitation-evaporation ratio of 60 per cent. The western boundary of this mixed grass association he placed approximately at the 100 meridian instead of near the 101 as indicated by Shantz. At the north, the situation was different, but Schaffner's study did not cover that area. Identified on the map by the nearest important towns, the eastern boundary of the mixed grass prairie plains at the north lay between Mitchell, and Sioux Falls, South Dakota, passed southward through Yankton, turning southwest past Lincoln, Nebraska, but about twenty miles west of that city, crossing Kansas just east of Mankato, Concordia, Minneapolis, McPherson, and Oklahoma just west of Enid, and El Reno. The western line of the mixed grass association was located by Schaffner east of the Marbut soil boundary by approximately one degree and in Kansas near the towns of Oberlin, Moorland, Ness City, and Dodge City instead of Marbut's (1923) line of North Platte, Nebraska, crossing Kansas through Atwood, Hoxie, Gove, Dighton, Cimarron, Liberal, and southward through the Texas Panhandle by way of Amarillo. The frequency and extent of the revision of soil maps (1935, 1938, 1941) suggests, however, that attempts at correlations must necessarily be doubtful until there is more agreement on the soil classifications themselves. Weaver and

Himmel (1931) insisted that, except for certain modifications in water relations, the type of soil has "a very minor effect upon the groupings of grassland dominants."

Aikman (1935), who was followed closely by Carpenter (1940), adopted a modified version of the mixed grass transition zone and narrowed its width making it less conspicuous in comparison with Schaffner's interpretation. Its eastern boundary lay mostly between the 98 and 99 meridians, except at the extreme north and south where it curved westward to or across the 100 meridian. The western boundary lay mostly between the 100 and 101 meridians, swung northwest at the north along the Missouri, thence to the Canadian line at the 105, while at the south end it angled southwest to about the 102 at the Rio Grande river. Aikman approached the problem of dividing the grassland into three types of grass associations from the standpoint of precipitation efficiency:

This extensive plant community may be divided throughout its length from north to south into two almost equal-sized units, corresponding to its naturally dominant types — the prairie on the east, and the short grass plains on the west. The two important factors which seem to account for this division are simply the amount of precipitation (rain and snowfall), and the proportion of this moisture that is left available to vegetation as against losses through evaporation, run-off, and colloidal absorption or binding of moisture by the soil. The tall or prairie grasses require a combination of these two factors (such as occurs to the eastward) which makes for a relatively high available moisture content in the soil for a relatively long time through spring and summer; whereas the short grasses are adapted to the more difficult moisture situation that exists to the west.

There is, of course, no definite dividing line between the two plant communities, but rather a north-south zone under which precipitation, decreasing generally from east to west, occurs in just the requisite marginal amounts to insure the growth of both tall and short grasses in association or mixture; this association of plants may occur as a commingling of individuals or of limited tracts or patches, as determined by local variations of soil.

In describing the moisture requirements of the western edge of the tall grass prairie, Aikman estimated that in the central latitudes a precipitation of 26 to 30 inches, allowing six inches for excess evaporation over northern North Dakota, was approximately the equivalent of 18 inches at the north. These grasses used all the available soil moisture by mid-summer on the average, but it was adequate to insure a permanent tall grass cover of sufficient density to prevent the invasion of short grasses. In central South Dakota and northeastern Nebraska, however, the heavy soils were a handicap that permitted some invasion of shorter grasses from the west. In respect to the mixed prairie, Aikman said that

The eastern boundary of the mixed prairie is a line west of which prairie grasses, which assume a bunch grass habit because of a shortage of available soil moisture, no longer entirely dominate the area but share dominance with short grasses which become permanently established.

In a related paper Zon (1935) marked on the map "a line of equal effectiveness of precipitation" running from the Canadian boundary to north central Texas, 15 inches at the 49 parallel being equivalent to 22 inches at about the 33 parallel. This highly irregular line straddled the 101 meridian most of the distance and lay almost wholly between the 100 and 102 meridians. This was somewhat west of Aikman's western boundary of the mixed grass zone:

The western boundary of the mixed prairie may be described as a line east of which sufficient rain falls and penetrates the soil to wet it periodically to a depth varying from 24 to 30 inches. Plants whose roots penetrate the soil to a depth of 24 inches or more obtain moisture for sustained growth for a period of at least $2\frac{1}{2}$ months each year, even in drought years....

On typical soil west of the area thus defined, a periodic shortage of water even at a depth of 18 inches precludes the maintenance of a definite permanent tall-grass population within the shortgrass cover.

Short grass plains

Shantz (1911) was the first to make a major ecological study of the short grass association, the middle latitudes being the center of his activities. In 1923 (and later versions 1924, 1938) he fixed the eastern boundary at approximately the 100 meridian instead of the 101 or west as indicated by Marbut's (1923) soil map as the dividing line between the black and dark brown. As a guide he suggested that when the lime accumulation zone was less than two feet from the surface only short grasses could grow, and they grew best when it was one or two feet below the surface. The shallow lime accumulation zone meant that between it and the water table below was a dry soil zone. When it was more than thirty inches or in the more humid region where there was no such lime zone, the tall grasses dominated. The western boundary of the short grass plains was indicated for the most part as the Rocky Mountains.

Aikman (1935) based his analysis of the short grass dominance on root and soil moisture relationships:

The adaptation of the grasses to the heavy soils and the light precipitation of this region (and to similar soils in the mixed prairie region) is quite definite. In the heavy, fine-textured soils, percolation of moisture is slow and limited in depth, and there is a high percentage of run-off, especially of the heavier rains, so that for months or years the subsoil may remain dry. The short grasses make the fullest use of the topsoil moisture. Their root systems, fibrous and shallow, almost completely cover the area, taking up the small amount of available water in a limited time, and their season's growth and reproductive processes are completed in a few weeks.

The most important dominant of the entire shortgrass plains community (association) (fig. 76) is blue grama, Bouteloua gracilis (H. B. K.) Lag. In the central and eastern part of the Plains, from southern South Dakota southward, buffalo grass, Buchloe dactyloides (Nutt.) Engelman, is associated with blue grama. In well-developed short-grass plains, the cover is short and has the smooth appearance of a closely grazed pasture. Taller grasses, as three-awn, Aristida spp., and herbs are more abundant on the eastern border of the Plains, but even here the role of the tall-grass constituents, because of the shallowness of moisture penetration, is not important and permanent enough for the community to be considered mixed prairie. Because of the complete occupation of the top few inches of soil by the fine, fibrous roots of the blue grama and buffalo grass, these grasses are shown to be the true dominants of the community.

Costello (1944) described the extent of the short-grass plains in Colorado as from the foothills eastward, and in Wyoming he designated as short grass approximately the eastern third of the state, distinguishing it from the semi-desert to the westward through the South Pass region. These western limits agreed approximately with Marbut's western boundary of the great plains determined by the zone dividing the brown from gray desert soils. Shantz (1923) identified the short grasses particularly with the brown soils. Not all animals observed this approximate boundary, but several typical plains species or sub-species did; the buffalo, the plains prairie dog, the white tailed deer, the plains jackrabbit. Grinnell (1923) made the 100 meridian the line beyond which the California burrowing rodents began to disappear or diminish; a range division that would tend to confirm the distinction between the short grass and mixed grass, and would indicate that the mixed grass zone was also a mixed animal zone as well.

The gramas were the principal dominants of the plains, and the blue grama (Bouteloua gracilis) was the aristocrat of the short grasses. Contrary to popular opinion, the buffalo grass (Buchloe dactyloides) was less hardy both as to moisture and temperature range than the blue grama. There was only one species of buffalo grass, while there were many species of grama, each with a particular range of adaptability. The buffalo grass required more moisture than blue grama, a fact which limited its growth to the relatively moist plains and the transition mixed grass region. Its northern range limit was southwestern North Dakota and its southern limit was southcentral Texas with the optimum toward the warmer southern end of its range limits. The tall grasses had a limited representation, except on sand hills as in north central Nebraska, and those of the upper valley of the plains streams such as the South Platte, Arkansas, and Cimarron, where the bluestems (Andropogon scoparius and others) were important. In the Colorado-Wyoming area, blue grama provided 50 per cent to 90 per cent of the forage on well managed ranges (Costello 1944). Other grasses, of the mid-grass class, might be present; red three-awn grass (Aristida longiseta), bluestem wheat grass (Agropyron smithi), needle-and-thread grass (Stipa comata), and sand drop seed (Sporobolus cryptandrus). Always present were pricklypear cacti (Opuntia spp.), more numerous after a series of dry years rather than as a result of overgrazing (Turner and Costello, 1942). Big sagebrush (Artemisia tridentata) and other species appear in parts of the short grass plains of Colorado and Wyoming and particularly in the semi-desert.

Northern and southern grasses

In the northern part of the grassland, in Canada and the northern United States the grass species differed substantially from those of the mid latitudes and followed individual distribution patterns. The boundaries between grass associations did not seem to be determined in the same manner as factors of climate and soil, or the operation of those influences were not adequately understood as they control northern vegetation. In the northern vegetation the grasses were most important, with composites second and legumes third. The differences among authors in characterizing the dominants was confusing. Shantz (1938) described the valley of the Red river of the north occupying a narrow strip in western Minnesota and eastern North and South Dakota as dominated by porcupine, June, and slender wheat grasses. Another narrow strip to the west with a crescent-shaped western front pointing to the northwest and southwest was needle-and-thread, June, and bluestem wheat grass. The Missouri river country of western North Dakota was blue grama, June, needle-and-thread grasses. Western South Dakota was blue grama, buffalo, and western wheat grass. Reitz and Morris (1939) listed the most important group in the plains division of Montana as bluestem wheat grass, blue grama, needleleaf, and other sedges, and plains blue grasses (Poa spp.).

Aikman (1940) treated the northern grasses in relation to true prairie, mixed prairie-plains, and short grass plains. In the tall grass portion of North Dakota and northern half of South Dakota, the chief dominant was designated as slender wheat grass with bluestem wheat grass invading from the southwest and porcupine and needle-and-thread grasses invading from the west. Grasses with occasional dominance were little bluestem, side oats grama, June grass, prairie drop-seed, and big bluestem. The mixed prairie-plains would have, therefore, the tall grasses in definite bunch form together with the mid-grass species designated above as western invaders and the short grasses, the gramas, especially the blue grama.

The southern end of the grassland presented also its problem of variations from the pattern of the mid-latitudes. Shantz (1938) designated a crescent-shaped area through central Texas from the Red river to the coast (approximately the western part of Clements's and Shelford's coastal prairie (1939) as "mesquite-desert grass savannah." The extreme southwest part of Texas, south of the range of blue grama and buffalo grass, as major dominants, and southern New Mexico was designated a mesquite (Hilaria spp.) grass desert merging farther west into the creosote desert. The northern interior desert was designated sagebrush as far east as the South Pass junction with the short grass plains. No area was designated as a complete sand desert. Chapline and Cooperrider (1941) designated as semi-desert-grass plains all of southern Texas south of the range of blue grama and buffalo grass as major dominants, southern New Mexico and Arizona, and adjacent Mexico except mountain areas. They described it as resembling short grass plains, but the principal short grasses were gramas and curly mesquite (Hilaria belangeri); in the depressions tobosa grass (Hilaria mutica) in bunch grass form; and in damp places, sacaton grass (Sporobolus wrighti) and some other tall grasses. Mixed with the grasses and grassland areas were desert vegetation, shrubs, mesquite trees, yucca and cacti, and as on deserts, during the rainy season there was a burst of ephemeral plants.

It is not the purpose of this study to deal extensively with the desert as such, but the problem of the relation of the desert to the grassland at one end of the range of environmental extremes is just as essential to the grassland problems as the transition to forest at the other end. At the outset it must be recognized that the problem of the desert-grassland boundary has not received as full an investigation by ecologists as that of the forest-grassland boundary, and therefore more definite conclusions must await fuller analysis and more general agreement. Sumner (1925) challenged the assumption, widely held, that the intensity of competition for existence was more severe in the desert than elsewhere. The problem of survival was different, rather than more severe, the desert placing the critical emphasis on environmental factors not so much stressed by other natural regions. Clements (1934) set a maximum of a five-inch rainfall as the test of a desert. Approaching the problem from a completely different set of concepts, Shantz (1923), followed Marbut's (1923) new soil classifications, limited the great plains grassland to the brown earth area, designating the gray soils as desert. More broadly, Shreve (1934) held that it is not always either possible or important to distinguish arid from semi-arid land, and that the term desert was relative, an adequate definition being a composite one.

The application of the Clements (Weaver and Clements, 1938; Clements and Shelford, 1939) standard reduced the desert within the United States to a small area, the lower Colorado river basin and southern Nevada, designated as a desert shrub climax. The two dominants of this formation were Larrea Mexicana (creosote bush) and Franseria Dumosa, but these shrubs extended eastward as far as western Texas although not as dominants. The conspicuous appearance of these shrubs in this eastward extension were attributed to disclimax resulting from overgrazing in a region where the true climax formation was short grass plains. The sage brush climax was limited by Clements to a small portion of the northern Great Basin area, but as a result of overgrazing, as a disclimax formation, sage brush extended more widely over grassland climax areas. Carpenter (1940) omitted all of central and southern Texas from the "Grassland Biome" defining it as grassland-desert ecotone and applying the Shantz (1938) label of "mesquite-desert grass savannah," and the extreme southwestern part of Texas and New Mexico as "mesquite grass desert." Carpenter pointed out also that on the basis of animal population these desert grass areas were related to the short grass areas. He omitted also the sage brush area from his Grassland Biome, pending more adequate data.

Following closely the Clements precedent, Whitfield and Anderson (1938) and Whitfield and Benter (1938) presented in detail aspects of desert vegetation. In the former paper, with a map, two types of desert-plains grassland formations were described and located. Southwestern Texas, the extreme southern portion of New Mexico, and southwestern Arizona were described as a Bouteloua-Hilaria faciation. A narrow band of country farther north was designated as a Hilaria-Bouteloua faciation. Northwestern New Mexico and northeastern Arizona were assigned to a woodland and montane association.

The leading authority on the North American desert is Forrest Shreve, long in charge of desert research under the sponsorship of the Carnege Institution. Unlike so many other specialists, he was not committed to extreme views, insisting always on the incompleteness of the data and the tentative character of most conclusions. The inadequacy of total annual rainfall as an index of desert conditions was emphasized by his study (1934) of rainfall, runoff, and soil moisture showing that there was little relation between total rainfall and soil moisture available to plants. Rains of less than 0.15 inches had no effect, and as most rains fell in torrents, causing high relative run-off, the additions to soil moisture were relatively small. The rainfall type was biseasonal, and drouth periods were frequently long. Drouth separated by one or more small rains was the same in effect as one continuous drouth. Surface rooted vegetation was closely related to rainfall variations, but deep-rooted plants (shrubs) were affected little. Another Shreve study (1934) reported that among the several continental deserts of the world their vegetations possessed little relationship. The desert vegetation of each of the continents was found to be more closely related to those of its own adjoining low-rainfall regions. Animal life had been less systematically studied, but the same conclusion seemed to apply. The understanding of deserts, from such a point of view, must proceed from the study of the processes by which living things were enabled to adapt themselves to diminishing moisture requirements.

The most comprehensive but brief description of the North American deserts was that of Shreve (1942). He explained three different kinds of definitions of the desert; that of the geographer, mostly in terms of physical and climatic characteristics; that of the biologist, "distinguished by the life that prevails there"; and that of the economists, emphasizing the requirements for human occupation such as irrigation. Shreve mapped four North American deserts; the Great Basin, the Mohave, the Sonoran, and the Chihuahuan, but pointed out that like great natural regions there were no sharp boundaries, only gradual transitions. The Great Basin desert included southeastern Oregon, southern Idaho (south of the Snake river), southwestern Wyoming, most of Utah, all of Nevada (except the southern tip), and a small area extending into north-central Arizona. The Mohave desert included the southern tip of Nevada and southeastern California. The Sonoran desert included the extreme southeastern corner of California, somewhat more than the southwestern third of Arizona and most of Mexico and lower California fronting on the Gulf of California. The Chihuahuan desert was located in the northern upland interior of Mexico, extending into the United States to include only a small area of the lower Pecos valley in Texas, and into south-central New Mexico up the Rio Grande valley.

The predominating feature of the North American deserts as a whole was shrubbery, and whatever other vegetation appeared it was mixed into this shrubbery matrix.

The shrubs consisted of two principal types, the woody types of which creosote bush (Larrea) was the most conspicuous, and the semi-shrubs of the sagebrush and salt bush type. The former were characteristic of the southern and the latter of the northern deserts. The northern boundary of the Mohave desert was designated as approximately the northern range limit of the creosote bush.

Shreve designated three types of vegetation structure, simple stands, mixed stands, and rich stands. The simple stands were composed of two or three dominants; in the north sagebrush associated with either rabbitbush (Chrysothamnus), or saltbush (Atriplex), or both; and to the southward creosote bush associated with two or three plants, white bursage (Franseria dumosa), Encelia (Encelia farinosa), or American tarbush (Flourensia cernua). A single species in each case determined the appearance of the community; the coverage varied from 8 per cent to 15 per cent, with the open spaces broader than the spread of the plants. These simple stands occupied about 12 per cent of the North American deserts, mostly in the United States. The mixed stands which were composed of four to twelve species of plants affording more variation in height and spacing, occupied about 60 per cent of the North American deserts, the areas more favorably situated with respect to moisture, soil, and topography than the areas occupied by simple stands. Most of the desert in the United States was of this type, and the desert studies of Shreve (1917) and Shreve-Hinckly (1937) at the desert laboratory near Tucson represented a sample of this desert. The rich stands were composed of more than twelve species and were found mostly in the Sonoran desert, and the Chihuahuan desert at an altitude of 5000 to 6000 feet.

It should be noted that the Shreve mapping of the deserts limited the southwestern deserts of the United States to such an extent as to exclude from that classification most of the higher northeastern and eastern portion of Arizona, nearly all of New Mexico, and all of Texas except the lower Pecos valley. The portions that were not mountainous were desert-grassland transition, but the boundaries differed substantially from the mapping of Weaver and Clements (1929, 1938), and Clements and Shelford (1939) and Whitfield and Anderson (1938). These differences were even more sharply emphasized by Shreve's challenge of the validity of the ecological concepts of succession and climax as applied to desert vegetation. This analysis of the desert and the exclusions emphasize that the southwestern United States, popularly designated as desert, is relatively moist in comparison with a dry desert such as the Sahara. Many varieties of cacti were conspicuous in the more moist portions of this desert and the adjoining grassland and as Maximov (1929) pointed out, cacti did not grow in a dry desert. Shreve's mapping of the Great Basin Desert was significant also, in giving it an extent nearly coincident with that area as a physiographic region, an interpretation at variance completely with the Weaver and Clements and the Clements and Shelford limitations.

If the historian may be permitted to add one more proposal to the many already offered, it is to map the grass associations on an east-west basis to recognize water relations and altitude, and on a north-south basis to recognize the latitudinal differences in light and in temperature. From north to south, east of the Rocky Mountains, there would be three divisions of grasses. At the north, the long day and short season habits would dominate as plant characteristics; and at the south, short day and long season habits, because these are essential to reproduction and perpetuation of the species in a natural environment. At both of these extreme ends of the grassland, the east-west distribution in relation to moisture is somewhat obscured by the requirements of the light and temperature factors. Only in the mid-latitudes, Kansas and southern Nebraska, do all the variations in response to the moisture and altitude factors appear in clearest differentiation. An acceptance of this approach to the problem would resolve much of the conflict that appears in the literature on the mapping of grass associations. Schaffner's (1926) treatment of the mid-latitudes is particularly significant in such a treatment, and he expressly excluded the application of his description from applying to the country north and south of Kansas and Nebraska.

In mapping these vegetation areas it would clarify matters to introduce the concept of nuclear and mixed associations. Under such an approach the grassland east of the Rocky Mountains would include four nuclear grass associations; the bluestem grass (eastern), the needle-wheat grass (northern), the mesquite-grama grass (southern), and the blue grama-buffalo grass (central great plains). Between the nuclear areas, mixed grass associations would combine in varying proportions the species of grasses and other vegetation transitional from one nuclear area to another.

In the Great Basin there are three major divisions of vegetation, excluding forests associated with mountain formations, each a nuclear area; the Palouse prairie at the north, the sagebrush desert in the mid-latitudes, and the creosote bush desert at the south. Of course, this applies primarily to the area within the United States. The mixed plant associations, shrub and grass, connect the grassland proper east of the Rocky Mountains with the desert brush associations of the Great Basin.

The concept of nuclear grass associations would apply only to those areas where the ecological factors combined to provide approximately optimum conditions for the dominant plants, and to a lesser degree, animal species. This would imply a relatively high degree of uniformity. The concept of mixed grass association would be applied to those areas of wide fluctuations in ecological conditions; areas in which optimum or near optimum conditions prevailed for a large enough proportion of seasons to permit the species to achieve the position of dominants, but within limits where unfavorable extremes, such as temperature or moisture or both, would not be so severe during other seasons but what the survival of the species in question would be assured sufficiently for them to recover and maintain codominance. This would imply a relatively high degree of diversity, but a diversity rather definitely restricted to a limited number of ecological patterns of species composition. Among other things such a treatment of northern and southern grass centers would tend to redress the balance relative to plant physiology by giving definite recognition to temperature and light along with water relations. Although the emphasis here is upon the grasses, such an approach would conform fairly closely also with prevailing knowledge of animal distribution.

Chapter Eight

THE GRASSLAND AND EARLY EXPLORATIONS: SOUTHERN

The best way to clarify the problem of the grassland is to assemble the facts that describe its character as seen by early observers. The limits of space permit only a sampling of the evidence and for that reason the selections chosen center upon the middle of the nineteenth century and apply to the central and southern portions where the topography invited travel in an east-west direction.

The 35 parallel or Canadian river route across the southcentral portion of the grassland was described in 1845 by Lt. J. W. Abert traveling from west to east, and by Captain R. B. Marcy, Lt. James H. Simpson, and Captain F. T. Dent, marching westward from Fort Smith, Arkansas, in 1849. Dent's report gave a sketchy description of the country, but without a clear consciousness of grass, as far west as the area between $96°$ and $97°$. Marcy and Simpson wrote independent reports on the same expedition, both based upon their journals. These reports were among the earliest written records of this region and represent the situation prior to the heavy emigration of 1849. As both expeditions followed the river closely for wood and water, high prairie and upland plains conditions were inadequately described. Furthermore, none of the writers knew much about the grasses, their names and characteristics. They were interested in grass primarily as forage, having only a vague idea of the genus groups and were definitely not species conscious. The trees and forbs received more attention, but Marcy was learning about grasses as is shown on his return through Texas. In fact, botanists did not yet possess much scientific data on these grasses. Buffalo grass, for example, was misnamed at that time, the female was not yet identified and the correct scientific description, classification, and naming did not occur until 1859 by George Engleman (cf. also Gray,1859).

In order to interpret the narratives of the explorers or other travelers in the grasslands, it is imperative that attention be directed to the meaning of terms as employed by each writer. For example, Marcy's description of the Llano Estacado table land where he crossed its northern extremity June 14, 1849 (quoted in full later in this chapter) used these terms — "desert prairie", "Zahara of North America", "desolate waste", and "barren plains", — all in one paragraph, but the herbage upon it was "a very short buffalo grass." He mentioned that it was treeless, twice compared it with the ocean with respect to the sense of space it imparted and the absence of landmarks to measure distance and direction, and also twice commented on absence of water. Just what made it a "desert" according to Marcy's standard of values? He compared it with the Sahara, yet it was not a dry, sand desert without vegetation. It is certain that the soil was covered with vegetation, "a very short buffalo grass", but in the same sentence the area was "barren plains." It seems clear that his subjective standard of measure was trees and water in the form of running streams or springs. When most early grassland observers wrote of the "barrenness" of the country, the context seems to indicate a similar set of standards, and it was only occasionally that the writers used language which indicated explicitly that "barrenness" meant literally absolutely no vegetation of any kind. Relative to Nebraska and Dakota of 1855, 1856, and 1857, Lt. G. K. Warren wrote of "the desert character of the country," but contrasted it with "the deserts on the Green River and the Snake River, west of the South Pass", because in the former "good grass will generally be found all over these plains", while in the latter "even a sufficiency for animals cannot be found." It should be noted that he used the word desert in both instances. Even in records dealing with trees in the grassland, it is well to note what Gleason (1922) found in examining original land survey records of the eastern prairie — that land was designated as prairie which had standing trees up to two feet in diameter. This may be a relatively extreme illustration, but nevertheless it emphasizes that in the minds of many forest men trees meant commercial saw timber. According to such a standard, most of the prairie streams to the westward were treeless. All these things must be given consideration in interpreting the records, and particular care must be exercised in interpreting isolated phrases, sentences, or even paragraphs from any individual observer, in fact, the only safe procedure is to study carefully the context of a whole description.

In writing of a prairie fire, October 17, 1845, Abert reported that "in the state of Missouri, and on the frontier where the grass is tall, they are extremely dangerous if the wind be high. The grass we now saw on fire was not more than a foot high." The location was between the forks of the Canadian river just west of $96°$, and it is evident the grass was a mid grass, not a tall grass. Farther west, however, about $97° 30'$, [near the present town of Norman,] the "low bottom of the Canadian river was covered with tall grass", and near the same place the prairie chickens concealed themselves "in the tall grass." In 1849, Captain Dent wrote of the country west of the present town of Ada on the south side of the Canadian, and between Ada and the 97 meridian: "I came out on emphatically 'The Plains' ..." on the ridge between the Wichita and the Canadian.

Across the river south of Norman, and just west of Purcell, the Indian guide warned Marcy (May 5, 1849), that this was the last chance to get hickory timber for wagon repairs. This was near the 97 meridian. Simpson thought, however, that good oak axles for wagons could be found nearly as far west as 98 on the stream, but he warned that the oak in the Cross Timbers was often unsound at the core. On the north side of the Canadian, about 98° 30', Abert (October 1, 1845) reported "the prairie land covered with a luxuriant growth of grass." On this side of the Canadian the Cross Timbers, varities of oak chiefly, extended as far west as 99°, but the south bank was treeless.

While yet in the vicinity of 97° 30' Marcy (May 15, 1849) recorded the wetting of the botanical collections, which he lamented could not be replaced as the expedition was near the "Grand Prairie", "where the character of the vegetation is entirely different from what we have passed." On the south side of the Canadian the Lower Cross Timbers extended as far east as 97° 30' and the Upper Cross Timbers to a point near, but west of 98°. During the day's journey west of 99° Abert (September 26, 1845), reported that they started great numbers of meadow larks "from the tufts of grass which covered" the small, level plain. Such an entry would indicate bunch grass, not a short grass sod. The Marcy expedition was following the divide most of the way through this region, making camps in the ravines near the streams. An emigrant with the Marcy expedition wrote May 27, 1849 from west of 99° (Foreman, 1939) "The grass here is kept short by the immense herds of buffaloes...." Two days later Simpson described the divide country as "a confused mass of small reddish-colored hills, scantily covered with grass...." On May 31, the Marcy expedition had reached Antelope Hills near the 100 meridian (the Texas-Oklahoma line) and Simpson remarked that we "will at once get into the region of buffalo grass, prairie dog villages, ponds, singly disposed and far remote from each other...." Marcy's entry for the following day read that a small lake had been reached and "the buffalo grass grows luxuriantly upon its banks. The grass is very short and thick; but the animals are very fond of it, and it is very nutritious." It should be noted here that he specified buffalo grass only on the banks of the lake. Abert's (1845) descriptions for the same area, eastern Panhandle of Texas, were indefinite, except for emphasis on sandy soil and that grass was found along the streams. Of Shady Creek (101° 40') Simpson wrote: "The cactus, which has from time to time been seen along the whole route, today appeared in all its glory."

Near the 102 meridian the Marcy expedition (1849) met the northern escarpment of the Staked Plains following a route (June 13) between the bluffs of 100 and 200 feet high and the Canadian river. On June 14 the route climbed to the top for 28 miles. This was west of 102°, in the present Oldham county, Texas, bordering New Mexico, of which Marcy wrote

When we were upon the high table-land, a view presented itself as boundless as the ocean. Not a tree, shrub, or any other object, either animate or inanimate, relieved the dreary monotony of the prospect; it was a vast— illimitable expanse of desert prairie— the dreaded 'Llano Estacado' of New Mexico; or, in other words, the great Zahara of North America. It is a region almost as vast and trackless as the ocean— a land where no man, either savage or civilized, permanently abides; it spreads forth into a treeless, desolate waste of uninhabited solitude, which always has been, and must continue, uninhabited forever; even the savages dare not venture to cross it except at two or three places, where they know water can be found. The only herbage upon these barren plains is a very short buffalo grass, and, on account of the scarcity of water, all animals appear to shun it.

On the north side of the Canadian, nearly opposite the Staked Plain, Abert (September 10, 1845) described the uplands as he proceeded eastward:

[we] entered a more desolate country than we had hitherto seen. The high and dry table lands were covered with but a few scattered plants, and were altogether desert-like.

Along this stretch of the river he commented that "the cactus and mesquet [sic][trees] were most abundant," and in the bottoms "the tall grass grew abundantly;" but "From the idea impressed by the barrenness of the country, it appeared incredible that so many Indians could obtain their subsistence from it." West of this, 102° 47', according to his calculations, Abert wrote September 6: "Among the plants which were most abundant, we noticed the artemisia [sage brush] and the 'cucurbita aurantia', which are characteristic of the plains since leaving Bent's Fort" [between La Junta, and Las Animas, Colorado]. About 103°, Abert was obliged to keep some distance north of the canyon rim on account of deep ravines: "The road, too, was rendered extremely rough by a species of dry grass, which grew in tussocks so hard it was like travelling over a road strewed with brickbats." Farther west, to a point opposite Tucumcari mountain, New Mexico, Abert emphasized cacti and mesquite trees on the prairie beyond the canyon rim.

On the south side of the Canadian, toward the Pecos in New Mexico, Marcy reported, June 22, 1849, that "the grass is short, the Mexicans having recently made this a pasture ground for large flocks of sheep.... The country in this vicinity is a miserable sandy plain, and fit for no purpose but for grazing sheep." In the vicinity of Santa Fé he found "the grass ... short, as it is everywhere in the vicinity of Santa Fé."

These day-by-day entries for 1849 were supplemented by general observations. Marcy commented that the Cross Timbers

bordering the great western plains, forms the boundary line between the woodlands and the prairies. East of this area the country was timbered and watered; whereas, in the west, it is an ocean of barren prairie, with but here and there a feeble stream and a few solitary trees.

It would seem as if the Creator had designated this as an immense barrier, beyond which agriculturists should not pass — leaving the great prairies for the savage to roam at will.

Simpson's summaries appeared in both his report and on his set of four maps. "In regard to wood, water and grass," he divided the trail into three segments, and for the eastern division

from Fort Smith to the antelope hills 100° we found these essential requisites generally abundant, always ample.... From Antelope Hills to Shady creek [101° 40'], of grass there was no deficiency; of water occasionally an abundance — always a sufficiency, except in two instances, which now, on account of present experience, may be obviated.... From Shady creek to Santa Fé the grass was rather scarce; it generally, however, proved sufficient, and it is believed that an ample supply may, in most cases, be obtained by a search for it along the ravines and streams aside of, and at not impracticable distances from the road.

Simpson's four maps divided the route differently at 97° 30', at 100° 30', and at 104°. No comment on grass or water seemed necessary on map one, but on the second the note stated "Prairie grass generally and mezquite sparcely as far as camp 41 [Antelope Hills] generally quite sufficient on the prairie[,]abundantly so upon the streams. Buffalo grass from camp 41." An inscription upon the face of the map itself between 100° and 100° 30' read: "Buffalo grass begins to appear." On the third map (100° 30' to 104°) the note read: "The water becoming more scarce, its location, where it appeared, in sufficient quantity, is indicated in writing. The grass, an alternation and sometimes a commixture of Buffalo and Prairie, with an occasional patch of Mesquite; begins to become rather scarce after reaching Shady Creek." On map four (104° to 107°), the note read: "Pasturage an alternation of buffalo and mesquite grass beyond Tucumcari creek rather scanty; a search for better generally rewarded."

The years 1844 and 1845 and 1848 and 1849 were wet years for the central plains region, so far as the generally unsatisfactory rainfall records of military posts are available. This is important to the interpretation of Abert's Journal of the fall of 1845 and the records of the Marcy expedition when the condition of vegetation should have been favorable. Abert described the ravines, the hard soil, and a hail and rain storm on the Canadian east of 104°, northeastern New Mexico, as of August 28 and September 1, 1845. The entry of September 1 ran:

We experienced some difficulty in crossing the streams which intersected our trail. Although now dry, yet they are deep, and the banks cut vertically by the immense bodies of water which rush through during the rainy season; for these little ravines carry off the drainage on a wide extent of prairie, whose surface baked in the hot sun, absorbs but little of the water which the clouds sometimes so boisterously shower upon them.

The storm record of August 28 read:

The hail stones had piled up so as to form dams between the tussocks of red grass, and the rain which accompanied them caused the hard ground to be covered ankle deep in water, while the hail which whitened the prairie was one and a half inches deep. Many of the hail stones did not melt until next morning.... The storm was over in less than an hour.... We had been looking for water, that we might encamp. Instead of dry ravines, we now found impassable torrents; and, as they bore along in their rapid currents huge fragments of fallen rock, we readily conceived the manner in which the deep channels had been excavated.

This erosion kept the most of the rivers of the grassland turbid and the waters of those flowing through the red beds of the south central plains, such as the Canadian, red in color. Simpson commented on both points September 10 and 24, and on the former occasion reported that the water was too muddy for bathing. The return to the Canadian river through the sand hills near the 100 meridian provided a theme for a description of wind erosion on that and other western rivers:

The strong wind keeps the sand upon the banks, forming constantly changing hillocks, and often advancing inwards like the dunes on the seacoast. The course of the streams may be traced for miles on a windy day, by the drifting clouds of sand which hang over their beds.

Some information upon the general distribution of animals came out in these reports. Abert noted the change from deer to buffalo, east and west of 97°, and prairie chickens east of that line. Simpson first noticed "buffalo wallows and bones" just west of this line, and Marcy commented on the same place as the eastern limits of the range of wild horses. Simpson mentioned prairie dog towns about the 100 meridian. Before drawing conclusions, however, it is imperative that the limitations of these records be emphasized. In passing through the country on a continuous line of march they could record only what they saw, while an intensive examination of the whole area through a year or a series of years might have revealed, and probably would have revealed, a somewhat different distribution of facts. The important conclusion is that the conditions described were so obvious, they were apparent under such circumstances as representing a generalized picture. Subsequent accumulations of knowledge have served to confirm largely these records, but with modifications in detail.

The influence of animals upon the vegetation was also a matter of comment. Abert described the buffalo paths converging at springs and the river crossing and the accompanying soil erosion. The relation of the buffalo to the grass was described by one of the emigrants under Marcy's escort, Dr. J. G. Candee (Foreman, 1939); "The vegetation in some places and at some seasons is luxuriant.... In some places, and at the proper season, the forage for miles in extent, hundreds of thousands of acres, is literally eaten to the turf by the immense herds of buffalo...."

The year 1849 was the first of peace after the annexation of Texas, the Mexican war and the annexation of the Spanish southwest. It was a year of the gold rush to California, and it was one of feverish interest in exploration of the new accessions of territory to take stock of what had been acquired. Marcy's return from Santa Fé was by way of Doña Ana, a new town just above El Paso, which by inadvertance in boundary making had been left in Mexico. From Doña Ana, Marcy's route lay eastward across the southern end of the Staked Plains and thence northeast to Fort Smith. This route, if it proved feasible, would be a shorter route than by Santa Fé by which to connect with Cook's Gila route to California.

Marcy started east September 1, 1849 through the San Augustine Pass of the Organ mountains, past the Sacramento range and the Guadelupe mountains to Delaware creek and the Pecos river. All the way he recorded several varieties of grama grasses, cacti, and on several camps the only fuel was mesquite bushes. As the Comanche guide said they could not cross the Staked Plains from that point, they followed the Pecos southeastward to near where it intersected the 103 meridian, the Horsehead Crossing. The water of the Pecos was muddy, like the Rio Grande. From this crossing the expedition headed east of northeast directly across a range of sand hills, "white drift sand ... destitute of soil, trees or herbage." On October 3 they reached the eastern escarpment of the Staked Plains at about 101° 45', reaching the Big Spring at about 101° 30' for camp. Here the mesquite trees were becoming larger, and the accompanying map carried the inscription "Mesquite Timber" as applying to the country along the road northeast to the Brazos Crossing at about the 98 meridian. Similarly the inscription descriptive of grasses indicated grama and mesquite grasses extending from the Organ mountains to a point near the 99 meridian. On October 5, (ca 101° 20', and 32° 55') was the first mention of the combination of mesquite trees and mesquite grass, and the following day they crossed a continuous prairie dog town of eleven miles, "covered with short buffalo grass," the first mention of that variety. Here also was the first sighting of turkeys. On October 11, (ca 33° 15', and 100° 20') occurred this interesting description: "We have been travelling through groves of mesquite timber, with a beautiful carpet of grama grass underneath, nearly all day." Crossing rolling prairies they were approaching from the western side the headwater of Clear fork of the Brazos. September 12 (ca 32° 25', and 100°),

over as beautiful a country for eight miles as I ever beheld. It was a perfectly level grassy glade, and covered with a growth of large mesquite trees at uniform distances, standing with great regularity, and presenting more the appearance of an immense peach orchard than a wilderness. The grass is of the short buffalo variety, and as uniform and even as new mown meadow, and the soil is as rich, and very similar to that in the Red river bottoms.

The Clear fork of the Brazos was reached October 20, the location being identified by Marcy as latitude 32° 40' and longitude 99° 30', finding Comanche and Kickapoo Indians encamped:

the valley being covered with several kinds of grass that remains green during the winter, they come here in autumn, graze and fatten their horses, and are ready for the buffalo on their winter migration to this region. We found the first pecan timber here that we have seen since we left the Creek Nation.

On the following day this significant entry indicated the transition country: "The mesquite wood and grass continue very abundant, and we occasionally see the grama grass." This was just southwest of the Brazos crossing above the mouth of Clear fork where Fort Belknap was soon to be established, (ca 33° 45', and 98° 40'), Northeast from the Brazos were mesquite and oak openings, with occasional prairies, and over the divide to the west forks of the Trinity river there was oak and other hard timber, and "there is grass in this valley which grows to the height of six or eight feet, with a round jointed stock, and a head upon the top with seed which our animals eat eagerly, and I think must be very nutritious." This was associated with the further entry that the mesquite was diminishing in favor of oak groves, and "the grass has also changed from grama to the mesquite variety. The divide in question which had marked this transition was about 98° 30'.

Four reconnaissances and reports of 1849 contributed to the printed record of Texas geography south of the 30 parallel, centering largely on San Antonio (near 29° 25', and 98° 40'); Michler, a military road from Corpus Christi to Fort Inge (near 29° 15', and 100°) on the headwaters of the Leona; Smith, a military road by the southern route from El Paso to San Antonio past Fort Inge; French, leading a military supply train west over the same route; and Whiting, a road from Fort Inge to the Rio Grande at Eagle Pass.

From Corpus Christi (ca latitude 27° 40', and longitude 97° 30') Lt. N.Michler proceeded up the Nueces river on the west side just outside the bends through a prairie "covered with fine Mesquite grass, and interspersed with mesquite trees and live oak moats," crossing the muddy Nueces river (ca 28° 20', and 98° 15'), "you rise to a mesquite flat" about a mile through, thence across "a beautiful valley, perfectly clear, and covered with fine nutritious mesquite grass", about eight miles to the Frio river. Again just outside the bends, the route followed up the west side of the Frio over "a perfect dead level prairie, covered with excellent grass and mesquite trees", but the bottoms were difficult to penetrate because of heavy timber and thick chaparral undergrowth, "together with mesquite and cactus of every description." Crossing the Leona (ca 99° 20', and 28° 45') the road, of necessity, followed up the ridge nearer the Frio than the Leona because of the greater quantity and density of the chaparral which was so thick "it is difficult to make your way through them." Michler included also a generalized description of the country that is highly significant:

As a general thing, I may here remark concerning the land both on the Frio and Leona, from these rivers back, that it may be divided into four parallel strips— the first, next to the river, consisting of heavy timber, and a heavy black soil; the second, a mesquite flat, of small width, and a soil of lighter nature, and very fertile; the third, a range of low hills, covered with loose stone, and thick chaparral; the fourth a wide open prairie, the soil generally very dry, but covered with excellent grass, the latter article being generally very scarce close to the rivers: and again you sometimes find a second line of chaparral hills beyond the prairie land. Each of these strips is distinct, and parallel to the general course of the river.

Lt. W. F. Smith reconnoitered for a road between San Antonio and El Paso. Chaparral cutting was indicated as necessary for the road west of the Pecos. He had gone west through south central Texas, but returned down the Rio Grande river by a southern route leaving the road near the river at about 30° 30', and 105°, crossing the Pecos, down that river and thence east to the headwaters of the San Pedro reaching the Wool road from the latter river over country "almost entirely the rolling mesquite prairie." Captain S.G. Cross made the journey from east to west in June with a quartermaster's train. His comment on the Pecos river was that "Its waters are turbid and bitter, and carry, in both mechanical mixture and chemical solution, more impurities than perhaps any other river in the south." Lt. Michler, who made a trip from the Red river to the Pecos by way of the Big Wichita and the upper Brazos river late in the year, was more extreme in describing the Pecos river as a "rolling mass of red mud." Of the grass between the two rivers Michler declared that it was "mostly the fine curly mesquite," except from the White Sand Hills westward which was "a thick growth of chaparral" and "the grass...indifferent and the soil poor and unproductive." Of the country east of El Paso, French thought it would afford a place for grazing cattle,... were water discovered in abundance,..."

The report of Lt. W. H. C. Whiting dealt with the Texas frontier defense problems and for southern Texas is particularly valuable for the country south of the San Antonio-El Paso road which ran south of the Brazos escarpment of the Edwards plateau. It was this outcropping limestone formation, mostly south of the 30 parallel, which he designated as the northern limit of future settlement. This formation turned north above San Antonio, between 98° and 99°, and is evidently lost in the tablelands of the Brazos river. The military road and the line of forts below this escarpment and along the headwaters of the numerous streams forming the Nueces river, had been chosen as a natural frontier defense and communications line. Fort Inge, one hundred miles west of San Antonio, on the military road was the post at the headwaters of the Leona, "one of the most important and desirable positions in Texas." Here the road to Eagle Pass on the Rio Grande turned southwestward, a circuitous road, following an old smuggling trail, which General Wool had established in his recent invasion of Mexico. Whiting laid out a straight road of fifty miles, some twenty miles shorter which he recommended be cut through the chaparral to Fort Duncan at Eagle Pass. Of the Leona site, he wrote:

The grazing in the vicinity, through the rich mesquite flats of the Leona, is unrivalled.... Shingles are readily procured from the cypress of the Sabinal and the Frio [where it was being manufactured].... The forests of mesquite which clothe the Leona bottom afford abundance of fuel.

The country between Fort Inge and Fort Duncan he compared unfavorably with that of Fort Inge: "Subject to almost continual drought, badly watered, clothed with cactus and thorn chaparral, it presents an aspect dreary and desolate in the extreme. This effect is increased as the Rio Grande is approached."
Hundreds of north-south Indian trails and the two main trails ran through the country near and mostly west of Fort Inge, and as he pointed out "no country is better adapted than this to the Indian in his purposes of depredation, escape, and concealment.... Once in the chaparral of the great plain between the Nueces and the Rio Grande, ... pursuit is wellnigh hopeless." This is the kind of country that forms the setting for the early Texas cattle business, which is described in J. F. Dobie, Vaquero of the Brush Country (1929).

In eastern Texas only in a few places had the frontier of settlement reached beyond a line followed by Lt. Michler in October 1849, on his way from San Antonio through Austin and Dallas to Fort Washita north of the Red river. This curved line of 380 miles ran from near 99° at the south to 98° near Austin, and about 97° near Dallas. He reported the first hundred miles as passing through fairly well-settled country, but farther north settlements occurred only at intervals of ten to fifteen miles. The principal crop was corn for local use, and near the Red river some cotton was produced. The line of frontier forts for Indian defense ran below (east of) the northern extension of the Balcones escarpment, a route described by Lt. Whiting from Fort Worth south and west to the headwaters of the Guadelupe river and the hill country northwest of San Antonio, approximately one degree of longitude west of the Michler route. Although only indirectly involved in this study of the grassland, Whiting's strategy for Indian defense has some importance to this description. He objected to the existing line of numerous forts with small garrisons because the frontier of settlement would soon reach them and the natural barrier and because such posts and forces could not deal effectively with the Indian problem. He recommended five large military posts in the high plains west of the natural frontier in a chain from the Little Wichita on the Red river by way of the old San Saba Fort to Presidio del Norte, the probable head of navigation below Fort Duncan. Manned with large forces of mounted troops the Indian could be controlled effectively on the open plains.

EARLY EXPLORATIONS: SOUTHERN   93

In addition to Marcy's route northeast-southwest across the plains from Doña Ana to Fort Smith, several other reconnaissances were made during the same year. Across the south central area of Texas, two expeditions set out from San Antonio for El Paso partly by different routes. Lt. W. F. Smith went northwest by way of Fredericksburg to the headwaters of the south fork of the San Saba river thence west and southwest. Lt. F.T.Bryan started by the same route but went north to Brady creek at a point east of 100°, then turned west past the headwaters, southern tributaries of the Concho regaining Smith's route to the Horse Head Crossing of the Pecos. Bryan reported his section of the route "covered with scattered mesquite and mesquite grass;" and farther west on the Concho July 3 the grass was burnt up, providing "little sustenance to our animals", and July 3, beyond the Concho, "grazing very indifferent." Smith reported his section as "thinly covered with mesquite tree" and he saw several clumps of live oak. Toward the west end of the joint route there was not sufficient water for a military road without digging wells.

In northwest Texas, between Marcy's northeast-southwest route of 1849 and the Red river two important exploring parties crossed the area, Lt. Michler in 1849, and Marcy in 1854. Michler started from Fort Washita, November 9, heading south of west crossing the Red river above the mouth of the Little Wichita, thence southwest up the Big Wichita river, thence somewhat west of south across the Brazos reaching the 100 meridian near the Clear fork at about 33°, thence southwest along the Marcy route to the Pecos. Marcy's expedition of 1854 to the headwaters of the Big Wichita and of the Brazos opened unexplored country on a line somewhat west of the Michler route. Of this expedition, there is a private account by William B. Parker in addition to the official report of Marcy, although it is evident that Parker leaned heavily upon Marcy's reports of this and an earlier Red river expedition of 1852.

As Lt. Michler was making his departure from Fort Washita, Marcy arrived from his expedition and was able to give important information. Michler's route took him through the Cross Timbers to a crossing of the Red river just above the mouth of the Little Wichita, 104 miles of travel. East of the Cross Timbers he reported:

The soil is of a sandy nature throughout the entire distance [19 miles]. The prairie was already very dry at this season, the species of gramma being most abundant; here and there spots of mezquite. Saw several varieties of cactus today.

Between Mud creek and the Red river the grass comment was

The further west we travelled, the better grazing we found— the
gramma, sedge and buffalo grass the most abundant, but the mezquite [grass?] constantly becoming more frequent.

After crossing the Red river the route was up the divide
between the Little and Big Wichitas about 86 miles, when
it turned south from the divide, ten miles to the Brazos
river:

Near the Red river the soil is slightly sandy, and you meet with
some few post-oak mots. It then becomes a fine mezquite country,
well timbered with mezquite, and for miles perfectly level, and
even when a rolling prairie, the elevations and depressions are
small. The grass at first is principally gramma, and the ordinary
sedge, and their species, but then come the fine early mezquite
and the winter mezquite.

The extent of the processes of erosion are described, as seen between the divide and the bed of the
Big Wichita:

... the ground is exceedingly rough and uneven: deep gullies had
washed through the clay and sand, and numerous small mounds had
been formed by the swift currents during the high freshets to which
this stream must be subject. From the amount of drift scattered
about, it must rise to a very great height, and its current become
remarkably swift.

South of the divide, toward the Brazos was "a
continuation of the mesquite range." The waters of the
Little Wichita had been clear, but the Big Wichita, and
the Brazos were red, although the latter was without mud.
The grasses of the narrow Brazos bottoms were "sedge and
water-grass, and on top of the bluffs again spread out
the mesquite flats." He must have crossed the Brazos
near the 100 meridian and struck Marcy's trail about 45
miles southeast and reached the Clear fork, according to
his calculations at about 33° and 100°, traveling 118
miles. "Day after day the country was almost perfectly
level," and except for a range of sand hills covered with
scrub oak, "the rest was either mesquite flats or a very
slightly rolling mesquite country.... The whole country
was well timbered with mesquite, but most of it had been
killed by prairie fires." The timber on the streams was
reported scanty at first, increasing farther south.
Across the divide between the Brazos and the Colorado he
reached Big Spring, located by Marcy about 32° and 102°
[the present town of Big Spring, Howard county, is north
and east of Marcy's location];

The country here undergoes a complete change. You now meet with
high rolling prairies, arid and destitute of timber, and scarcely

EARLY EXPLORATIONS: SOUTHERN

any grass but the most miserable kind. Occasionally you cross low sand hills, containing some low cedar and scrubby oak.

The Big Spring vicinity, a favorite Indian camp, was described: "The soil is chiefly sand; the grass is poor; no timber but young mesquite and cedar; some scrubby elm borders the stream." Between the Big Spring and the Pecos:

Our road lay now over a high arid plain, perfectly destitute of timber— scarcely even a sprig of mesquite, except in the neighborhood of water holes.... It seemed destitute of all growth of any kind, and nothing to be seen upon it excepting the antelope, the wolf and prairie dog town. The grass was scattering, and miserably poor; occasionally a small spot of mesquite was found.

Marcy's expedition of 1854 for the headwaters of the Brazos started July 15 from Fort Belknap, established in 1851 about ten miles below the Dõna Ana road crossing of the Brazos (Foreman 1937) passing northward to the Little Wichita "over a rolling country covered with groves of mesquite trees," and "a dense coating of verdure;" thence westward on the Big Wichita. Parker was more specific on the second day, July 16, saying that the rolling country was "covered with buffalo grass and mesquite timber...." and on the Little Wichita two days later, that the plain was "covered with mesquite grass." On July 19, the word picture presented a "wide prairie with its yellow coating of [dry] buffalo grass, studded with the pale green mesquite, a beautiful combination of a landscape painting." On account of scanty prospects for water, the wagon train was sent to camp upon the Brazos and Marcy proceeded, July 29, to the headwaters of the Big Wichita with pack animals, and July 31 turned south to cross the divide to the Brazos river. On the upper waters of the Big Wichita the only timber was of dwarf red cedar and mesquite.

On July 30, the day before reaching the headwaters, Marcy recorded that:

The portion of the valley over which we have been passing for the last forty miles is barren and sandy, and the only wood land is upon the bluffs, which are covered with dwarf cedar, with an occasional lonely cottonwood or mesquite in the valley. Here and there may be seen a small patch of wild rye or gramma grass, but the principal herbage in the valley is a coarse variety of grass unsuited to the palates of our animals.

The following day from a high point on the headwaters overlooking the valley of the Big Wichita, Marcy concluded that it is, in almost every respect, the most uninteresting and forbidding land I have ever visited. A barren and parsimonious soil, affording little but weeds and coarse unwholesome grass, with an admixture of cacti, of most uncomely and grotesque shapes, studded with a formidable armour of thorns which defies the approach of man or beast, added to the fact already alluded to, of the scarcity of wood or good water, would seem to render it probable that this section was not designed by the Creator for occupation, and I question if the next century will see it populated by civilized man. Even the Indians shun this country....

Parker's account again gave more variety and detail. On July 29, starting up the Big Wichita with only the pack animals, he described the plain as "covered with buffalo grass and mesquite timber;" then a dry lake bed "covered with luxuriant green grass, making an oasis in the comparative desert;" then, coming to a tributary crossing their route, there was an abrupt change,

as far as the eye could reach, was a barren and desolate waste, broken and torn into ravines, mounds, gullies and defiles, the soil a bright red clay, and not a tree or a shrub, except a white dwarf cedar, to be seen; ...

later a meadow a mile wide for many miles, flooded during the rainy season, and covered with grass very thin and coarse, like in salt marshes. On July 30

we found some land good enough to grow trees of a considerable size, but the most part was a barren waste, covered with gypsum, with here and there the low stunted white cedar and patches of very thin coarse grass. In the fertile spots grew China tree, the live oak and the mesquite....

It was during the day of July 31, that the little expedition turned southward toward the divide when Marcy's record ran: "On leaving the Wichita, we travelled south towards the Brazos for six miles through mesquite groves...," while Parker wrote that they "entered an extensive plain covered with thin coarse grass and stunted mesquite timber" and on arriving at the Brazos side of the divide camped on a tributary with "plenty of grass and wild rye for our animals."

On the next day, August 1, proceeding southwest in the direction of the objective, the headwaters of the Brazos, Parker's record of the day included prairie dog villages, then "the most barren, rugged and broken country we had yet met with, covered with stunted mesquite trees and dwarf cedar, the ground one mass of broken rocks;" and later "a plain covered with a singular growth of dwarf oaks ... but the highest not more than two feet high," but bearing acorns. August 3 provided more variety; a fertile plain "covered with a rich growth of buffalo grass and very large mesquite trees;" then a prairie dog town for ten miles with holes about seven feet apart; and finally arrived at a spur of the Llano Estacado, ascended and proceeded for six miles on a "broad level plain ... covered with buffalo grass and mesquite trees, and extending as far as the eye could reach in a perfect level towards the dim, cloudlike mountains at the head of the Brazos."

On account of the illness of the Indian agent Neighbors, and all suffering from gypsum water, they decided not to risk further the safety of the expedition. Returning to the edge of the escarpment Marcy wrote:

> Towards the east from this elevation nothing could be seen but one continuous mesquite flat, dotted here and there with small patches of open prairie, while in the opposite direction ... we discovered the elevated mountains beyond the head of the Brazos.

Parker's word picture was more vivid:

> The view was the most extensive and glowing in sunset, the most striking that we had enjoyed during the whole trip, combining the grandeur of immense space— the plain extending to the horizon on every side from our point of view— with the beauty of the contrast between the golden carpet of buffalo grass and the pale green of the mesquite trees dotted its surface.

Proceeding eastward down the Brazos on the north side to rejoin the wagon train, Marcy described the country the next day, August 4, as gently undulating, covered with mesquite trees, the soil very rich, and "producing several varieties of gramma and mesquite grasses...." At the Flat Rock camp, they recuperated. This was about 99° 30' west longitude on the south side of the Brazos on a creek, where "a grove of stately elms lined the banks of the stream for a quarter of a mile. Under the trees grew a rich growth of wild rye." Heading southward on August 9, four day's travel bringing them to the Clear fork of the Brazos of which Marcy wrote: "A change takes place in the physiognomy of the country in passing from the Main or Salt fork, to the beautiful Clear fork of the Brazos, which seems almost magical," and "all, within the space of a day's travel." The waters of the Brazos were red, muddy, and bitter from gypsum, running through a valley almost destitute of timber, while the Clear fork had clear water, running over a limestone bed, pure and sweet, and banks heavily timbered.

Marcy took the opportunity to summarize the status of the mesquite tree in the southern grassland as "it covered a great portion of the country over which we travelled...." The first public notice of the tree, in a scientific sense, was by Dr. Edwin James in connection with Long's expedition of 1819. Blooming in May, it produced long pods of beans which ripened in September. As a tree, it was short, scrubby, thorny, four to fifteen inches in diameter and not over twenty feet high. Torrey classified it first as Prosopis glandulosi, but others differed with him. Although incompletely known, the principal range of the species was set by Marcy as between 26° and 36° north latitude, and 97° and 103° west longitude. North of 33° the tree became smaller, farther north it was a bush, and north of 36°, or north of the Canadian river, it disappeared. To the west, it was found between the Rocky Mountains and the Pacific coast, but flourished better in the Gila valley than anywhere else west of the Rio Grande:

In the journeys I had made before upon the plains, I had observed the mesquite tree extending over vast tracts of country, and I had noticed some of its useful properties, such as its durability and its adaptation for fuel, but I was never so fully impressed with its many valuable qualities as during the past summer.

* * *

It is only indigenous to the grass plains of the west and south, extending far beyond the limits of most other varieties of trees, and it would seem from its locality to have been planted by an all-wise Providence with special reference to the wants of the occupants of a section of the country suitable to the growth of no other tree.

Although the mesquite tree was adapted to dry climate and grew on the upland away from streams, Marcy reported that settlers thought of it as an indicator of fertile soil and competed for good mesquite land. The manner of growth and distribution on the ground attracted the attention of all observers of which Marcy was typical: "The trees stand at wide intervals, upon ground covered with a dense carpet of verdure, and a stranger approaching one of the groves cannot resist the impression that he has a peach orchard before him...." It is this characteristic habit of growth that explains the descriptions already noted of mesquite trees with an understory of either buffalo grass or mesquite grass, according to local circumstances.

Because of its many possible uses, Marcy thought the mesquite tree would be highly important to the future occupants of the region. He said that the wood burned freely, even when green, making a hot fire second only to hickory; it was durable "much used for building in southern Texas and Mexico," being well preserved in old ruins. The beans were eaten by wild horses, deer, antelope, and turkeys, and being highly saccharine and nutritious was used for food by the plains Indians, and the desert Indians pulverized them and made cakes. In crossing the desert, emigrants depended upon them for livestock forage

The trees produced a gum which was thought to be equal to gum arabic.

In surveying the Caddo Indian reservation during mid-September, 1854, Marcy's camp was located south of the Brazos about fifteen miles from Fort Belknap, in what is now western Palo Pinto or eastern Stevens county, and Parker described in some detail the grass situation:

Near our camp I found large quantities of the black mesquite grass, a very favorite grass with all who have tried it, and I collected a stock of the seed, which I trust may stand our climate, as from the avidity with which our animals eat it, I am sure it would be a great addition to our northern crops, either for pasture or fodder. It grows about as high as timothy, and has a head on it like wheat. The grasses met with are the white gramma, the blue gramma, three varieties of the sedge, the buffalo grass, the bearded mesquite and the black mesquite.

Of these, the buffalo grass would make a beautiful sod for lawns, as its growth is very short and velvety, appearing more like the thickest kind of moss than grass. I observed that our horses eat it in preference to any other, even when it was quite dry, and green succulent grass in its vicinity. I could not procure any seed.

It may be too much to assume that Parker was accurate in his nomenclature, but if so, the mesquites he was describing belonged to the genus Hilaria, and it is clear that he was distinguishing them from the gramas and the buffalo grass.

To those who still think of the grassland as monotonously uniform and uninteresting, Parker's comment based on his journal of July 19, is recommended:

My wonder has been throughout my journey that so few if any of our artists ever join expeditions to the plains. A portfolio could soon be filled with novelties, compared with which the hackneyed subjects universally to be found on sale or exhibition sink into mediocrity. Every variety can be found there, hill, dale, lake, valley, mountain, river and plain, whilst color, tint, light and shade are constant in quantity and quality. Let but the experiment be tried, and prairie scenery will become a valued gem in the gallery.

Why is it that no one returns from the plains disappointed. It is because their anticipations have been doubly realized. This fact is to my mind conclusive, that visits of artists to the plains would not only end in adorning the art, but give a better impression of that comparatively 'terra incognita'. I say a better, not a full, impression, for to be fully realized it must be seen and passed over.

Between the Canadian river, described by Abert, Marcy, and Simpson (1845 and 1849), and the Big Wichita and Brazos rivers described by Michler, Marcy, and Parker

(1849, 1854), lay the watershed of the upper Red river, the Texas Panhandle, a terra incognita until 1852, when Marcy made its description a matter of record as far west as the Staked Plains. This expedition started May 16 officially from Cache creek, a northern tributary of the Red river, just east of 99°. On this stream there was much good timber; pecan, elm, hackberry, ash, wild China, and especially overcup oak (Quercus macrocarpa), some trees as large as four feet in diameter. Ascending the North Fork past the Wichita mountains they saw sand hills ten to thirty feet high, blown up by the winds, which supported "a very spare vegetation of weeds, grape-vines, and plum-bushes." Just west of 100° they passed a post-oak grove of 400 to 500 acres and an "extensive tract of mesquite woodland." On June 5, in the bend of the river on the south side

the country we traversed was exceedingly monotonous and uninteresting, being a continuous succession of barren sand-hills, producing no other herbage than the artemisia, and a dense growth of oak bushes, about eighteen inches high, which seem to have attained their full maturity, and bear an abundance of small acorns.

On June 7, at about 100° 30' Marcy recorded: "The grass, however, as we found it everywhere on the Red river and its tributaries, is of a very superior quality, consisting of several varieties of grama and mesquite." Marcy followed a tributary on the north side of the Red river June 9, just east of 101 and found it "fringed by large cottonwood trees, ... and grass in the valley, ... consisting of mesquite and wild rye,...." Further along on the Red river to the south of them was "a range of sand hills extending back about five miles upon each side of the river." On June 14, reaching about 101° 30' they recorded "fine mesquite and grama grass," and cottonwood trees, "a never failing resort when the grass is gone." The following day he recorded that "The herbage for the last twenty miles of our march has suffered much from drought, and the grass in many places upon the elevated lands is entirely burnt up. We, however, continue to find excellent grass in the valleys near the borders of the small streams, and upon the river itself."

From June 17-19, Marcy crossed from the Red to the Canadian river just east of 102° which provided the occasion for a description of the grass over the Staked Plains as "generally a very short variety of mesquite, called buffalo grass, from one to two inches in length, and gives the plains the appearance of an interminable meadow that has been recently mown very close to the earth." On Sandy creek of the Canadian, however, the grass was "wild rye and mesquite."

In passing south from the north fork to the main fork of the Red river, just east of 102°, and below the

Staked Plains escarpment, Marcy passed a variety of country; especially noticeable were prairie dog towns and sand hills — at one place a labyrinth of barren sand hills, for fourteen miles. On June 26, all day was spent passing through a prairie dog town twenty-five miles across where the holes were about 20 yards apart. The Red river valley was particularly attractive to the Kiowa Indians, he wrote, because "the exuberant and rich grama grasses ... everywhere abound in the river bottoms...," in fact "several varieties of the grama," and mesquite groves.

The use of the name mesquite grass by Marcy requires caution, because in one instance, in 1852, he called buffalo grass a mesquite grass. In his report of 1854, on the Brazos expedition, there is no apparent confusion. He was clear in his treatment of the grama grasses:

> The range of the grama grass, so far as my observations have extended, is bounded on the north by near the parallel of $36°$ north latitude, and on the east by about the meridian of $98°$ west longitude. It extends south and west, so far as I have travelled; it appears, however, to flourish better in about the latitude of $33°$ than in any other. As there is generally a drought on these prairies, from about the 1st of May to the middle of August, it would appear that the particular varieties of grasses that grow here do not require much moisture to sustain them.

The plants collected on this expedition by Dr. G. G. Shumard included twenty-nine grasses, which were classified by Dr. John Torrey. Other plants of interest were the sage (Artemisia filifolia) from the upper Red river and two sedges, Carex Muhlenbergii and festucacea, from the headwaters of the Trinity river of Texas. It is of interest also, to note that the mid grasses on the list were collected mostly on the Trinity and Washita rivers or that general vicinity, or about the 98 meridian, or in the Wichita mountains east of the 100 meridian.

In trying to explain the barrenness of the plains, Marcy was influenced partially by the idea of the sterility of the soil, and the gypsum was thought of as one means of its restoration. On the whole, however, Marcy recognized the significance of climate in a determining rôle, the prevailing rainfall pattern, and the survival and vigor of buffalo grass and the grama grasses because of their adaptation to that environment. He pointed to the growth of timber to near the 99 meridian,

where the road emerges from the woodlands and enters the great plains, where but little timber is seen except directly along the borders of the water courses. The soil soon becomes thin and sandy, and, owing to the periodical droughts of the summer season, would require artificial irrigation to make it available for cultivation.

Chapter Nine

THE GRASSLAND AND EARLY EXPLORATIONS: CENTRAL

The country between the Canadian and the Platte rivers has a history quite different from that to the southward. The Santa Fé and the Oregon Trails introduced disturbances into the natural environment, and the emigrant Indian tribes, settled in the region east of the 98 meridian after the general removal act of 1830, introduced substantial changes into the tall grass prairie. A view of the original grass condition requires, therefore, an early body of records, when scientific knowledge in general was less advanced, and less widely disseminated, and the plant families of the grassland were quite strange even to those who were scientifically inclined. F. A. Wislizenus, the German explorer, traveled through the South Pass in 1839, making some general scientific observations. Leaving the north bank of the Kansas river at its mouth in May, he proceeded northwest to the Platte river at Grand Island where he first observed buffalo grass, then known as Sesleria dactyloides. He had passed over the tall grass prairie without a mention of the kind of grass growing there. It was the novelty of the buffalo grass that inspired this first specific grass species entry in his record. Past Chimney Rock on the North Platte, he noted that the country was "very sandy, the vegetation scant," also that "we observed very many bitter herbs, especially wormwood [Sage, Artemisia]; also Pomme Blanche (Psoralea esculenta), whose knobby root contains much starch, has a pleasant taste, and is gathered by the Indians." Leaving Fort Laramie on the road west to the crossing of the North Platte, he noted June 15, "the change in vegetation," and the conspicuous landscape feature, the sage and cacti: "This Artemisia is found on both sides of the Rocky Mountains, in sandy soil, where the grass grows sparcely or not at all." It varied in size from a foot to the height of a man, and the large forms made good fuel. Around the southern point of the Wind river mountains on the Sweetwater the plain became hilly, sandy, and had a "little grass, but the more sage brush, and quantities of buffalo," but cacti were becoming rarer, and mosses more frequent.

John C. Fremont's first expedition, the one to the Wind river mountains through South Pass was made in 1842, the last year in which reasonably natural conditions could be observed on the immediate vicinity of the trail to Oregon, because in 1843, the big emigration began, and with it, disturbance of nature along the main trail. As in the case of so many explorers with some botanical interests, Frémont was not grass conscious in the sense of discriminating with respect to species and the composition of the grass cover. An incident in the Laramie vicinity illustrated his point of view vividly: On making camp July 13, some of the party went to kill buffalo, others to gather buffalo chips, but "I amused myself with hunting plants among the grass." Not a word appeared concerning the species of grass among which he was hunting "plants." He did not recognize buffalo grass on his first expedition and seemed unacquainted with the grama grasses on both the first and the second.

The expedition started June 10 from Cyprian Chouteau's trading house near the mouth of the Kansas river, by way of the Santa Fé trail, thence the Oregon trail past the site of Lawrence, reaching the southeast corner of the present Marshal county, Kansas (about 96° 15'), June 19, when he recorded that the rise from a 700 foot to a 1400 foot altitude "appeared already to have some slight influence upon the vegetation." The plant that struck him as most conspicuous was the false indigo, and near the 40 parallel, in crossing broken limestone country, June 20, he recorded that

> In these exposed situations grew but few plants; though, whenever the soil was good and protected from the winds, in the creek bottoms and ravines, and on the slopes, they flourished abundantly; among them the <u>amorpha canescens</u> [false indigo with purple flowers] still retaining its characteristic place.

On the dry ridge which the trail followed he noted the same day the wild rose everywhere, "the most beautiful of the prairie flowers," and "The artemisia, absinthe, or prairie sage, ... is increasing in size, and glitters like silver...." On June 23 he crossed Sandy creek, the country becoming very sandy, "and the plants less varied and abundant, with the exception of the <u>amorpha</u>, which rivals the grass in quantity...." The same day he noted cactus, and later thistles. By this time the expedition was south of the present Hastings, Nebraska, somewhat west of 98°. On June 25, "the road led across a high level prairie ridge, where were but few plants, and those principally thistle, and a kind of dwarf artemisia [sage]." The night camp was on the Platte river near Grand Island (98° 45').

Below the forks of the Platte river (about 100°) June 30, Frémont found himself in "the midst of the buffalo, swarming in immense numbers over the plains, where they left scarcely a blade of grass standing.... Clouds of dust rose in the air from various parts of the band, each the scene of some obstinate fight." This is a revealing entry relative to the effect of the buffalo on grass, and this and subsequent entries July 4, 8, emphasize the thinness of the vegetational cover and the dust rising from the soil pulverized by thousands of hoofs. Hunting buffalo the next day Frémont noted that "a thick cloud of dust hung upon their rear, which filled my mouth and eyes, and nearly smothered me. In the midst of this I could see nothing, and the buffalo were not distinguishable until within thirty feet." Among other things, the herd crossed a prairie dog town for nearly two miles, and this suggests further the scattering over the whole area of the loose dirt dug up by the prairie dogs. The holes were spaced three or four to every twenty yards square.

Above the forks along the South Platte, July 4, about 101° Frémont described vividly the relation between plains streams and the distribution of soil materials by water:

Leaving camp, our road soon approached the hills, in which strata of a marl like that of the Chimney Rock, hereafter described, make their appearance. It is probably of this rock that the hills on the right bank of the Platte, a little below the junction, are composed, and which are worked by the winds and rains into sharp peaks and cones, giving them, in contrast to the surrounding level region, something of a picturesque appearance. We crossed this morning numerous beds of the small creeks which, in time of rains and melting snow, pour down from the ridge, bringing down with them always great quantities of sand and gravel, which have gradually raised their beds four to ten feet above the level of the prairie, which they cross, making each one of them a miniature Po. Raised in this way above the surrounding prairie, without any bank, the long yellow and winding line of their beds resemble a causeway from the hills to the river. Many spots in the prairie are yellow with sunflowers (helianthus).

By July 6 he had reached Lodge Pole creek where "a few willows on the banks strike pleasantly the eye, by their greenness, in the midst of the hot and barren sands." In this region "the amorpha was frequent among the ravines, but the sunflower (helianthus) was the characteristic [plant]. The impression of the country travelled over today was one of dry and barren sands." This was about 103° west and he crossed from the South to the North Platte west of 104°. In the entry for July 13 he attempted an explanation of the barrenness of the plains at this elevation of 5440 feet:

the constituents of the soil in these regions are good, and every day served to strengthen the impression on my mind, confirmed by subsequent observation, that the barren appearance of the country is due almost entirely to the extreme dryness of the climate. Along our route, the country had seemed to increase constantly in elevation....

EARLY EXPLORATIONS: CENTRAL 105

On the same day the expedition rode "over a plain covered with innumerable quantities of cacti," buffalo were found in the ravines "which always afford good pastures." This was a distinction repeated frequently, grass sufficient for grazing only in the stream bottoms. Frémont reached the North Platte some thirteen miles below Fort Laramie on July 15. After moving somewhat west up the river, he summed up his impressions July 22:

With the change in the geological formation on leaving Fort Laramie, the whole face of the country has entirely altered its appearance. Eastward of that meridian, the principal objects which strike the eye of the traveler are the absence of timber, and the immense expanse of prairie, covered with the verdure of rich grasses, and highly adapted for pasturage.... Westward of Laramie river, the region is sandy, and apparently sterile, and the place of grass is usurped by the <u>artemisia</u> and other odoriferous plants, to whose growth the sandy soil and dry air of this elevated region seem highly favorable.

One of the prominent characteristics in the face of the country is the extraordinary abundance of the <u>artemisia</u>. They grow every where— on the hills, and over the river bottoms, in tough, twisted, wiry clumps; and, wherever the beaten track was left, they rendered the progress of carts rough and slow. As the country increased in elevation on our advance to the west, they increased in size; and the whole air is strongly impregnated and saturated with the odor of camphor and spirits of turpentine which belongs to this plant.

The following day he commented upon the unparalleled drouth that year which had made it impossible for the fur traders to float their catch down either the North or the South Platte: "Everywhere the soil looked parched and burnt; the scanty yellow grass crisped under foot, and even the hardiest plants were destroyed by want of moisture." To rapid evaporation at this high altitude without protection of timber "should be attributed much of the sterile appearance of the country, ... and the numerous saline efflorescences which covered the ground." The Indians were having difficulty in finding grass, a condition revealed by a camp where cottonwood trees had been cut for their horses, and failing to find noon grass, Frémont did likewise— "Usually, Indians use cottonwood only in winter." On July 27, "we travelled later than usual, having spent some time in searching for grass, crossing and recrossing the river before we could find a sufficient quantity for our animals. Toward dusk, we camped among some artemisia bushes, two or three feet high, where some scattered patches of tough grass afforded a scanty supply." Summing up the country between Fort Laramie and the crossing of the Platte July 28, Frémont wrote:

The face of the country cannot with propriety be called hilly. It is a succession of long ridges, made by the numerous streams which come down from the neighboring mountain range. The ridges have an undulating surface, with some such appearance as the ocean presents in an ordinary breeze.

It will be remembered that wagons pass this road only once or twice a year, which is by no means sufficient to break down the stubborn roots of the innumerable artemisia bushes. A partial absence of these is often the only indication of the track; and the roughness produced by their roots in many places gives the road the character of one newly opened in a wooded country.

It is evident that there was no disturbance of nature by man that could account for the dominance of sage brush, and the scarcity of grass, in this region. Hearing from Indians of worse conditions in the Sweetwater valley, Frémont decided to travel light and cached his wagons and supplies "in the sand, which had been blown up into waves among the willows." According to his calculations this position was $106° 30'$, and $42° 50' 53$."

In the Sweetwater valley above the Devil's Gate "the upland part of the valley, ... is overgrown with artemisia," and later "our fires tonight were made principally of ... artemisia, which covered the slopes." Near the South Pass "a variety of asters may now be numbered among the characteristic plants, and the artemisia continues in full glory; but cacti have become rare, and mosses begin to dispute the hills with them." The last part of this comment is very like that of Wislizenus. Across the South Pass on the Little Sandy, there was still artemisia in full bloom, "and, numerous as they are, give much gayety to the landscape of the plains." West of the South Pass there had been no buffalo, but on returning August 19 they again enjoyed roasted ribs for dinner.

The second Fremont expedition, 1843-1844, covered but a part of the same territory as the first, but in both going and returning he was covering similar country which provided variations of the first experience. The record is more important than the first in many respects because he had gained experience both in recording natural history and in acquaintance with the territory covered. The route westward was up the Kansas river, and the Republican river, to a point near the northwest corner of the present state of Kansas, and across the Republican to the South Platte river at about $103° 30'$. The return was down the Arkansas to Bent's fort, thence northeast to the headwaters of the Smoky Hill river in eastern Colorado and down that stream to the bend south of the present site of Salina, thence east to the Santa Fe Trail.

Frémont made his start May 29, following the Oregon Trail to the crossing, but then deviated by following the Kansas river up the south side. On June 4, he observed Kansas Indian women digging prairie potatoes psoralea esculenta). Arriving at the Smoky Hill, June 8, he crossed on rafts, then traveled for several days along the south bank of the Republican through a well-watered and timbered country. Occasional elk and antelope were seen and later they became more frequent. At the Big Timber the party divided because of the heavy load on the wagons, Fremont and small party, June 16, crossing the Republican and the north fork of the Solomon. This was about 98°, probably between Concordia and Beloit, Kansas. Along the Republican to the Big Timber "the country was everywhere covered with a considerable variety of grasses — occasionally poor and thin, but far more frequently luxuriant and rich." The record of this much of the route was most unsatisfactory. One day's travel from the Big Timber brought the first notice of bunch grass (festuca) and buffalo grass (sesleria dactyloides): "Amorpha canescens (lead plant) continued the characteristic plant of the country, and a narrow-leaved lathyrus occurred during the morning in beautiful patches. Sida coccinea occurred frequently, with a psoralea ... and a number of plants not hitherto met.... Fremont travelled near the source of the Solomon tributaries. Just east of 99°, where he crossed the Pawnee Indian trail, "Amorpha, with the same psoralea, and a dwarf species of lupinus, are the characteristic plants." Near the Pawnee trail prairie dogs were first seen, "Sida coccinea was a characteristic on the creek bottoms, and buffalo grass is becoming abundant on the higher part of the ridges." Why didn't he name the grasses on the slope levels and low prairies? Two days later, June 21, on the divide between the Solomon and the Republican, "plants were few; and with the sward of the buffalo grass, which now prevailed everywhere, giving to the prairies a smooth and mossy appearance, were mingled frequent patches of a beautiful red grass, (aristida pallens,) which had made its appearance within the last few days." According to Frémont's calculations this was about 99° 45', probably in present Norton county, Kansas. Two days earlier, on the 99 meridian the first buffalo were sighted and only at this time was the first one killed, an old bull. On June 23 he travelled along a stream, "which was populous with prairie dogs, (the bottoms being entirely occupied with their villages,).... We gave to this stream the name of Prairie dog river." On June 25 there were "buffalo in great numbers, absolutely covering the face of the country," and that evening he camped on a stream near the Republican where the artemisia filifolia was first seen. This was at the northwest corner of Kansas between the state line and the Republican river in Nebraska. On the morning of June 26, they "found suddenly that the nature of the country had entirely changed. Bare sand hills everywhere surrounded us ...; and the plants peculiar to a sandy soil made their appearance in abundance." It is evident that by plants, he meant forbs, not grasses. These sand hills were a prelude to the Republican river,

> whose shallow waters, with a depth of only a few inches, were spread out over a bed of yellowish white sand 600 yards wide.... The features of the country assumed a desert character, with which the broad river, struggling for existence among the quicksands along the treeless banks, was strikingly in keeping. On the opposite side, the broken ridges assumed almost a mountainous appearance; and, fording the stream, we continued on our course among these ridges.... We traveled now for several days through broken and dry, sandy region, about 4,000 feet above the sea, where there were no running streams; and some anxiety was constantly felt on account of the uncertainty of water, which was only to be found in small lakes that occurred occasionally among the hills.... The soil of bare and hot sands supported a varied and exuberant growth of plants....

Continuing in a general west and northwest direction Frémont struck the South Platte on June 30 west and north of 103° and 40°. During the next ten days Frémont explored the country between the Platte and the Arkansas river, and July 11, at a 7,000 foot altitude, occupied "a piney elevation, into which the prairies are gathered, and from which the waters flow, in almost every direction, to the Arkansas, Platte, and Kansas rivers.... [east of 105°, and north of 39°]. With occasional exceptions, comparatively so very small as not to require mention, these prairies are everywhere covered with a close and vigorous growth of a great variety of grasses, among which the most abundant is the buffalo grass, (<u>sesleria dactyloides</u>)." He was so impressed with this top of the plains that he made a drawing showing Pike's Peak 40 miles in the distance (opposite p. 114). It is evident, also, that he did not mean that all these grasses were mixed together uniformly, but that from place to place, soil to soil, the plains represented a variagated pattern of these grasses. It is almost certain also that Frémont was confusing the grama grasses, especially the blue grama, with buffalo and calling them all by the latter name. As most of the botanical specimens were destroyed on the return trip, the grass record is missing from Dr. Torrey's report on the collection.

The next segment of Frémont's journey was begun July 26 from St. Vrain's Fort, northwest, up the Câche-a-la-Poudre river. As he crossed the divide between the headwater of the stream and the Laramie river August 1 and 2, artemisia reappeared and by the afternoon of August 2, "although the road was not rendered bad by the nature of the ground, it was made extremely rough by the stiff tough bushes of <u>artemisia tridentata</u>, in this country commonly called sage. This shrub now began to make its appearance in compact fields; and we were about to quit for a long time this country of excellent pasturage and brilliant flowers." From here for the next few days the principal vegetational theme was <u>artemisia</u> as they proceeded around the Medicine Bow mountains and turned west to the headwaters of the Medicine Bow river and then to the North Platte river. On August 4, "with the exception of some thin grasses, the sandy soil here was occupied almost exclusively by artemisia...." On August 6, nothing [was] to be seen but artemisia bushes; and, in the evening, found a grassy spot among the hills, ..." This was just south of the Sweetwater river, near the South Pass, but east of the divide, and August 7,

Our road the next day was through a continued and dense field of <u>artemisia</u>, which now entirely covered the country in such luxuriant growth that it was difficult and laborious for a man on foot to force his way through, and nearly impracticable for our light carriages.... during the day there had been but very little grass, except in some green spots where it had collected around springs or shallow lakes.

On August 15, the expedition had reached the Green river by way of South Pass, with artemisia still a companion.
On Frémont's return from the Pacific coast in 1844 he descended the Arkansas river to Bent's fort and then crossed in a northeasterly direction to the headwaters of the Smoky Hill river west of 102°, descending that stream to a point probably somewhat west of Hayes, Kansas, about midway between 100° and 99°, when under a date line of August 17, he summarized the landscape quite clearly as of three types, short grass, mixed grass, and tall grass:

The country through which we had been travelling since leaving the Arkansas river, for a distance of 260 miles, presented to the eye only a succession of far-reaching green prairies, covered with the unbroken verdure of the buffalo grass, and sparingly wooded along the streams with straggling trees and occasional groves of cottonwood; but here the country began to change its character, becoming a more fertile, wooded, and beautiful region, covered with a profusion of grasses, and watered with innumerable little streams which were wooded with oak, large elms, and the usual varieties of timber common to the lower course of the Kansas river.
As we advanced, the country steadily improved, gradually assimilating itself in appearance to the northwestern part of the State of Missouri. The beautiful sward of buffalo grass, which is regarded as the best and most nutritious found on the prairies, appeared now only in patches, being replaced by a longer and coarser grass, which covered the face of the country luxuriantly.

In 1849, Major Osborne Cross led a regiment of dragoons from Fort Leavenworth to Oregon, leaving the fort May 20. For six seasons, the Oregon emigration had been moving over the trail, and in 1849 an even greater volume was bound for California; 4000 wagons accompanied by some 50,000 animals were ahead of Cross's dragoons on the south side of the Platte. The situation provides an opportunity for comparison with descriptions of pre-emigration days. The third day out, he commented on the change taking place in the country, trees to grass, "an endless prairie country,... very beautiful at first sight, but becomes tiresome beyond description after the novelty has worn off.... Nothing from day to day but the broad canopy of heaven above, and the greensward below." Of the small streams he crossed, Wolf creek, the Big and the Little Nemaha, the Vermillion, the Big Sandy, and the Little Blue, and many others, he made the significant comment that they were not fed from a permanent source, but were to be looked upon only as "drains to the prairies." On the third day out above Fort Kearney, June 4, the valley of the Platte was as destitute of timber as the adjacent prairie;

although it is large, it is but a drain for the melting snows from the mountains, and can only be remarkable for possessing more sand bars, less depth of water, and more islands half covered with useless timber, than any other stream of its size in the country. It is not navigable, nor can it be made so, and, in a commercial point of view, has very little to recommend it.

Back from the bluffs on June 7, the entry ran,

nothing could be seen but large buffalo trails; the deep ravines were much trodden and torn up, forming wallows, which are resorted to by them when these places are partially filled with water.

Two days later Cross recorded, "The road today much cut up by gullies, which are the natural drains from the highlands to the river...."

The influence of the flood disturbance and of the water erosion was revealed by the entry for June 19, the first camp past Chimney Rock:

wood, as usual, was scarce, but we obtained enough in the valley for our use, that had been swept down from the hills by the heavy rains which frequently fall during the summer. What was found, principally consisted of dwarf cedar and pine. We had but very little for our horses at this encampment and the grass began to change as rapidly as the face of the country.

An example of the flash floods of the plains occurred on July 25, west of Fort Laramie; rain and hail:

It lasted but a short time, and was very partial, as the rear division got none of it. The water came in torrents from the hills....

The ravines, which a few minutes before were dry, soon became filled....

On June 28, six days west of Fort Laramie, "the hills and valleys ... were entirely destitute of anything like vegetation, except artemisia." The camp of June 29 was at the mouth of Deer creek, near the Platte river crossing:

Our road today passed over a dreary and uninteresting route— more so than any since leaving Fort Laramie. The hills are not so high as you approach the Platte, but entirely barren. Nothing was to be seen but the artemisia, or wild sage, which is extremely uninteresting, having neither beauty nor usefulness to recommend it, and its odor by no means pleasant. We are now destined to travel a very long distance where this shrub was constantly seen, and in greater quantity than had already been met with, for it may be said that we had just entered it, and it was not very plenty or large, compared with what we afterwards met with on the route.

There must be something in the composition of the earth particularly adapted to its growth, for, whenever grass was scarce, we invariably found it in great quantities. I have travelled for days, before reaching the Columbia river, where nothing could be seen on the highlands and plains but artemisia, which for miles looked as if the whole country had been cleared of all vegetation to make room for it.

The crossing of the Platte occurred on July 5, grass was scarce and the divisions were ordered to travel one day apart, "The face of the country having entirely changed since leaving Fort Laramie, it was only at certain points in our day's marches hereafter that grass could be procured, and even then in limited quantities."

The following day the theme was continual, comparing the grass then available with that between Fort Laramie and Fort Kearney:

there was no portion of the route but what grazing could be had at any moment, though much better in some places than at others, but such is the formation of the soil, and its extreme sterility, that you are compelled to travel sometimes a whole day before getting to a spot where you can find the least quantity, and these places this spring have been so frequented that the grass had been entirely consumed.

West of the continental divide on the Bear river, beyond the Green, Cross called attention, July 27, to another change of scene:

Since arriving at Green river, I observed a great change in the soil among the mountains and ravines. We were now getting to where a fine, short grass was to be found on the sides of the hills and ravines; although not very thick, it was considered very nutritious, which I presume must be the case, as our animals would leave the bottoms and climb to the top of the highest hills to hunt for it.

At Steamboat Springs, the roads forked, the California gold emigrants turning off in that direction, few continuing to Oregon, so Cross found the grass good, August 2, the first day west of that point. The Snake river country was ahead which "was entirely destitute of grass to the Cascade mountains, a distance of 700 miles." On August 13, five days out from Fort Hall, "the scenery for the last two days was much the same, the picture being made up of distant hills, barren wastes, and wild sage, with not a tree to intercept the view," and near Solomon Springs two days later, "the road was so pulverized, that, by every revolution of the wheels, it would fall off in perfect clouds."

The entry of August 19 related that during the preceding night a norther had struck with such force that the "wagon covers were torn to pieces, and our tents blown down over us, and in the morning we were completely buried alive in sand, which had drifted on the tents as they lay over us. The morning continued windy, raising clouds of dust so thick that the wagons descending the hills, were completely enveloped...." After passing Fort Boise, the country changed and the "hills were well covered with bunch grass, which was very strengthening and much sought after by the mules, and we were fortunate in getting it for them through to the Grande Ronde."

The Santa Fé trail had been travelled by traders since the early 1820s, but the trade did not become big business until after the outbreak of the Mexican war in 1846. Disturbance to nature could not have been very extensive along its course until after that date. Three reports on conditions are chosen here as a basis of description as of that year.

Dr. Wislizenus, of St. Louis, attached himself to the wagon train of A. Speyer, which started from the frontier near Independence, Missouri, May 22, 1846. He was equipped for making scientific collections and observations. He took the tall-grass prairie for granted, making little description of its vegetational cover. On the first full day on the road, May 23, his comment was that

> The grass had all the freshness of spring, and the whole plain was so covered with flowers, principally with the sky-blue Tradescantia virginiea [spiderwort] and the light red Phlox aristata, that it resembled a vast carpet of green, interwoven with the most brilliant colors.

At council Grave he commented on the timber of hickory, oak, walnut, elm, and ash, and generalized on the relation of this landmark to the country east and west:

The vegetation is quite luxuriant, and the soil is very fertile For agriculture, as well as raising stock, the place would be excellent.

Council Grove forms, as it were, a dividing point in the character of the country east and west of it. The country east of it is formed of prairie, with slight ascents and descents — constant undulations, ... resembling the waves of the ocean.... This eastern portion is well watered, and along the water courses sufficiently timbered to sustain settlements.

A short distance west, the country rises suddenly to the elevation of 1500 feet, and ascends gradually towards the Arkansas to 2000 and more feet above the sea. The intermediate country yet exhibits sometimes the short, wavelike form of the eastern portion, but oftener it resembles already the plateaux of high plains between the Arkansas and the Cimarron, those representatives of the calm, immense high seas, where the horizon extends further, the soil becomes drier and more sandy, the vegetation scantier, timber and water more rare. The country between Council Grove and the Arkansas form the transition to the sandy plains on the other side of the Arkansas [100°]; the soil is generally less fertile than in the eastern portion, but all along its water courses.... settlements might succeed, though they would have to depend more upon stock-raising than agriculture.

Near Diamond Springs, the following day, he saw the first antelope; on the Little Arkansas four days farther west, but still east of 98°, the first prickly pear cactus (Opuntia vulgaris). The first mention of buffalo grass was west of Pawnee fork of the Arkansas, but he had been in the buffalo grass country for several days. Although this was only June 6, "the short buffalo grass is rather dry as everywhere else now." Sixteen miles farther, the vicinity of the present Kinsley, Kansas, 99° 30', a dry camp was made "with tolerable grass." The Cimarron crossing of the Arkansas near the present Dodge City, 100°, was made June 10. Ahead of them was sixty-six miles of trail to the Cimarron river without dependable water. From the river:

Our road led through deep sand. Grass was very scanty, but there was quite an abundance of sand-plants, and the ground was covered with the most variegated flowers, especially the gay Gaillardia pulchella, that it looked more like an immense flower garden than a sandy desert.

The first camp had "poor, dry grass," the buffalo had entirely disappeared,

not even buffalo chips, ... were to be seen. The high plain between the Arkansas and the Cimarron, whose elevation ... is about

3000 feet, is the most desolate part of the whole Santa Fe road.... The soil is generally dry and hard; the vegetation poor; scarcely anything grows there but short and parched buffalo grass and some cacti....

Above the lower springs of the Cimarron, June 14,

the soil has now become entirely sandy; different species of artemisia ... cover the whole plain; horn frogs, lizards, and rattlesnakes find a comfortable abode in the warm sand; thousands of grasshoppers occupy all shrubs and plants; mosquitos and buffalo gnats the air; — what a great place for settlements this would be!

A dry camp June 14 had "tolerable grass, considering we were on the Cimarron," and the following day the Cimarron was crossed, finding running water for the first time in the river and grass was better:

In looking back from here towards the Arkansas, it is hardly necessary to remark, that this whole country, from the crossing of the Arkansas to the crossing of the Cimarron, will never be settled, from the scantiness of grass, the scarcity of water, and the entire want of wood.

Brigadier General William H. Emory made the trip to Santa Fé and other points in the Spanish southwest as an officer in the Army of the West which marched against the Mexicans during the summer of 1846, about a month later than Wislizenus. Emory divided the country between Fort Leavenworth and Santa Fé into three parts, the breaks being indicated as the Pawnee Fork of the Arkansas river near 99°, and Bent's Fort on the Arkansas river west of 103°. The analysis which followed revealed, however, a quite different interpretation of the country traversed and one that recognized four kinds of landscape and vegetation. He made no detailed description of the segment from Fort Leavenworth to the Pawnee Fork because it had been so often traversed. Nevertheless, he gave a general description:

Trees are to be seen only along the margins of the streams, and the general appearance of the country is that of vast rolling fields, enclosed with colossal hedges. The growth along these streams, as they approach the eastern part of the section under consideration, consists of the ash, burr oak, black walnut, chestnut oak, black oak, long-leaved willow, sycamore, buckeye, American elm, pig-nut hickory, hackberry, and sumack; towards the west, as you approach the 99th meridian of longitude, the growth along the streams becomes exclusively cottonwood. Council Grove creek forms an exception to this, as most of the trees enumerated above flourish in its vicinity, and render it, for that reason, a well known halting place for caravans.... On the uplands the grass is luxuriant, and occasionally is found the wild tea ... and pilot weed.

At this point in his description the substance is that the region from about 98° to 100° was a transition country between the prairie and the plains. This fact is made clear, merely by changing the sequence of his sentences:

The transition is marked by the occurrence of cacti and other spinose plants, the first of which we saw in longitude 98°. Near the same meridian the buffalo grass was seen in small quantities, and, about noon, our party was cheered for the first time by the sight of a small band of buffalo.... The next day immense herds of buffalo were seen.

Then the earlier sentences in the description makes the true sequence of ideas:

As you draw near the meridian of Pawnee Fork, 99°, west of Greenwich, the country changes, almost imperceptibly, until it merges into the arid, barren wastes described under that section....

A later sentence concluded the sequence:

The section of the country embraced between this point and Bent's fort is totally different in character from that just described, but the change is gradual, and may be anticipated from what has been said in reference to the appearance of the country so far east as the 98th degree, or even the 97th meridian.

Emory's characterization of the country between Pawnee Fork and Bent's fort (west of 103°) pointed out that the Arkansas bottom was

generally covered with good nutritious grass. Beyond this [bottom] the ground rises by gentle slopes into a wilderness of sand hills on the south and into the prairie on the north.... as you approach Bent's Fort, the hills generally roll in more boldly on the river, and the bottoms become narrower, and the grass more precious.

The eye wanders in vain over these immense wastes in search of trees. Not one is to be seen. The principal growth is the buffalo grass, cacti in endless variety, and rarely that wonderful plant, the Ipomea leptophylla ..., man root, ...[which] serves to sustain human life in some of the many vicissitudes ... to which men ... are subjected.

At the Big Sandy (about 102° 25') yucca appeared "and marked a new change in the soil and vegetation of the prairies." By judicious selection and distribution of camps near Bent's Fort the grass in the Arkansas bottom served the Army of the West while it was recuperating for the next stage of the long march. Thereafter the fuel was often buffalo dung and sage brush. Wild game was scarce, except buffalo, within this range, but their range was uncertain. This particular year it was 98° to 101°. Emory warned against an army trying to live off the plains.

Lt. J. W. Abert's journal of the Canadian river expedition of 1845 had been reviewed already. He had started from Bent's Fort and had kept a rather detailed account of the early part of his trip. He was with the Army of the West in their march in 1846 and kept notes on natural history. As in the case of Frémont and Marcy, he seemed to have learned from his first expedition. His report of the expedition of 1846 showed that he had been doing some thinking about his data, and whatever the stimulus or source of inspiration, he had arrived at some rather well organized ecological conclusions recorded for the first day out from Fort Leavenworth, June 27:

The ground is what is called 'rolling prairie', of gentle curves one swell melting into another. The soil around is extremely rich; the whole country is verdant with a rank growth of the 'tall grass', as it is called by way of eminence, when compared with that which grows beyond the region of the walnut and the hickory. Here are many varieties of useful timber: the hickory, the walnut, the linden, the ash, the hornbeam, the maple, the birch, and the beech, also cottonwoods; but beyond the limits of the 'tall grass', there is the cottonwood only.

As was customary with naturalists of the period, being forest men, Abert placed his emphasis on trees and forbs, rather than upon the grasses and their species. Unlike most of them, however, he observed the relations of animals, the pocket gopher, to the tall-grass land as well as the prairie dogs to the short-grass land. On the second and the sixth day (east of 110 mile creek on the Santa Fé trail) he described their operations:

Whenever we rode to the side of the road we noticed that our horses would frequently sink to the fetlock, and saw on the ground little piles of loose earth, like small ant hills, being about 5 inches high and 10 or 12 inches in diameter at the base, and without any opening; they are formed by the sand rats, or gophers, (pseudostoma bursarius) and although their habitations cover the prairies, there are few persons I have met with who have seen them. (June 28).

The mounds made by the gopher, or sand rats, were more abundant than heretofore, and in several places a number of these mounds had been made so close together that the distinctness of each was completely lost in the mass, covering an area of five or six feet. (July 2).

Grasshoppers grew in abundance, and the same varieties of timber at 110 creek as on the Kansas crossing below the mouth of the Wakarusa, but he thought the season dry because the creek, which was full of water in 1845, had only "a few scattered pools" in 1846. "Independence creek had the first running water since the Wakarusa, and had elm, cottonwood, hickory and oak timber." The camp on Big John creek, close to Council Grove, had walnut, oak, and sycamore. It was here about 96° to 96° 30' that the hickory and the oak disappeared from his list of trees, and if his tall-grass-hickory-oak theory was consistent, this was near the edge of the tall-grass prairie.

Somewhat west of 97° was the Turkey creek branch of the Little Arkansas, the camp of July 8:

> We had now reached the short grass, that is not more than four or five inches in length, and we saw little patches of the true buffalo grass, (<u>sesleria</u> <u>dactyloides</u>) a short and curly grass, so unique in its general character that it at once catches the eye of the traveller.

At the same time, Abert referred again to animal relations to the grass:

> On either side of us we observed little circular spots marking the places where the buffalo once wallowed.... These old wallows are now overgrown with plants that grow more luxuriantly than on other portions of the prairie.
>
> It is seldom, now, that the buffalo range this far east; no signs of old excrements are to be seen; and the bleached bones left upon the plains by the hunters have long since mouldered away

Two factors contributed to the renewed vigor of the vegetational cover; the disturbance of the soil from the wallowing of the buffalo, and the fact that the wallows caught the rain. The first prairie dog town to be mentioned was July 11, west of Cow creek (present Rice county) just west of 98°, and the next day on the Great Bend of the Arkansas, cactus (<u>Opuntia</u>) and "on all sides the little mounds of loose earth thrown up by the pocket gophers." In the vicinity of Pawnee Rock (99°) "we ... entered upon the vast plains of buffalo grass;" here the buffalo covered the plains, and that night the wolves, that followed the buffalo herds, provided them with a serenade.

The expedition had reached Pawnee fork of the Arkansas, just west of 99°, which was flooded. While waiting for the waters to recede, Abert tried to dig out a prairie convolvulus (<u>Ipomea</u> <u>leptophylla</u>) or man root. After following its stem for about twelve inches, it suddenly expanded to about 21 inches in circumference and about two feet long. The tea or lead plant, false indigo (<u>Amorpha</u> <u>canescens</u>), "is in some places so abundant as to displace almost every other herb," and prairie indigo (<u>baptisia</u> <u>leucantha</u>) was plentiful. Both were of the legume family. On July 16 they were on Coon creek, in the vicinity of the present Kinsley, where "the neighborhood is generally destitute of grass." The position was the 99° 30' meridian, and the following day's entry contained several items of outstanding ecological interest:

> We have now entered that portion of the prairie that well deserves to be considered part of the great desert. The short, curly buffalo grass (<u>sesleria dactyloides</u>) is seen in all directions; the plain is dotted with cacti and thistle (<u>carduus lanceolatus</u>) while only in buffalo wallows one meets the silver margined <u>euphorbia</u>; and in the prairie dog villages, a species of <u>asclepias</u> [milkweed] with truncated leaves.

Here occurred the only note on wild horses, and the "buffaloes seemed as if trying to surround us. We saw scarcely anything else far or near. The whole horizon was lined with them." And where there were buffalo there were dusky wolves (<u>canis nubilis</u>) in large numbers, an association that was so much a part of nature that the buffalo paid no attention to them.

As Abert became seriously ill July 21, he discontinued his notes for the remainder of the trip. He arrived at Bent's Fort on July 29. In summary, however, he pointed out that "As one approaches Bent't fort, he meets with many varieties of artemisia, with the <u>abione canescens</u>, and ... <u>yucca angustifolia</u>" (soap root, Adam's needle, Spanish bayonet, or the Mexican name palmillo).

In the absence of an Abert journal for 1846 covering the region south of Bent's Fort, that of 1845 is used here. Starting down the Arkansas river August 16 on the south side of the site of the present city of Las Animas, Colorado, at the mouth of the Purgatory or Las Animas river, a stream twenty yards wide and "highly charged with sedimentary matter." This right bank was

> one continued series of hills and sand plains. We noticed a profusion of prairie sage (<u>artemisia tridentata</u>) being about the only shrub that grows in these sandy regions. This plant seems to love a dry and arid soil, covering, as it does, millions of acres of the great desert at the eastern base of the Rocky mountains. In some places it grew so luxuriantly that the stalks might be used for fuel.

Up the west side of the Purgatory river for three days, then crossing over the Timpus creek, "cactus were numerous", and the divide between the streams was "a succession of sandy rolls covered with artemisia, yucca angustifolia, and a species of cactus — cactus Peruviana." The Indians found it necessary to have soles on their mocassins. "In this country they are obliged to sole them to protect the feet against the numerous cacti. The soles are of 'par flêche' — inside edge cut perfectly straight, and the toe pointed." In the camp of August 19, in the Timpus valley about twenty-five miles by air-line southwest from Bent's Fort, "As there was no timber, we were forced to use the artemisia as a substitute, which grew so luxuriantly as to be almost impenetrable. We found it answered very well, burning with a slightly crackling and clear flame." Two days later the entry recorded that "the artemisia is literally alive with a large species of gray hare, lepus Americana."

Either cactus species were multiplying, or Abert made a more careful entry August 20 higher up towards the divide: "The cactus were very numerous; among which we saw the cactus openetia [opuntia], a kind resembling a small canteloupe half hidden in the ground, C. melocactus, and the C. Peruviana...." He commented on the effect of the great drouth but the bluffs toward the mountains were green with grass and trees. This was a characteristic of the country south of Raton Pass, a desert-adapted vegetation on the plains, and the green of the pines, cedars and grass on the mountain elevations.

These descriptions of the grassland at mid-latitude tend to confirm the views of those ecologists who hold to the three-way division of grass associations; tall grass, mixed grass, and short grass. They emphasize that disturbance was not only the normal condition in nature, but that disturbance was a positive contribution to the well-being of vegetation and soil. Conservation theory should not be oriented in terms of prevention of erosion and indiscriminate destruction of wild animals and so-called weeds, but in terms of control measures derived from appreciation of the ecology of the grassland in its natural condition. Furthermore, the distribution of modern agricultural crops is found to be in accordance, broadly, with the ecological divisions found among the grasses. Corn and soft winter wheat in the mid-latitudes extend westward as far as the tall grass and the oak and hickory, about $97°$ in Kansas. West of that line, are the hard winter wheats, and the dry land grain sorghums. The outcrop of Flint Hills limestones and sandstones in the Bluestem Pasture Region sharpen this line of division, but the broader combination of ecological factors determine it as a matter of agronomic experience. The line of the 100 meridian, approximately, marks another transition in which the hard winter wheat must divide attention with range livestock and grain sorghums.

Chapter Ten

FACTORS IN GRASSLAND EQUILIBRIUM

Plant relations

A land of uniformity and monotony was the first reaction of forest men to the grassland, and to many it was not only their first but their lasting impression. Nothing could have been more erroneous. The deficiency was in the mind of the forest man and not in the grassland. Because of the absence of trees, the grassland possessed in its structure fewer vegetational layers and a lesser range of height of the layers and lesser spread between them. The composition of the vegetation presented, nevertheless, a wide range of variety of grasses, forbs, and woody plants. All were in intense competition with each other and with invader plants. The tall grass prairie possessed more layers and height than the mixed prairie; and that in turn than the short grass plains, but the desert-grass transition again introduced increased numbers of layers, height, and spread. Not only were there more layers and greater height at the two extremities of the grassland, the forest borders and the desert borders, but the proportion of woody plants was likewise greater in those borderlands.

The layering of vegetation was an aspect of biological equilibrium in the competition of nature; some plants thrived on light, others were destroyed by it; some thrived on shade and others were destroyed by it. In the long periods of time involved in the evolution of the grassland formation, the equilibriums were worked out which established vigorous plants in each category in the places in which they possessed the qualities of survival. Invader plants could not survive unless they possessed characteristics and vigor which fit them into a niche in the complex system. In few places were there pure stands of any one species. In fact, under most circumstances, pure stands of any vegetation were signs of weakness rather than strength. The fullest equilibrium in nature was attained over long periods of time and stresses where species of varied characteristics provided the most complete interchange of compensations with each other, with animals and with soil. In cultivated fields of controlled pure stands, the farmer provided those compensations artificially and periodically; by rotation of crops, addition of fertilizers, and machine tillage. In nature, these processes operated simultaneously and continually as a consequence of the variety in the forms of plant and animal life. The grasses constituted the principal portion of the vegetation, the forbs of the composite family were usually second, and the legumes third in rank. The literature has not made clear the rôle of the composites and was not sufficiently specific about the contribution of the legumes to the fertility of the grasslands. The popular opinion was to call them all weeds, and as they had little or no forage value, to wish to kill out all such weeds. The presence of the legumes was vital, however, as nitrogen fixers of varying efficiency and there is need of studies describing exactly and quantitatively the place of each of the major legumes, the varying combinations of them, and their over-all significance.

In collecting plants in the grasslands, the earliest explorers seemed more interested in what they called plants rather than grasses. These forbs were a conspicuous feature of the landscape and appeared prominently in the herbaria even though many of the collectors lost large parts of their findings as the result of accidents of travel. Nicollet's area of collecting was between the Mississippi and the Missouri rivers, partly forest, partly grass (1836-1840), and he listed 82 composites and 33 legumes, as well as species of other families, and 42 grasses. He lost about half of his specimens. Frémont (1842) lost part of his collection but, the catalogue of those preserved included 93 composites, 33 legumes, and 18 grasses. Abert's (1846) list included 19 composites, 18 legumes, and 5 grasses. Emory's (1846) southwestern desert collection included 11 specimens of grasses which were sufficiently complete for identification, and 9 not identified as to species, at least 2 legumes, and 4 composites. Marcy's (1852) Red river expedition yielded 27 each of composites and legumes, and 29 grasses. John Torrey was the principal authority who made the classification for most of these collectors.

Within the unstable equilibrium of plant competition for light, water, and nutrients, seasonal distribution was an important factor. Many small annuals matured and seeded in the early spring before the grasses and other perennial plants made their growth. The characteristic grasses tended to make their growth in the late spring and early summer, becoming dry by mid-summer, but even among them there was a seasonal succession. Many prominent tall forbs made their principal growth during the summer and fall.

Below the ground surface there was competition in many respects more significant (if one aspect of biology could be more significant than another) to the problems peculiar to the survival of vegetation in the grassland climate than above the surface. That the most of the grassland vegetation was under the ground was no mere figure of speech. The roots of the grassland plants presented a wide variety of forms and habits. The roots of the grasses were fibrous, and of the short grasses finely fibrous. Those of the forbs mostly were either branched or tap roots. As pictured diagrammatically by Weaver and Fitzpatrick (1934, p. 123) for the tall grass prairie of eastern Nebraska, there were roughly three levels of roots; prairie June grass (Koeleria cristata) in the first fifteen inches, little bluestem (Andropogon scoparius) extending to five feet, and the Psoralea floribunda, a forb, more than five feet. By individual species, the depths were more varied so that the idea of three levels tended to be minimized, except for the basic fact that different species tended to occupy different levels of the soil and thus did not compete with each other for water and nutrients so much as the individual plants of a species competed with each other. The big bluestem (Andropogon furcatus) roots reached depths of five to seven feet and the Amorpha canescens, a nitrogen fixer, as much as twelve to sixteen feet. Roots of different species responded differently to the drouth of the 1930s in the vicinity of Hays, Kansas (Weaver and Albertson, 1943). The depths of buffalo grass roots before and after the eight year drouth were four and one-half feet and two feet respectively; blue gramma five and two feet. On a different plot in the same vicinity growing the taller grasses, big bluestem prior to the drouth reached five feet, and seven years later, six feet; bluestem wheat grass (Agropyron smithii) depths were six to seven and seven to eight feet respectively for the two periods, and sideoats grama (Bouteloua curtipendula) five and six respectively.

The plant that captured Frémont's imagination from the start of his expedition in 1842 as characteristic of the tall grass prairie was the Amorpha canescens (lead or tea plant or false indigo). His interest in it was excited by its outward appearance, but by unforseen coincidence the emphasis the plant received in his journal contributed in an important manner to better understanding of the mechanism of biological equilibrium. Science was later to establish the rôle of certain legumes as nitrogen fixers and this one fell into that class (Weaver and Fitzpatrick, 1934). In his journal describing the natural history between Fort Leavenworth and Bent's Fort in 1846, Abert mentioned the same plant and also the prairie indigo (Baptista leucantha), as conspicuous. Wislizenus (1839) and Frémont (1843) commented on the Psoralea esculenta (various common names, Pomme Blanche, Pomme de prairie, and prairie potato), and all travelers in the southern grassland wrote of the mesquite trees, both legumes. Marcy (1854) commented that settlers recognized mesquite land as particularly valuable and competed for its possession. It seems possible that there was more reason for this choice than just the tradition, usually unfounded, that forest land was more fertile than land without trees. It is an error to assert, as has been done (Shelford 1944), that it was cattle that spread the mesquite tree from south central Texas northward to Oklahoma. All the explorers reviewed, who covered the grassland from the Canadian river southward between 1845 and 1854, testified to the presence of the mesquite prior to any cattle drives through that region (Chapter 8).

Among the various means by which the grassland plants survived the severe fluctuations of climate not the least were those below ground. The several varieties of _Ipomea_ and of _Psoralea_ presented in exaggerated form, in their large roots, a reservoir for storage of water and food. The plant food reserves served also as food for rodents and for Indians. Few of the grasses most characteristic of the drier portions of the grassland produced any substantial crop of viable seed. Severe drouth, heat, and periodic evergrazing tended to reduce the probabilities of seed production and prairie fires the possibilities of seed survival. In the age-long process of evolution of grassland plants, those had survived and established dominance that could most successfully propagate themselves, and except for the buffalo grass which spread by stolons (runners), the most of them depended primarily upon underground parts, rhizomes for the most part, but some upon corms, bulbs, or tubers. It should be emphasized also that such underground parts were particularly important as defenses against the temperature extremes of heat and cold. The plant tops died down each year and might be burned off, the seeds might not mature or might be destroyed, but the underground protection made survival possible. Thus in the ecology of the grassland, the vegetation had attained a stability against the hazards of light, water and temperature.

Animal relations

Animal ecologists were behind the plant ecologists in their study of the grasslands and plant ecologists ignored largely the animal factor, yet scientific information on the influence of animals upon soil formation and upon plant succession and climax was essential to the understanding of the grassland as a natural region. Grinnell's (1923) study of California rodents is one of the most significant American works available. He estimated the burrowing rodent population as constituting one-half of the whole number of mammals in the state. These burrowing rodents extended east to about the 100 meridian, beyond which they began to disappear. The rodent relations to soil and vegetation were summarized under nine heads: the substratum was weathered by the opening up of deep holes; substratum material was brought to the surface, scattered, and subjected to weathering; wind and water distributed this loose soil; rainfall was absorbed through the rodent runways, run-off minimized and evaporation retarded; a more vigorous vegetational cover was promoted by this conservation of moisture; soil fertility was improved; buried vegetation was incorporated into the soil; runways, galleries, and holes counteracted the packing effect of hoofed animals; and the rodents as grass eaters competed with grazing animals in consuming and converting vegetation which contributed to soil fertility, and to restriction on growth, the extent depending on varying numbers. Ants which were characteristic also of grasslands, made important and similar contributions as had been elaborately explained years earlier by Charles Darwin.

Formosov (1928) summarized his own and the research of other Russian scientists on the subject of the grasslands of central Eurasia, regions little disturbed by the activities of modern civilization. He tied his paper directly into that of Grinnell by restating the latter's conclusion that "On wild land the burrowing rodent is one of the necessary factors in the system of natural well being." The Russian observations confirmed fully all that Grinnell had said and elaborated and emphasized it, but there were important respects in which Formosov went further. Among the former points was the conclusion confirming the importance of loess formation by wind-transported dust from rodent mounds. Of the additional conclusions, a most significant one was the effect of raw soil thrown out upon the surface in retarding vegetational succession. Until such soil materials had been subjected to the long term soil forming process they promoted the growth of a vegetation different from that prevailing upon fully formed soil. The final effect, therefore, of such animal influences was to "contribute towards maintaining a more stable existence for the dominant vegetation...."

The role of the hoofed animals, antelope, mountain sheep and wild asses, was explained by Formosov. In ungrazed areas, excessive growth of grasses, especially the taller and bulkier type, the feathergrass (Stipa capillata), suppressed partially or wholly the weaker plants such as Festuca ovina, Koeleria gracilis, and Poa bulbosa. Furthermore, the excess of dead cover smothered the dominant grass itself, resulting in replacement by weeds and a new succession sequence before reestablishment of the original dominant. Where hoofed animals grazed, however, they removed excess leafage, and by tramping broke loose dead stalks preparing the way for new growth of grass. Grazing and tramping down of grass, also reduced water losses through transpiration and evaporation during dry periods, and tramping promoted the natural reseeding process. In extended spaces the grass was sometimes tramped out altogether, especially at resting places and around lakes and watering places. The whole complex of activities contributed to the well-being of the animals and promoted the equilibrium of soil-vegetation relationships through the process of natural tillage. As a conclusion to this summary of Formosov's study, the fact should be pointed out that some of the dominant grasses involved were species of the same genera that occupied so largely the northern portion of the North American grassland, and the animal population had performed in North America a similar function. With settlement the original large wild animals were largely exterminated, but many of the small ones still occupied the grasslands where agricultural operations did not interfere. Probably the insect population remained more completely, but in somewhat changed proportions, and some new ones had been introduced. One tentative reservation should be made to this generalization, however, until more is known of the regional distribution of microorganisms of the soil.

The rôle of insects was an important component in natural processes. Hayes (1927) demonstrated that in Riley county, Kansas, a large part of the insect population burrowed into the top thirty inches of soil for the winter. If they died there their remains enriched the soil, but as most of them emerged the next season they channelled and areated the soil both in entering and in leaving. It is important further that grasshoppers and other insects were sensitive to heat. As pointed out by Ball (1937), among the effective means of controlling grasshoppers in the grasslands was to graze off the grass in an infested area, another example of the relative stabilization of equilibrium through natural biological controls.

The Clements (1916, etc., and Weaver and Clements, 1929, 1938) theory of climax and disclimax as applied to the different subdivisions of the North American grassland was based on the contention (1936; Yearbook No. 39, 1940) that in the natural state, before the coming of the white man, the influence of wild animals, fires, and Indians on the vegetation was negligible. Thus he denied that in the true prairie the bluestems were real dominants, that on the plains the gramas and buffalo grasses were real dominants, and that on the desert the mesquite shrubs, creosote bush, and sagebrush were real dominants. In all cases they were apparent dominants because of overgrazing, fires, rodents, and human influences, and by protection could be restored to nature. One of the strangest aspects of Clement's (Yearbook No. 32, 1933; No. 36, 1937) argument was in attributing greater vigor and capacity of resistance to these plants, which he called false dominants, than to the real dominants. Obviously, his definition of dominant capacity was one that recognized as true dominants only those grasses that thrived under perpetual optimum conditions. Even under Clements's climate definition of the grassland, this seemed inconsistent, because climate analysis revealed the grasslands as areas of fluctuating climatic extremes rather than uniform climatic optimum. It seemed inconsistent also with his whole theory of dynamic ecology to base the concept of dominants upon requirement of an optimum which must necessarily be substantially static. In this as well as later, in declaring (Yearbook No. 39, 1940) that at their maximum, the wild animals exerted "only a transient effect upon the [grass] climax," his views seem to conflict with those expressed elsewhere (Clements and Shelford, 1939) insisting that plant and animal relations should always be studied together as bio-ecology.

A field experiment with a buffalo herd in the Wichita National Forest, Oklahoma, was carried out in the summer of 1933 (Yearbook No, 32, 1933). The contention of Clements was that this proved that a buffalo herd straggled like cattle when grazing on the open range; that buffalo did not graze grass clean and uniformly; that a grazed area recovered its growth of mid-grasses when fully protected; that the mid-grasses were the true dominants; and that it was demonstrated conclusively that the buffalo herds did not produce the short grass plains disclimax. Irrespective of whether or not his conclusions might be correct, his summer's experiment proved nothing. It is an example of the too-frequent assumption that a laboratory experiment is proof of what would happen in nature. A small herd of semi-domesticated, fence-broke buffalo would not necessarily graze in the same manner as the mass-herds of wild animals in their annual migration in open space, and their effect upon grass would not necessarily be in any manner comparable. Furthermore, the influence upon grass and soil would not be determined by one season, but would have to be considered on the long-term basis which would allow for climate fluctuations, frequent prairie fires, prolonged drouths, and the pulverizing of denuded soil by thousands of hoofs.

In his dicta concerning the buffalo experiment, Clements did not consider the problem of population numbers, increase potential, die-ups resulting from drouth, overgrazing, severe winters, and disease. In fact, elsewhere, he (Clements and Shelford, 1939) denied the relevance of these factors attributing control of numbers of wild animals in general to sunspot cycles.

The most effective statement of the historical point of view as opposed to the theoretical hypothesis of Clements, and of Weaver and Clements (1929, 1938), was a brilliant paper by Larson (1940) applying particularly to the northern plains. He contended that the short grass plains was a true climax and that the survival influence was a natural and integral part of it. The argument was directed particularly at Weaver's and Clements's three types of evidence; the reappearance of the taller grasses under protection, the reestablishment of the taller grasses during wet years, and the photographs taken by the Hayden expedition of 1870. The first point was met by quotations from historical records indicating that buffalo were probably as numerous as cattle and that the overgrazed grassland was a natural condition. The second point was answered by appeal to Taylor's restatement of the Liebig law of minimum as the test, not the maximum (in this case rainfall) as the test of plant behavior. The issue of the photographs was met by pointing out the location of the sites photographed as non-typical plains and that the interference with the buffalo herds provided unnatural conditions. A final point was Larson's insistence that the "marked ability of the short grass dominants to withstand overgrazing," indicated an adaptation of long duration.

Vestal (1931), in a review, likewise challenged the Weaver and Clements (1929) theory of the true short grass climax of the southwestern desert. Agreeing that overgrazing existed and caused to some extent the results described in the prominence of desert shrubs, he pointed out that in this region as in South Africa there was reason to assume that overgrazing took place under natural conditions and that the characteristics of desert shrub-short grass had prevailed for many thousand years. Carpenter (1940) recognized also the validity of the animal relations, and although he cited little specific historical evidence, he quoted an early description by J. Hildreth of the country near the 98 meridian in Oklahoma, between the site of Oklahoma City and the Wichita mountains, in which taller grasses, short grasses cropped close by buffalo, bushes, and cactus all appeared.

Irrespective of the controversial questions of cause, there were fluctuations in the numbers of wild animals in nature. At the minimum extreme in numbers there would tend to be more food than the animals would consume, and at the maximum end there would not be enough. Over a term of years that would span a series of such fluctuations there seems to be good reason to believe that a relatively stabilized equilibrium would be maintained and that regardless of increase potential, one plant and one animal of each of the component climax species would survive. Of course, if that were to be interpreted literally it would mean a static end-product condition, which is not valid. The ideas of change and succession mean that there would be a long term drift or tendency for the composition of the life forms to change. Except for intervention of some unusual disturbing factor, however, such change would be slow. As respects the numbers and food supply for both plants and animals the numbers would tend to approach the danger zone of survival. The biological principle has become so generally recognized as a fact that the burden of proof lies against anyone challenging its validity. As a matter of adequate scientific proof, as applying to Clements's theories of climax and disclimax in the grasslands, the verdict for the present must be rendered, not proven.

Buffalo numbers, in what is usually called the state of nature, would carry the story back to the jurisdiction of the anthropologist or at least as far as the historical geographer as outlined by Sauer (1941), but within the time-span of written records they were subject to disturbance from two directions, from the Spanish with their horse culture from the south, and from the English and their appropriation of the land which drove the Indians west. The direct pressure of the Spanish upon the buffalo range was relatively slight, but indirectly their northward march through Mexico as far as Santa Fé in the sixteenth and seventeenth centuries influenced all animal relations, and also introduced domesticated stock, sheep, cattle, and horses, to graze the grass of the occupied land. More fundamental to the buffalo problem, however, was the acquisition by the Indian of the grassland of the horse culture. The traditional dates and explanation of the acquisition by the Indian of horses was the escape of horses from either the Coronado or De Soto expeditions, or both, after about 1540. The discovery by Aiton (1939) of the Coronado muster-roll destroyed completely this convenient hypothesis when it was revealed that, of the horses on that expedition, there was only one mare and there was no record of her escape. Furthermore, the chances of survival of a few horses and their progeny in a wild country was virtually zero, even if there were proof that one or more pairs of stallions and mares escaped. The spread of the horse culture among the plains Indians, from the centers of Spanish influence in New Mexico, must not have occurred until late in the seventeenth and early eighteenth centuries as a relatively slow process. The Indian on horseback undoubtedly became a major disturbance factor to all animal life of the areas affected, because it gave to the Indian for the first time a mobility that revolutionized his culture, including his methods of hunting, particularly the hunting of the buffalo, and the following of buffalo migrations. Supplies of meat for food, and skins for clothing and shelter, must have modified the Indian population in relation to the buffalo population, and to the population of all animals directly or indirectly affected by hunting on horseback, or made accessible by the new potentials of Indian migration in horse pursuit of sources of food supplies. In the long run the whole biological equilibrium was affected by the introduction of the horse factor.

GRASSLAND EQUILIBRIUM

In the English advance from the east, the forest game was first affected by settlement, which eventually displaced both game and Indians in the regions actually occupied, as well as by systematic commercial trapping for furs by the French and by the English in the early seventeenth century. The buffalo in the interior were disturbed directly by the encroachment of white settlement west of the Appalachian mountains in the last quarter of the eighteenth century. By the time of the first Wislizenus (1839) and Frémont (1842) expeditions, it was a well established conviction among mountain men such as Fitzpatrick, Bridger, and Carson that the buffalo numbers had been already rapidly depleted. Plains Indian raids into Mexico were attributed (Report of Secretary of War, T. S. Jesup, 1850) to the driving pressure of this increasing scarcity of game. The progressive change in Indian-buffalo relations reached other Indian-game relations, and in turn was reflected in the animal-grass relations. Clements (1928) advanced the theory of a buffalo concentration zone resulting from the pressure of the advancing white frontier upon the buffalo range. There does not appear to be any historical basis for such a theory, however, as the problem of diminishing buffalo numbers had become a serious Indian food supply issue before the middle of the nineteenth century and was forcing rapidly, prior to 1850, a new Indian policy of reservations and government annuities to the disturbed Indians (Malin, 1921). Incidentally, also, all this occurred prior to the invention and use of the revolver, the breech-loading Sharps rifle, and the repeating rifles usually associated with the wholesale slaughter of the buffalo.

Any discussion of the problem of buffalo numbers in the pre-horse culture era, or the changes in numbers of buffalo and Indians during the century or so of the horse culture prior to the Anglo-American contacts, must necessarily be primarily theoretical. Attempts to reduce the problem of buffalo numbers to a mathematical basis become somewhat absurd when subjected to analysis. Within the period of Anglo-American records some evidence can be assembled tending to show a relation of buffalo numbers to grass, water, and winter storms which resembles closely the experience of the domestic range cattle industry. The Abert and Emory reports of 1846 represented relatively good moisture and grass conditions of the plains and the buffalo migration route was well to the west, Emory said, between $98°$ and $101°$. His report was focused on the matter of subsisting troops sent across the plains, and he warned that "their [the buffalo] range is very uncertain." He pointed out that "the buffaloes are sometime driven by the severity of the winter, which is here intense for the latitude, to ... feed upon the cottonwood." Kansas newspapers reported that during the drouth period of the 1860s spring flood water of the Kansas river carried masses of carcasses of buffalo that had died on the plains during the severe winter of scarce grass and storms. Charles Goodnight (Haley, 1936) reported large scale die-ups in the Texas Panhandle and in 1872 the T. P. Roberts upper Missouri river report mentioned that "Dead buffaloes were quite numerous on the plains about the falls." As a result of the severe drouth of 1860 Kansas newspapers reported that one effect had been to drive the buffalo eastward earlier than usual and many were then east of the Republican river. That meant east of $98°$. At this time the encroachment of settlement upon the buffalo range was reducing the buffalo to a limited space which interfered with free movement in the natural adjustment of their range to fluctuating climatic and grass conditions. This deals with freedom of range, rather than concentration of numbers, but it is only in this sense that there could be any validity to Clements's theory of concentration, and that is not his version of it.

The study of the natural history of the buffalo has important bearings on contemporary great plains policies. It was a part of the cattleman's propaganda to argue that there would have been no great plains problem if the farmer had been excluded and the area left to the cattle. The behavior of the buffalo disproves that propaganda conclusively, as their survival depended upon freedom in dry years to vary their migration range as far east as the tall grass prairie. The buffalo, in open space, did what the government was called upon to do as emergency relief in the 1930s in shipping great plains cattle east to the tall grass country for feed.

Furthermore, as a normal system of conducting the cattle business, the great plains and the desert southwest were not independent regions. After the cattle tick controls were inaugurated, it became standard procedure to breed cattle on the range, then ship them east to pasture and feed lot for maturing and finishing for slaughter. The Kansas-Oklahoma Bluestem Pastures, east of $97°$ became the largest single eastern pasture area in this livestock economy of regional interdependence (Malin, 1942).

The influence of animals in the grassland was in many respects in the nature of a natural tillage. Abert (1846) described the old buffalo wallows east of $98°$ that had become covered over with vegetation, and farther west, some in which that process apparently had not been completed. On the former he said that the plants "grow more luxuriantly than on other portions of the prairie," and on the latter that "only in the buffalo wallows one meets the silver margined euphorbia." The prairie dog towns also received the comment that there appeared "a species of esclepias, with truncated leaves." The influence of the pocket gopher was multiple as indicated by the fact that the horses' feet sank fetlock deep in the loose earth, and the earth from the subsoil was scattered over the vegetational cover. From the standpoint of the immediate effect on plants alone, the wallowing of the buffalo, and the digging of holes by the gophers, prairie dogs, badgers, kit foxes, etc., as well as the Indian digging roots for food, was disturbance. From the standpoint of succession and climax theory, the disturbance set back the plant succession causing a repetition of greater or lesser extent of plant succession to reestablish the climax. The processes were not occasional, but continuous. A third aspect of the action of these animals was that they contributed to the penetration of rainfall into the loosened soil and into the holes, diminished or prevented run-off into the streams, and conserved that moisture against rapid evaporation from the loosened soil. The digging of the soil and the subsequent mixing by rain, wind, and animals of that soil with vegetation and animal droppings was the final stage in the process of natural tillage performed by these animals as a part of the natural processes of perpetuating the long term vigor of the vegetation of the grasslands. In this long-time sense, and as a completed operation, the actions of the wild animals were not plant disturbance in a negative or destructive sense, but rather a positive process of natural renewal.

Climate relations: wind, water, and soil

In undertaking the description of climatic relations of the grassland from the standpoint of biological equilibrium, the historian is embarrassed by the volume of ill-advised propaganda of the conservation movement. Much of this is in the form of publications of private organizations, interested in some manner in the problem; much has been written by government officials, and issued by commercial publishers; but the greatest volume of this material is governmental publications. Nothing in this discussion is to be interpreted as opposition to conservation, but it is intended as a protest against misinformation and misuse of information in an excess of zeal to sell conservation policies to the public. Some of these policies were themselves unsound. Furthermore, all policies, sound or unsound, should be subjected to the test of full public discussion. The leading book in its field, Soil Conservation (1939), by H. H. Bennett, an official in the federal soil service, opened with a discussion of the 'virgin land' in which he pictured a nearly perfect biotic balance and removal of soil no faster than new soil was formed in nature. The water in the rivers was clear, he maintained, except under flood conditions which sometimes muddied the Mississippi and the Missouri. Another example is found in a governmental publication, Little Waters: Their use and relation to the land (1935, Revised 1936), by H. S. Person, in which was presented a diagram, (figure 29) representing erosion: for 1492 the surface lines were horizontal denoting complete absence of erosion, and for 1935 the lines showed differing degrees of denudation and gullying. This diagram and diagrams from other sources of similar validity were used in school books, fixing in the minds of students visually such totally false impressions. It should be recorded, however, that Person's text was carefully written, for the most part, the figure being the offending aspect of the publication.

Among the standard forest conservation arguments, dating from the early stages of that movement, was the contention that cutting forests caused floods and erosion. A challenge to this is found in Person, Little Waters, which questioned whether cutting timber modified much the run-off of water, as undergrowth sprung up which might possess greater density. The point of emphasis was that erosion was promoted only when the land was cleared off, and kept clear of new growth, but not by cutting the commercial timber. Another kind of argument was that floods were caused by careless farming which had caused the loss of the top-soil (Sears, 1936 and Meyerhoff reply, 1936). Similar charges were made with respect to plowing or grazing the grassland.

The cause of floods could not be attributed to such origins, as every school boy knew who read the spectacular story of the capture, during a flood, of Fort Vincennes by George Rogers Clark in the American Revolution, when neither the trees nor the top soil of the Ohio valley had been disturbed. In the history of Missouri river floods, that of 1844 seems to have set the all-time record at Boonville, Missouri, of 32.8 feet. Even the flood of June 1947 reached a crest at only 32 feet. Although several qualifying factors should be considered in interpreting these records of over a century, the simple fact presented by the figures is sufficient challenge to the propaganda that severe floods are man-made and of recent origin.

The Lewis and Clark expedition found the Missouri muddy in 1804 (Elliot Coues, Editor), and the geologist G. C. Swallow, in his Kansas report of 1866, reviewed the deposits of varying thickness in the Kansas and Missouri river bottoms, with emphasis on 1844:

That from the flood of 1844 is very conspicuous throughout the length of the Missouri and Kansas bottoms in this state. It is sometimes six or eight feet thick, particularly in low bottoms, so heavily timbered as to obstruct the current.

In 1854 an observer said of Missouri river water that "a common drinking glass full of it, allowed to settle, deposits a sediment at least half an inch thick," (New York Tribune, June 22, 1854). The Roberts engineering report of 1872 declared that "The water of the Missouri River, from the mouth of the Muscleshell down, never, even in the lowest stage, becomes clear...."

In the accounts of the Texas explorations of 1849 only a few clear streams were mentioned, and in plains geography the occasional appearance on the maps of a stream named Clear creek is a matter of significance in emphasizing that a clear stream was the exception. The clear headwaters streams were mostly limited to the limestone outcrop along the escarpment of east-central Texas, or similar outcrops in other parts of the grassland, but such clear beginnings did not flow much distance until they became muddy. Abert (1845) noted the muddy waters of the Canadian, on occasion too muddy for a bath. He described how the violent rainstorms eroded the ravines. Similar descriptions were recorded by Frémont and by Cross. The last named related also, how he gathered fuel in the upper Platte valley from the drift on the upland washed down by the floods.

The character of the streams of the drier grassland has been insufficiently emphasized and appreciated in connection with the history of the area. Except for the beds of the Missouri, Canadian, and Pecos, few rivers of the plains lie below the ground water level (Webb, 1931). The typical plains streams lie above the ground water level and often the beds are higher than the surrounding country. This was noted by some early travelers, Frémont (1842) in particular, describing the Platte tributaries. Geologists did not notice and describe this peculiarity until later, Warren (1859) describing the Niobrara as running along a ridge and having few tributaries. Merrill (1924) excused Warren's lack of understanding of the explanation, because he was not a geologist.

A second characteristic of the streams of the grassland, and this applies to those of most of the region, is the one stressed by Cross (1849) that they were merely "drains of the prairies" carrying off the local precipitation. To put it in the negative; they were not fed in volume by deepseated perpetual springs or glaciers. Cross made the point also, that the Platte was really only a drain of the mountain snows which melted into spring floods, and a drain of the prairies. When the melting of the snows and the spring rains syncronized, the floods were frequently disastrous. Both the rains and the melting of the snows were highly seasonal, so after the spring floods, the stream beds were dry, or nearly so, most of the remainder of the year.

The muddy character of the Mississippi river in 1834 so impressed the geologist G. W. Featherstonhaugh, while on an expedition to the Ozark mountain region, that he suggested experiments to determine the annual load of sediment carried by the river. By this procedure he proposed to calculate the age of the river and the rate of the retreat of the ocean as the result of silt deposits. Merrill (1924) credited him with being the first American geologist to make such a suggestion for estimating geological time.

The excesses of political agitation, the sensationalism of various types of social agitators, and the lack of historical perspective of the 1930s, planted in the public mind erroneous ideas of causes of wind erosion, and dust storms. The film, The Plow that broke the Plains, distributed by the federal government, was only the most conspicuous of the devices around which was crystallized the idea that the plow was the cause of dust storms, and second only to the plow, allegedly, was overgrazing. The legend was built up assiduously that the western country was becoming a desert. Rexford Tugwell, undersecretary of agriculture, delivered an address May 15, 1935, in which he pictured the doom of the Trans-Mississippi West as of 2235 A. D., 300 years hence, unless something like the Tennessee Valley Authority was adopted for that area. It was presented in the form of a journal of an exploring expedition investigating the Great Desert:

This week we have crossed the Mississippi River and have journeyed in our high-wheel motors deep into the great desert. Our dust masks have been useful, for without them we should be unable to travel for more than an hour or two after dawn. The Mississippi was nearly dry so that our pontoons sufficed for the crossing.

Our records show that at the junction of this with another river, the Missouri, there was once a considerable city and that this was a country devoted to the cultivation of grain. There now are only moving pieces of dust for hundreds of miles. Of the city little remains except some skeletons of twisted steel. It is not recommended that excavations be carried out at this point, since everything of historical value was moved to the eastward as the desert encroached.

The cause of these desert conditions is different from that which ruined the civilization to the east. Here it was the exposure of the plains to the wind. There it was the destruction of trees and the washing away of hills by the characteristic torrential rains of summer. Today we are camped on the bank of a river which falls over an escarpment evidently built of masonry. We think it must have been intended to dam up a canyon and form a lake to furnish continuous power. We assume that the lake filled with silt and that the power or irrigation venture failed because the river runs only during the spring floods.

We have seen no living thing since leaving the Tennessee Valley. We expect to return soon for the study of the records which have concentrated there as civilization has disappeared elsewhere on the continent. We expect to spend a few more weeks analyzing the soil of this desert, measuring, as well as we can, the climatic changes since vegetation disappeared, and collecting specimens of various remains.

Of course this address was not representative of the serious work of the regular staff of the department of agriculture, but unfortunately the political influence of such extremes left its mark upon the myth built up about the great plains.

In a spirit similar to Tugwell's desert speech, Paul B. Sears, a well-known botanist, published a book, Desert on the March (1935). The latter part of the book contained much information on sound soil conservation practices, but it was set in a historical framework indicated by the title. In reviewing the history of civilization, the destiny of different peoples was represented as being determined by soil destruction — man-made deserts were pictured as marking the seats of once-great empires. From the standpoint of historical methodology, Sears did not state his question in a form that was subject to proof — it was a meaningless question. From the standpoint of historical evidence, he did not present documented facts that could be shown to possess the alleged cause-and-effect relationship between civilizations and deserts. In any case, this supposititious historical introduction did not prove that the plains, or any other part of the United States, was undergoing a man-made change into a desert.

In 1935 the Tugwell directed Resettlement Administration began production of the documentary film, The Plow that broke the Plains (3 reels, 30 minutes). As the Literary Digest (May 16, 1936) described it, they entered Montana in September, "worked into Wyoming, on the wings of a blizzard, shot scenes in Colorado, Western Kansas, and the Texas Panhandle, from which they were blown by high winds and choking dusts." There were only 700 words in the story, the remainder being told by the camera, with music "as an explanation and emotional accompaniment." The picture sequence was grass, cattle, the plow, drouth, duststorms, and desert, supposedly representing fifty years of change. At the close, some of the projects of the Resettlement Administration were shown by which some of the human damage might be repaired, but it was said that damage to forty million acres could never be undone.

In May 1936, when the film was given the first public showings, much was made of the fact that the film was not taken by the regular distributors. A few candid observers pointed out that the film length of thirty minutes was too long for a short, and too short for a regular feature, and therefore it could not be programmed without disrupting theatre schedules. Also, it put city theatre audiences to sleep. Some theatres in New York tried the device of advertising it as "the picture no one dared to show," but without significant results. Of course, that was fraudulent advertising! The Survey Graphic predicted that the film would be in great demand among educational institutions, and that was correct.

As history the film was indefensible. To be sure, each separate photograph, of which the 2700 feet of film was composed, was an authentic picture as of the year 1935, but the film as a whole was not photographic history. The sequence of the photographs and the cause-and-effect relationships were not a camera record of historical change, they were arrangements of photographs of a single date pieced together to produce the illusion of a time sequence. The historical effect of change from grass to man-made desert was solely the result of the artificial design or purpose in the minds of the producers. In other words, if run in reverse order, the same pictures could be made to show the transition in fifty years from desert to grassland. It was all a matter of arrangement.

After three years of protests from the plains states, an Associated Press dispatch of April 19, 1939 quoted Representative Mundt of South Dakota as saying that the National Emergency Council had agreed to make no additional commitments to exhibit the film until it could be changed to meet the criticisms of the residents of the plains. The changes were not made, and the film was withdrawn from circulation. Copies of it, already in educational film libraries, however, were still available.

Photography is a medium of communication particularly adaptable to the uses of propaganda, or of mere sensationalism in the news. Faked drouth pictures, or photographs with misleading captions, became an issue during the latter part of 1936. The historian is indebted particularly to the Fargo (North Dakota) Forum for its aggressive attack upon such practices. This campaign succeeded in attracting national attention, was reported in August and September 1936 by the Associated Press and other news services, and thus became a matter of public record in the newspapers served by those news agencies.

One example was that of a bleached bovine skull which appeared in photographs, in different settings, designed to depict drouth damage. According to the evidence brought out in the controversy, the pictures were taken in May, 1936 in Pennington county, South Dakota. At that time of the year there was no drouth damage. Obviously, as the skull was conspicuously weathered, it was from an animal that had been dead a long time and from causes quite unknown to the photographer. According to the Associated Press report, August 29, 1936, the official photographer admitted moving the skull from place to place, but he denied that such a procedure was faking a drouth damage photograph. Another example, apparently merely sensationalism, was pointed out by the Fargo Forum, and involved a news reel agency which was charged with taking pictures of a farm family in Stutsman county, near Jamestown, North Dakota, supposedly fleeing from the drouth. The whole episode was said to have been "staged" for the moving picture camera and the family in question was paid for services rendered, after which they returned to their home.

John Steinbeck's novel, Grapes of Wrath, a bestseller of 1939 (film version 1940), exploited the drouth in the great plains as well as the migratory labor question in California. As applied to the plains, it's theme was the effect of drouth in the "Dust Bowl" and of tractor farming, as the causes of human degradation. As a literary work, undoubtedly it was an effective piece of writing. It was not the first time that fiction had been used for social propaganda purposes. As a piece of reporting of the contemporary scene, it was grossly inaccurate. The location selected, Sequoyah county, borders on Arkansas. It is on the western flank of the Ozark region, a hill country, not the plains. Its climate is of the high rainfall type, and climate and soil make it an area suitable for highly diversified farming, not mechanized wheat, cotton, or corn farming. The most effective criticism of the book is to be found in Elmer T. Peterson, Forward to the land (1942). Even so far as Steinbeck's argument could have been applied to the plains country, it misrepresented the factor of mechanization in agriculture, which was a constructive step in the agricultural revolution— significant to the agriculture everywhere in the twentieth century and nowhere else more so than to the plains.

A large part of the surface of the plains region had been formed by soil materials transported from the mountains and deposited over the parent rock. Water sorted the materials in the process of depositing them, and winds blew the finer particles up and down the plains area. A substantial portion of the plains soils, developed from transported materials, were classified and mapped by the federal and state soil agencies as wind-blown soils long prior to the dust storms controversies of the 1930s. Possibly part of that dust storm process of soil material distribution occurred prior to the establishment of the general grass cover, but it is certain to the historian that much of it continued throughout the whole span of time since the country was known to the white man. To be crystal clear, it is true that dust storms did not arise from a soil covered with vegetation, but the point at issue is that always some parts of the plains were bare or relatively so as a result of the activities of animals, especially rodents, and large portions were exposed to the wind from time to time through the action of fire, drouth, and wild animals. As Grinnell and Formosov pointed out, these influences were continually disturbing some part of the grasslands, delaying or destroying a theoretically normal succession series and forestalling any achievement of a uniform theoretical climax. It was only in periods of prolonged climatic fluctuations on the drouth side that these conditions became general and serious. The review of the exploring expeditions as sampled in Chapters 8-9 afford a conspicuous and reliable documentation of the soundness of this point of view.

Archeologists have shown from evidence that seems all but conclusive that during archeological time there were recurring periods of extensive soil blowing (Seltzer, 1940; Steward, 1940; Strong, 1940; Van Royan, 1937; Wedel, 1940, 1941, 1947). Excavations of Indian village sites show successive occupations of the same location, sometimes three or more, separated by thick deposits of wind blown material. There would seem to be little room for difference of opinion relative to this basic situation, although views may vary as to the length of the periods of soil blowing, as to why the sites were abandoned from time to time, and as to what became of the Indians during the intervals.

An explicit example of dust storms on the plains during historical time is afforded by the journal of Isaac McCoy, October-November 1830, recording an expedition engaged in surveying the boundary of an Indian reservation north of the Kansas river. The entry for October 18 was written near the southern boundary of the present Nemaha county, Kansas, about $96°$ west longitude:

>Had a little rain last night— the country is exceedingly parched with drought. When we got on to the prairies, the ashes from the recently burned prairies, and the dust and sand raised so by the wind that it annoyed us much, the wind raising. I found that the dust was so scattered that it became impossible to perceive the trail of the surveyors, who had gone a few hours ahead of the horses. While conversing with Calvin about the course we should go, we discovered the atmosphere ahead darkening, & as it had become cloudy, we fancied that a misting rain was coming upon us, and made some inquiry respecting the security of our packs. A few minutes taught us that what we had fancied to be rain, was an increase of the rising dust, sand, and ashes of the burnt grass, rising so much and so generally that the air was much darkened, and it appeared on the open prairies as though the clouds had united with the earth. Our eyes were so distressed that we could scarcely see to proceed.... The wind blew incessantly and excessively severe.... Was about to select a camping ground, when we met a man whom the Doctor [sent] to inform me that he could not proceed with his work, & that they waited for us in a wood a mile ahead. It being very difficult for me to look at my pocket compass I told the soldier ... to lead us back

He set off with great confidence that he could find his way back and in a few minutes was leading us north instead of west.... On finding the surveyors, we encamped for the residue of the day. Even in this wood, and after the wind had somewhat abated, the black ashes fell on us considerably.

Wind and dust accompanied the expedition farther west and on October 26 the Republican valley was reached: "Wind very high, scarcely allowing us to pass." October 27:

.... Today we reached the Republican,... and to our great disappointment we found it more destitute of grass than any place we had seen where wood was to be found. The river runs over a bed of sand— the banks low, and all the bottom lands are a bed of sand white and fine, and now as dry as powder ought to be. I never saw a river along which we might not find some rich alluvial moist bottoms, on which, at this season of the year, could not be found green grass. But here there is, in a manner none.

We examined along the river for grass until satisfied that none could be found and then turned back to a creek we had passed five miles back.... The scarcity of wood on the river and the sandiness & poverty of the bottoms, greatly discouraged me as to the country— While the great scarcity of food for our horses made us fear that we should not be able to proceed much further.

The entry of November 5 represents the country about the 98 meridian and read:

Completed the line of the outlet to 150 miles, and stopped. For some days we have discovered that our horses were failing so fast, that we must soon return, or lose them all.... We are beyond all Indian villages, and 50 miles, or more, into the country of Buffaloes—.

* * *

After we completed our survey, we turned on to a creek, and were looking for an encampment— the day calm & fair— when suddenly the atmosphere became darkened by a cloud of dust and ashes from the recently burnt Prairies occasioned by a sudden wind from the north. It was not three minutes after I had discovered its approach, before the sun was concealed, and the darkness so great, that I could not distinguish objects more than three or four times the length of my horse. The dust, sand, & ashes, were so dense that one appeared in danger of suffocation. The wind driving into ones eyes seemed like destroying them....

The storm commenced sun three quarters of an hour high in the evening, and blew tremendously all night. It had abated a little by morning. The dust was most annoying at the commencement. There was no clouds over us.

Another early example illustrates the conditions in the east central part of the present state of South Dakota, east of the Missouri river. I. N. Nicollet described his experience of July 1839: "as the growth [of grass] is too scant to prevent the dust from being raised by the almost incessant winds that blow over them [the plains], the traveller is very much inconvenienced." This entry was for the mid-summer when the winds were least severe and the summer growth of grass would have provided the fullest cover. It would be still more informative if continuous records were available on both of the preceding localities for the windy spring months of March, and April of the next year when the country was most barren of vegetation; when drouth, grazing, the tramping of hoofed animals, and prairie fires had most completely exposed the soil to the action of these winds. Furthermore, these examples deal with the country east of the 99 meridian, not the more critical high plains. Another example of a dust storm was related for the Snake river sagebrush country in connection with Major Osborne Cross's expedition to Oregon in 1849 (Chapter 9).

During the early years of Kansas settlement, the dust storms in 1855 evoked the comment that "this annoyance, however, will not be so great when the surrounding country is brought under cultivation, and the prairies cease to be burned" (Malin, 1946). During the 1930s the argument was just the opposite, that dust storms were caused by "The plow that broke the plains."

In the very nature of the case, the records are too incomplete to determine either the extent or the intensity of early dust storms as compared with the 1930s, but they are sufficiently comprehensive to establish the fact that on repeated occasions dust storms were extensive, and severe. These samples are only a few from the present author's collection of the records of a century of dust storms (Malin, 1946). To complete the record, attention is called to the fact that the National Archives has collected and organized for research purposes all the weather records of agencies of the federal government. They are described in a National Archives Special List No. 1, "List of climatological records in the National Archives" (1942). These materials have not yet been examined for information on dust storms. Sand and dust storm reporting was inaugurated during the 1870s, and these manuscript reports should contain a revealing story which would expand the one told by the present author from the printed sources (1946).

On the basis of experience in plains agriculture farmers came to the conclusion that the blowing of the soil, of the water, or of the wind transported types, did not necessarily damage the productivity of the fields concerned. On occasion it was pointed out that a field blown to the depth of plowing might produce better crops than formerly. To the eastern soil conservationist such views were not only incomprehensible, they were vicious. These views of the practical plainsman received the endorsement of scientific men in a "Symposium on Loess" (Elias, Editor, 1945) in which Elias said the same thing with respect to Nebraska loess and explained that the chemical breakdown of the constituent rock released supplies of necessary plant nutrients in a form available for roots. Of course, such an explanation on the part of the farmer or of the scientist was not to be interpreted as opposition to soil conservation, but it did call attention forcibly to the prevailing misinformation and misconceptions with respect to the soil problems of the plains.

A large part of the western soils of mid and low-rainfall areas have been formed from parent materials transported either by glacier, by water, by wind, or by combinations of them. They had been deposited to various depths over earlier geological formations. In some places the underlying rock protruded through the covering of transported material, or erosion had removed the covering material in part, exposing the underlying rock. In any case, the problems presented by transported soils were quite different from those of residual soils of high rainfall areas. In was the latter that presented the critical problem in equilibrium — the formation of new soil underneath by decomposition of parent rock as fast as erosion removed the top soil. If top soil were removed more rapidly than new soil formation replaced it, then the bare rock would eventually be exposed. It was this fact that made erosion, of even a small amount of the normal soil profile, a matter of serious concern, and any large proportion a major disaster. The problem of western transported soils was different, except for those spots where the apron of soil materials was thin or the underlying rock formations were exposed, and even there the fact of lesser rainfall resulted in differences.

The theory of soil maturity as held by the Russian-American school seems open to question as applied to the transported soils of the mid and low-rainfall region. Agricultural experience has already called attention to the fact that some sandy and some loessal soils could be blown from a field as deep as it had been plowed and the fertility and productivity remain unimpaired, or nearly so. Or possibly the picture would be kept more accurately in focus by bearing in mind the quips of western newspapers that sometimes admonished readers not to worry if their belongings were seen going north today, because when the wind changed they would be blown back. It is clear from practical experience that soil fertility and productivity are not necessarily related to the development, or to the preservation, of the theoretically mature profile. Even Marbut's Plate 6, <u>Atlas of American Agriculture</u>, shows much of the most productive land of the grassland area as having soils without profiles, or with only imperfect profiles.

The perspective of agricultural experience affords a demonstration of the difficulties in which the soil scientists became involved in their Soil erosion report and map (1935), in Part V of the Supplementary report of the Land Planning Committee of the National Resources Board. The reconnaissance survey of soil erosion for the entire United States was made by the Soil Conservation Service, using the services of "115 trained soil-erosion specialists, who visited every county in the United States and prepared an erosion map on the basis of actual reconnaissance,... Since the survey was made in 2 months, ending October 15, 1934, it was impossible to indicate more than generalized or predominant conditions." In defining the different erosion classes and mapping their distribution, reference was made frequently to the percentage of the top soil removed by erosive action of wind and water. No standards of measurement were presented. How was top soil defined and how was its original depth determined? What date was used as a basing point for original depth? According to the dust storm and water erosion record, such a date would be important and would have to be established, along with the record of the soil profile at that time. It is essential to an understanding of the unsatisfactory character of the report to make clear that there were no fixed markers against which top soil could be measured, nor "original" profile records that could be used for comparison. There is no indication of the extent of the acquaintance of the soil scientists mentioned with the counties which they surveyed, but it is obvious that they could not have possessed intimate knowledge of the physiographic history of those counties because no such historical records exist. Their estimates were necessarily guesswork of the most superficial kind, and based upon a soil theory that did not recognize adequately the fundamentals of the particular environment and its relation to soils.

The best test of the validity of this soil erosion survey is the production record of the following years. With the return of favorable weather conditions, 1942-1946, the hard winter wheat region experienced an unbroken succession of five good crops over most of the area, for some counties six, 1941-1946, a record not equalled by any other period since the country was settled.

For reporting purposes the state of Kansas is divided into nine divisions, by thirds in both directions. The eastern third, three divisions north to south, is traditionally a soft winter wheat area (wheat is not the leading crop), and is omitted from this discussion. It extends west to about 97°, the western limit of the tall grass-oak-hickory-walnut native vegetation. The middle third, three divisions north to south, extends approximately from 97° to 100°, the native mixed grass region, and a hard winter wheat region. The western third, three divisions north to south, is the short grass high plains, a hard winter wheat region. The accompanying table gives comparative yields for the six hard winter wheat divisions, 1945, 1946, and estimated for 1947.

Hard Winter Wheat Yields: Kansas

Division	1945 final (bu. per acre)	1946 final (bu. per acre)	1947 June 1 estimate (bu. per acre)
Central third			
North central	13.3	15.0	17.0
Central	11.3	16.4	18.0
South central	14.1	17.1	17.0
Western third			
Northwestern	22.3	21.3	21.5
Central	19.1	15.3	20.5
Southwestern (Dust Bowl)	17.4	13.1	22.0
State Average	15.5	16.2	19.0

In each erosion class over 25 per cent of the land of an area was said to have been affected, and on the map each class was colored separately and numbered. A large part of the western third of Kansas was marked with the symbols 4 and 5. This meant that for areas marked as erosion class 4, over 25 per cent of the land had been subject to "slight soil drifting" and that "small amounts of surface soil are removed." This class of land constituted a large part of the west central division of Kansas. Class 5 was similar to class 4 only more severely eroded, "where the soils have been blown off to depths ranging from 1 to 6 inches. The productive use of the land has been materially lessened, and the tendency is for its condition to become increasingly worse." Most of the southwestern division (the so-called Dust Bowl) was marked class 5.

In 1945, nineteen of the twenty highest-yielding counties were in the western third of the state, west of 100°. Stanton county, in the southwestern corner, led with an average yield of 31 bushels; Hamilton county, in the western central division, was second with 30.4; Cheyenne county, in the northwestern corner, was third with 30.1; and Wallace, and Kearney counties, in the central western division, tied for fourth place with 28 bushels each.

The eight counties in the northwestern corner of Kansas were marked with the map symbols 4, 27, 37, and 38 for the erosion classes and their distribution. Erosion class 4 indicated "slight soil drifting" and the removal of "small amounts of surface soil." Under the explanation of class 27 appears the statement: "Twenty-five to seventy-five percent of the surface soil removed," and "crop yields and farm efficiency have been reduced by soil losses,..." Number 37 represents a situation of "over 75 percent of all surface soil removed.... Loss of surface soil is severe and generally complete...." For class 38, the statement is that "Over 75 percent of the surface was lost.... Large areas of these lands are essentially ruined for cultivated crops. These conditions have resulted in abandonment of large areas."

The wheat crops of 1945 and 1946 were the climax of the period for the eight northwestern counties described in the preceding paragraph. The yield of wheat in 1946 was less than in 1945 but was 21.3 bushels per acre from 1,401,000 acres harvested, and was the highest yield for any similar area in the state. The state average yield was 16.2 bushels per acre. Of the ten highest county yields in the state, five are found in this group of eight northwestern counties. Only one county, Graham, fell below the state average. The three top yielding counties of the state were the three extreme northwestern corner counties, Cheyenne (24.7 bu.), Sherman (24.3 bu.), and Rawlins (23.5 bu.) lying in the area colored on the soil erosion map as of classes 4, 27, 37, and 38. In conclusion, these figures are all above the long term averages for either the state as a whole or for the particular counties named. If averages for the term of years 1942-1946 were used a similar general result would be obtained. If other groups of counties were used, conclusions in the same general direction would be obtained, the details varying with the area and the year. Obviously the soil was still highly productive, regardless of the soil erosion report of 1934.

The phenominal hard winter wheat crop of 1947 accents further the soil problem in relation to erosion history, fertility and productivity. At planting time in September 1946, there was a moisture deficiency, remedied by rains in the southwestern division of Kansas in early October, and in other divisions later. With the coming of rains, the wheat made a good growth affording wheat pasture for large numbers of cattle and sheep until heavy snows and severe winter weather forced some curtailments. On January 1, 1947 over a million sheep and lambs were on feed in Kansas, mostly on wheat pasture. Soil moisture tests, October 14-20, 1946, revealed moisture to a depth of 44.3 inches over the western two-thirds of the state— the hard winter wheat belt— summer-fallowed land having only a slight advantage over continuously cropped land. In April, 1947, soil moisture conditions in the western third of the state were the most favorable on record, 47.4 inches. A moist, cool, late spring provided the background for the unprecedented June 1 winter wheat yield estimate of 19 bushels per acre for the state, second only to 1942, (includes soft winter wheat of eastern third), or a total production of 277,761,000 bushels, 25,995,000 bushels above the previous record established in 1931. The 14,619,000 acres for harvest in 1947 was the largest of record.

The 1947 crop prospects were too good to be true. The crop season was abnormal on the favorable side, and before the season ended, on the unfavorable side, a long overdue fluctuation in weather to redress the averages. In April, tornadoes appeared in the grassland from the Mississippi river states westward, the climax occurring in the Panhandles of Texas and Oklahoma, destroying a large part of Woodward, Oklahoma, April 9-10. They continued into June. Violent electrical storms occurred also, with driving winds and rain which tangled the rank, heavy-headed wheat. Hail beat paths through many counties, one to three times. On May 29, occurred a freeze in the northwestern counties, and in Colorado and Nebraska; and again, June 11, snow and freezing weather hit the last named states. In the southwestern counties of Kansas hot winds came in June, especially on June 9, which blasted many fields. Other hazards were orange leaf rust, green bugs, Hessian fly, and root rot in scattered areas. And finally, in harvest time, a shortage of combines to harvest the grain, and a shortage of gasoline for combines and trucks, caused heavy losses, and there were some fires in ripened fields. Even weevils appeared in some grain as it came from the combines. The crop estimate of June 1 discounted only a part of these final hazards, but the most serious damage was revealed after June 1, or occurred after that date. With this fabulous crop made, and almost harvested, and a price of almost two dollars per bushel assured, the hard winter wheat region almost lived wheat twenty-four hours of the day. The July 1 estimate revised downward somewhat the figures of June 1, but the lateness of the harvest left uncertain the actual quantity of grain saved.

The Kingman, Kansas, *Journal*, July 11, 1947 reported that four hail storms had hit the county,

The harvest this year has been a headache for wheat farmers, caused by excessive rain, hail, and a shortage of machines.... With all the difficulties which have beset this harvest, farmers are not discouraged and are making arrangements to put in another crop and will gamble that 'next year may be better'. It is characteristic of the traditional spirit of the wheat farmer, if officers ever decide to enforce the anti-gambling law, the wheat farmer will be put out of business.

In Rush, Lane, and Wichita counties, reports of 30 bushel wheat were common and some fields made 40 and

50 bushels per acre. In Rooks county, yields of 45 bushels per acre were reported in the southern part of the county, while frost damaged fields in the northern part of the county were reduced to 5 to 10 bushels per acre. Phillips county had much the same story, 5 to 30 bushel yields: "The five bushel fields look just as good as the 30 bushel fields — from the road," lamented the <u>Phillips County Review</u>, July 10, 1947, but, "with yields vastly reduced by the freezing weather, and by a long path of hail, Phillips county appears to have almost a normal wheat crop anyway." The Norton County <u>Champion</u>, July 3, 1947 reported:

Estimates of the 1947 wheat crop in Norton county have been drastically reduced the past week or two by bad hail storms, and increasingly more evident frost damage, plus other factors, until it is now estimated that somewhere near 30,000 acres will not be harvested. What looked like a two million bushel crop — or better — a few weeks ago, now appears to be closer the million mark — with a resulting loss to Norton county of around two million dollars, a severe blow. [A million bushel loss at two dollars per bushel was the basis for this estimate].

    There are some farmers who will harvest no wheat whatever, and others will harvest a bumper crop. In between lie the vast majority with a lot of 10 to 15 bushel wheat....

    It has been one of those years that just naturally can't be classed as a good crop year from any angle. Everything in the book happened to the 1947 crop — but still there will be a lot of wheat harvested, compared with what we used to consider a 'normal' year in the 1930s.

    Decatur county, adjoining Norton on the west, started off the harvest with a report, according to the Oberlin <u>Herald</u>, July 10, of wheat on the Bremer estate yielding 50 bushels per acre: "That is mighty good wheat in any language, and it comes as a quieting note after the many reports during recent weeks of damage from green bugs, freeze, etc." The following week a 60 acre field made 61 bushels per acre of 63 pound wheat, and the price at the local elevator was $2.05 per bushel.

    These particulars have been presented in some detail to insure beyond the shadow of a doubt a record of the main issues. Wheat buyers reported the 1947 crop as showing one of the lowest protein averages on record. At Dodge City and Garden City percentages of 9.8 to 10 were recorded. These samples were lower than the great 1931 crop. In other respects, the 1947 grain was of the finest quality, testing up to 65 pounds per bushel. Agronomists had experimented, and debated the factors determining protein content, the most recent conclusions listing soil as the determinant, except climate in some instances, and variety. As a matter of historical record, high protein percentages were associated with unfavorable crop years such as the middle 1930s, and low percentages with favorable crop years such as 1931 and 1947. The claim that soil was the limiting factor did not meet the issue of why the same variety in the same field varied from year to year. What is meant by a favorable crop year? A crop year that is usually considered favorable for wheat in terms of weather conditions producing high bushel yields of grain is not necessarily favorable for high protein content. Which weather factor, moisture, sunshine, or temperature, or combination of them with soil and variety, are necessary to insure any particularly desired quantity or quality of product?

Nitrogen deficiency in the winter wheat region was a menace to the continued production of crops, according to some agronomists. The argument was advanced that in the higher rainfall areas nitrogen could be restored by legumes, manure, and crop residues as well as by commercial fertilizers, but that in the low-rainfall areas these factors were deficient, especially, varieties of legumes suitable for crop rotation programs. Experimentation revealed that in low rainfall climates commercial nitrogen fertilizers gave unsatisfactory results. Again it is in order to raise the question whether the soil scientists have arrived at a satisfactory theory of soil fertility and productivity for low rainfall climates. In the native vegetation the numbers of legumes, both of varieties and individuals, had been large in the tall grass country, but diminished to the westward in the mixed grass and short grass country. Descriptions by early explorers suggest that some of the heavily sodded grama-buffalo grass areas approached the status of simple stands, with very few legumes. How did the short grass country maintain its nitrogen supply? Studies of nitrogen supply under strictly native grass conditions are needed and might provide some perspective on the nitrogen question, and on the larger issues of soil fertility and productivity, and on the problems of soil conservation in low rainfall climates.

Evidence may be piled up without end to demonstrate that the soil erosion survey of 1934 did not deal with fundamentals. In total production, the crop of 1947 exceeded the record production of 1931, which preceded the drouth and dust storm years of the 1930s. The protein factor in the drouth years ran high and in 1947 low, as in 1931. In yields per acre, the crops of 1945, 1946, and 1947 exceeded those of 1931 and earlier. The Stanton county record yield of 31 bushels in 1945 was topped again by the estimates for 1947. When the final figures are available on the 1947 crop, they may fall below those given in the preceding table as estimated for June 1, but those reduced yields are not to be charged against nitrogen deficiency, loss of fertility, or of productivity, nor against wind erosion of soil. They were clearly the result of hazards intervening on the eve of harvest and after the record yielding crop was virtually made.

The crop of 1947 and two dollar wheat was a curse to the hard winter wheat region, in providing the final fillip to the wheat boom; high priced land, and large-scale speculative wheat production, which had already gone beyond reasonable proportions. The largest production reported under one management was 78 square miles in some seven counties in Kansas and Colorado. Many other instances of large acreage were on record. With the fluctuation in the weather to the unfavorable side, long overdue, such speculative mass-production operations create situations where it is a physical impossibility to supervise adequately such acreages under drouth and wind — dust storm — conditions. Unnoticed during the high tension over saving the fabulous wheat crop of 1947 was the announcement that after thirteen years the receivers for the Wheat Farming Corporation were closing the books on that unfortunate large-scale venture of the earlier wheat boom era. A 25 to 60 bushel yield, with wheat at two dollars per bushel, meant fifty to one hundred twenty dollars per acre gross income from a single acre and a single crop. What was such land worth, land that could not have been sold at ten dollars per acre, or possibly at any price, only ten years earlier?

Desert equilibrium

In dealing with the desert it is traditional to assume that there the struggle for existence was more intense than in other regions. This point of view has been challenged and needs revaluation, especially in a background of the recent findings in the field of developmental physiology. It would probably be valid to maintain that the problem of survival is different, for the several regions, but not necessarily more severe. On a long-term basis, stabilized plant and animal life appears to have been limited to replacement plant for plant and animal for animal. Thus in each region, irrespective of climate factors, the increase potential was offset by resistance factors of some kind which eliminated all but one survivor on the long term average. In a temperate climate these adverse factors may be intense competition between plants, or insects versus plants, or animals versus plants, or combinations of them, rather than drouth or heat. There would seem to be no difference as respects severity of the struggle so long as all increase is eliminated except replacement of individual for individual.

A second tradition about the desert placed the emphasis almost exclusively upon the water relation of plants and animals, with little or no attention to temperature, light, and soil. Possibly this second proposition possessed more valifity than the first, but even at that, there is doubt whether the assumption should be made without more complete information. The most comprehensive studies made thus far have been those under the sponsorship of the Carnegie Institution of Washington and they have dealt mostly with water relations of plants, and less intensively with temperature, but the photosynthesis studies sponsored by the Smithsonian Institution may prove even more important as explaining plant and animal life irrespective of climatic regions. The inadequacy of the studies of animal relations on the desert makes an attempt at a general summary of the factors of desert equilibrium peculiarly unsatisfactory.

Observations carried out on completely protected plots in the Arizona desert near Tucson have provided some interesting data on the problem of the composition of desert vegetation (Shreve 1929, Shreve and Hinckly 1937). The oldest plots were brought under protection in 1906, and others in 1910 and 1928. In thirty years the large perennials scarcely changed in number, but the small woody perennials and the grasses increased greatly. The grasses which were negligible in 1906 covered 2.7 per cent of the area in 1903, but the large perennial plants determined the appearance of the desert. In conclusion, Shreve and Hinckley emphasized that there was no common trend discoverable other than the increase in population of the small plants.

The limitations of the above experiment must be stressed in order not to leave erroneous impressions of its bearing on the controverted question of the nature of the desert climax. The areas were completely protected from grazing, and the relations of other animals to soil and plants were not considered in the report. The results can have, therefore, but a restricted bearing on the question of what the original wild state might have been. The limited area covered by grasses at the end of thirty years is worthy of note, but even more interesting is the fact that one large woody scrub, and no more, replaced each one that died during the thirty year interval. Although not constituting proof, these experiments tended to cast doubt upon, rather than support, the theory that the shrubs constituted a disclimax.

The sagebrush problem appears clearer, from the records of the early explorers, than the southern desert brush problem. Spanish influence had not extended much into the sagebrush country except at its southern extension, but the clear dominance of the sage was so conspicuous as to induce nearly every early traveler to comment explicitly upon it. In an earlier chapter (Chapter 9) Abert's (1845) description of the dominance of sagebrush, with its accompaniment of cacti was a conspicuous feature of his account of the country south of Bent's Fort, the present southeastern Colordao. Simpson's (1849) description of northwestern New Mexico and north eastern Arizona, the northern Spanish pueblo and Navajo country, deserves special attention. Unacclimated to the desert, Simpson was hypersensitive to its peculiarities and wrote with feeling of his trip westward from Santa Fé:

And, commensurate with this section, _arroyas_, cañons, _Mesas_, with their well-defined crests and escarpments; plateau and hemispherical mounds, intermitting dirty, clay-colored rills, dignified by the name of _rios_, (rivers) and an all-pervading dull, yellow, dirty, buff-colored soil, — have, in their respective magnitudes and relations, characterized the face of the landscape....

In regard to the fertility or productive qualities of the soil for the whole area traversed this side [west] of Santa Fé, saving the inconsiderable exceptions which have from time to time been noted in my journal, the country is one extended _naked_, barren waste, sparsely covered with cedar and pine of a scrub growth, and thickly sprinkled with the wild sage, or artemisia, the color of domestic sage, suggesting very appropriately the dead, lifeless color of the wild.

In the western part of the country explored, Simpson stressed, September 5, that "The artemisia, as usual, has been the chief, and almost the only, plant, especially upon the uplands," and again September 6, "The artemisia has been the chief _flores_. The cactus which has been seen but seldom, today was more prevalent. The return route eastward was by way of the pueblos of Zuni, and Laguna along which the vegetation was more plentiful..

Frémont's descriptions of the north-central prairie-plains have been summarized in an earlier chapter (Chapter 9) but for certain purposes a letter to the botanist John Torrey relative to the first expedition of 1842 emphasized more sharply what he thought of as characteristic plants of each subregion (Rodgers 1942). For the Kansas river valley uplands, it was Amorphia; for the lower Platte valley to the forks, it was Aster; for the return trip in September these valleys were yellow with Sunflowers; but from the Laramie fork to the continental divide at the South Pass, it was Artemisia which occupied "the place of the grasses." Frémont's second expedition, following a different route to South Pass and into the interior basin, emphasized artemisia again as the characteristic vegetation. For the country west of the South Pass, Frémont, in 1843, found sagebrush, and the account of Major Osborn Cross agreed with Frémont in emphasizing the sagebrush.

These records emphasize so strongly and in so many ways the prominence of the sagebrush for this perio prior to white occupation, except in the Spanish section where white occupation reached further back in time, that there seems little ground for the theory that the modern sagebrush characteristic was a disclimax resulting from overgrazing by domestic animals. In order not to be misunderstood, there is no denial here that there had been overgrazing and abuse of the grazing value of the sagebrush country. The purpose is simply to emphasize that the historical descriptions do not seem to justify the contention that the brush dominance and the scarcity of grass was of recent origin.

Conservation

The study of range management made great advances during the twentieth century (Agricultural History, 1944). Among the influences not sufficiently appreciated were the operations of rodents in the grassed areas as differentiated from the cultivated fields. The farmer tilled his land with machines and made war on rodents in his fields, but he carried this policy of exterminating rodents to the grassland without the accompaniment of artificial tillage. One or the other of these policies was a mistake. Range management experiments were carried out with contour furrowing using different widths, depths, and spacings for the conservation of moisture and control of erosion. This was artificial tillage and disturbance of the grasses which influenced succession and composition of the vegetation. This procedure was commended by Clements (Carnegie Yearbooks Nos. 37, 39, 40, 1938, 1940, 1941). Others advocated permitting rodents to operate, by keeping them under biological controls, thus permitting the burrowing animals to engage in natural tillage (Shelford, 1944). It is not the purpose of this discussion to pass judgment on the merits of the policies involved, only to call attention explicitly to the issues. If adequately administered natural biological controls can accomplish the desired end, why resort to the elaborate and expensive system of artificial range conservation; or would a combination of the two accomplish more effective results than either one separately? From the standpoint of conservation as a whole, the issues are broader by far than matters of soil, of grazing, or of forestry, and should include wild life resources and protection of fur-bearing animals as an integral part of a sound approach to conservation policies based upon an application of ecological principles. Programs for extermination of any particular species of wild animals should recognize the complexity of biological controls, and the fact that more damage than good may result from unsound policies (Hall, 1930).

One of the broadest and most significant discussions of nature and public policy was that of Shantz (1940), and one of his several points was the significance of wild land to man — roughly half of the world's land area or nearly the same proportion of the United States was wild land, which he predicted "will probably continue to be the back-log of civilization." Much had been done for conservation of forests, but not for wild land. What were the essentials of a policy for wild land? It was from a different approach, but in a similar broad spirit, that Shreve (1934) recommended the revival of the viewpoint of the old-fashioned naturalist in seeing nature as a whole. Such approaches challenged a reconsideration of that much abused term submarginal land — submarginal for what?

From the study of the ecology of the grassland in its natural state (Chapters 8 and 9) some important conclusions may be drawn, and one of the most important is that there was no such thing as an undisturbed grassland in the conventional sense. Man's turning over of the sod with the plow is only a more complete process of cultivation of the soil than took place continuously in nature. To say that a piece of grassland had never been plowed is a misstatement of the facts. It can mean only that the area had never been plowed artificially with man's agricultural tools.

New terminology might clarify much of the discussion of so-called native vegetation, because much of the disagreement relative to such questions stems from inaccurately or inadequately stated questions. Custom of long standing has settled upon the term "virgin" as applied to forests, or grassland, or soil, that has not been disturbed by civilized man. The term is a misnomer in the first place, because it is a sex term that has no proper applicability to vegetation, or to soil. In the hands of conservation propagandists, the term rape of the virgin continent, or rape of the earth, has often been introduced, carrying over into the discussion of nature the idea of sex crime. Nature does not offer any such parallel or analogy, and all such terms should be eliminated.

A new set of terms is offered here as a basis of discussion. As applied to the grassland, for instance, the term "native grassland" may be restricted to the condition prior to invasion by the white man's civilization. The term "domesticated native grassland" may be applied to areas after the white man's civilization has modified the vegetational cover somewhat and destroyed many or most of the native animals, especially burrowing animals, that influence vegetation and soil. The term "plowed native grassland" may be applied to land growing native grass species, but which has once been plowed, or artificially disturbed to an equivalent degree.

According to such a set of terms, it is clear that there exists in the United States little, if any, "native grassland", because the original animal populations have been largely destroyed, and because the composition of the vegetation has been modified substantially. In over-grazed areas, the grass composition has not only been modified, but weeds (forbs) have greatly increased, often new species have been introduced. In carefully conserved areas, such as those reserved regularly for prairie hay, or in which systematic "weed" killing campaigns have been followed, the original forbs, both composites and legumes, have been largely or altogether eliminated, and burrowing animals destroyed. These are "domesticated" areas because the native grasses and other vegetation found there are growing under a greater or lesser degree of artificial environment. The use of the term "plowed native grassland" is necessary to distinguish regressed areas from those not so disturbed, although the actual composition of the vegetation may not be substantially different from the other. The major difference may be only in the history of the treatment to which it has been subjected.

The importance of such considerations is emphasized by the conservation practiced by some land owners in the Bluestem Pasture Region of Kansas. Long years of use or misuse or ill-advised conservation measures, destroyed the legumes, and to compensate for this loss, lespedeza has been planted. This legume attains its greatest growth during the mid-summer after the bluestems have past their prime, thus affording a longer pasturing period, but the lespedeza is also an efficient nitrogen fixer, which rapidly contributes to the reconditioning of the grassland soil. In part, man is doing, artificially, with this introduced plant, what nature once did in the original "native grassland."

Idea of regional adequacy

Western civilization, which developed during the later medieval and the modern periods, was in some measure the product of its environment, a temperate and humid climate, a region mostly forested, and a soil more or less acid. The ideologies of this culture were fitted into this framework as a product of the historical process of adjustment. The region came to be accepted as right, adequate, and complete, and served as the standard of values by which all other regions were measured. Out of this background to both Europe and to eastern North America, among other things, grew the dogma of a geographical determinism that identified superiority in civilization with this particular set of standards. Closely allied with it was the idea of closed space resulting from the occupation of the temperate zone by the expanding westernized civilization and the consequent corallary of the struggle for control of the most valuable space.

In its relations with other regions, the tendency has been for this deciduous forestman to attempt to impose his particular culture upon them irrespective of its adaptability and to expect a uniformity of results in all parts of the world. Anything that resisted that pattern was assumed to be inadequate, inferior or deficient, even nature had blundered. The invasion of the grasslands of the world by this machine civilization began during the early nineteenth century, but the major phase of that occupation came only in the later part of the century. The grassland was described in terms of deficiency; treeless, sub-humid, and sterile. The assumption, that soil that did not grow trees was worthless for agriculture, proved to be not only false, but it turned out that forest soils were less productive. The natural neutral or alkaline reaction of the soil of the grasslands also proved its versatility in crop production over the acid forest soil which required constant treatments of lime for some crops. The forest region that grew trees, not grass, did not possess grass for livestock except as it was sown as a field crop — tame grass meadow and hay. On invading the grasslands, forest man refused to recognize native grass as the vegetation best adapted to the region, plowed up the native grass, and attempted to grow eastern tame grasses. When these grasses did not grow, he condemned the country. It took many years to learn that native grasses would perpetuate themselves and provide pasturage and hay for an unlimited time if only man would give nature a fair chance — which consisted mostly in just letting it alone (Malin, 1942, 1944). The grass man, if he took his region as the standard of values, could point the finger of scorn at the deficiencies of the forest land; grassless, wet, with an acid, leached, infertile soil. This study concentrates on the North American grassland, but study of an individual grassland is only the introduction to a comparative study of all.

It is not the intention of this statement of the case for the grasslands to imply that the forest lands were deficient. That would be to fall into the same error that afflicted forest man. There were differences as among the world's several natural regions and not a question of superiority of one over the other. The main purpose of this discussion is to reorient the prevailing point of view, and to insist that not only the grassland and the forest lands, but also the other regions such as the arctic, the desert, the tropics, and the oceans, were biologically complete products of nature, each expressed in its own manner and in terms of normal factors of equilibrium. The degree of success in the occupation by man of any of these land regions could be measured in terms of his ability to fit his culture into conformity with the requirements of maintaining rather than disrupting environmental equilibrium. The differences among the several regions did not represent deficiencies, rather, each difference represented an advantage useful to other regions by which they supplemented each other, and by which each made its own unique contribution to the world. There was no major land region to which some branch of the human race, together with plants and animals, had not proven adaptable and developed cultures as diverse and distinctive as the environment. If man found himself unable to cope with more than one of these several kinds of environment, it was man and not nature in that region that was deficient.

Other than the grassland, probably the arctic has been more maligned than any other region, although possibly Vilhjalmur Stefansson might challenge the assumption that the grassland has received the worst treatment at the hands of the deciduous forest man. The arctic land is not a wasteland nor the arctic sea a barren sea, and his emphasis on the capacity of plant growth under the influence of the light of the 24-hour day pointed to the importance of systematic and intensive research in the biological problems of the arctic. Raup (1941) emphasized also the lack of biological knowledge of the arctic and urged a program of research. The advent of air communications thrust both the grasslands and the arctic, and especially the latter, into the orbit of the new twentieth century struggle for power. With the challenge of the sea power theory of history, the grasslands became of greater world significance because of the land mass concept of power that was coming so largely to dominate twentieth century thinking, but also because of the agricultural productivity of their soils. The arctic gained significance because of the reorientation of outlook in political geography in terms of air power which follows great circle routes between the circumpolar land masses. The people who most effectively meet the problems of these two regions may lead the world in the next era.

Chapter Eleven

SCIENCE AND SOCIAL THEORY

During the early stages of World War II, the National Science Fund asked scientists to specify what they considered the outstanding problems for post-war research. The first item on the list was the analysis and study of human behavior (Robbins, 1944). Apparently the focus of interest was the relation of science to society, and under such circumstances there is a peculiar interest for the historian to examine the scientists' ideas on the relations between science and social theory. But the sciences are of interest to the historian in their own right as a portion of the material of history, and furthermore they have developed methodology and procedures of precise thinking that constitute a challenge to the historian and to the social scientist.

Ecology

Ecology has been defined variously without either botanists or zoologists arriving at substantial agreement, except in a general way, usually stressing that ecology deals with groups or assemblages of living organisms in relation to environment. In plant ecology there was a sort of hierarchy of names applied to groups of plants, beginning with the smallest unit; communities, associations, and formations. The principle of succession was fundamental also to ecological thought; the principle in botany held that on any particular piece of bare ground in the grassland, the plant population left to itself would follow a definite sequence beginning with weeds and ending in a relatively stabilized combination of grasses and forbs determined by environment. In a forest region, the equilibrium vegetation would result in a particular kind of forest. The term climax was applied to this equilibrium stage. Similar principles were applied to animals.

In its simplest classification, the theory that lay behind such attempts at definition was usually divided into three kinds; organismic (organistic), individualistic, and an intermediate group of ideas. Clements was the leader of the organismic group, announcing in 1905 his interpretation of units of vegetation as complex organisms passing through stages of development, youth, maturity, etc., similar to an individual plant.

These ideas were fully developed in his concept of climax (1916) as a complex organism shaped by the influence of climate (1934). The term biome was introduced also in 1916 to lay stress upon the relations between plants and animals, but excluding habitat, and involved the concept of biome as organism determined by climate. Clements had many followers as well as opponents in the United States. In South Africa, John Phillips championed a similar group of concepts, but with the additional association of Jan Smuts's philosophy of holism. Ecologists have discussed these matters in some detail in the reviews of such books as Clements and Shelford, Bioecology (1939) and Weaver and Clements, Plant Ecology (1929, 1938), and in many articles devoted to the analysis of terms and concepts of ecology. In such evaluations disagreements with Clements predominated with respect to his extremes of both theory and terminology.

The intermediate group of ecologists used the basic terminology of ecology such as communities, associations, formations, succession, and climax, but without committing themselves to organismic theories. There were all shades of difference, rather than any sharp line dividing many of them from the Clements point of view, and many repudiated outright the theoretical implications of organism. A moderate version of this intermediate group, such as that of Tansley (1920), the English botanist, was willing to use the organismic concept as an analogy that might serve a useful purpose. Cowles's view, emphasized by Cooper (1935) might be cited, but Nichols (1917) was more often indicated as representing best the intermediate position.

The most explicit and clear-cut expression of a different approach to the problem of concepts in ecology was that of Gleason (1917, 1926, 1939) in his individualistic concept of the plant association. In reviewing Gleason's last restatement of his views, the English ecologist Godwin (1940) said that, in the twenty-four years intervening since these views were first announced, for the most part they had been unanswered. Gleason emphasized the ever-changing character of vegetation, both as to space and time, that defied any system of rigid classification. No two spots in space were alike in composition of vegetation and changes were gradual, usually, even imperceptible often, until spots at considerable distances were compared, and only then did the extent of the change become apparent. The same concept of change applied to the time factor, both intra and interseasonal. Each individual plant or species stood by itself as respects survival and reproduction. Under no circumstances did an association migrate or reproduce as a unit. Of course, such down-to-earth realism excluded completely any idea of vegetation units as organism, even in the sense of analogy. Although not subscribing to the

Gleason definition as such, Transeau expressed much the same view in his progress report on the survey of the Prairie Peninsula (1929).

Climax as a general condition, extending with any degree of uniformity over a large area, is quite out of line with reality. The descriptions recorded by Abert in 1846, of the vegetation across Kansas, were sufficiently complete to establish the fact that disturbance, succession processes, and approach to, if not attainment of, climax conditions were continuous and no two spots of vegetation were necessarily in the same stage, in fact the probabilities were to the contrary. Intervention of new disturbances did not await attainment of climax because pocket gophers, prairie dogs, badgers, buffalo, and other animals, as well as hazards of weather, fire, and floods, all kept the vegetational cover in a state of variegation. It might be that forest vegetation, with long-lived trees, rather than grassland, approached more nearly to conformity with the theoretical ideal of climax. On such a view as this, however, Braun (1935) would seem to enter an objection in holding that the forest climax ideas had been associated with the younger forested areas of the post-glacial period. When the eastern grassland was called tall grass prairie; the central portion, the mixed grass prairie-plains; and the western part, the short grass plains; it was only in a general over-all sense that these names could apply to the predominating characteristics of the highly variegated, rather than uniform, landscape.

Vegetation as organism carried with it the implication of the climax stage as an end product beyond which there could be only decay or degradation. There was implied also something of teleology. All this went beyond the facts. To the present writer the use of the terms dynamic and climax seem paradoxical. Succession was dynamic, but the climax became essentially static, or if the organismic concept was used the disclimax of old age entered. A consistent use of dynamics and climax would seem to be along the lines of Gleason's individualistic concept of plant associations — the open system of endless variety of change in both time and space. The idea of the plant association or formation, as being the product of climate, was challenged by Tansley (1920, 1926, restated 1935), who held that all the physical factors of habitat, as well as climate, entered into the determination of the vegetational and animal characteristics of a region. Such units isolated for study, he called ecosystems. He warned against premature attempts to explain why until more facts were available. This would approach one significant step further in the direction pointed out by Sauer in his advocacy of historical geography as a study of regions from the standpoint of physiography and cultures comprehending the time-span beginning with the primitive culture of the anthropologist down to the present.

Incidental to his description of desert vegetation, Shreve (1942) questioned "the widely accepted view that there were certain 'fundamental units' of vegetation;" and the prevailing theory of plant succession, which was merely sequential and not genetic. He advocated that types of vegetation should be defined on the basis of their intrinsic characteristics. In applying these ideas to the deserts which grew simple stands of vegetation he pointed out that the uniformity in height and density obscured the differences in age of the individual plants and over a period of years there was little change. If the creosote bush was removed from an area dominated by that type of vegetation, seedlings would spring up, the excess numbers be eliminated and the face of the desert would assume its original character. This was the first and only stage in succession. He insisted "that the succession concept would never have been developed in a study of the vegetation of arid regions." In the desert, the vegetation made little impression on habitat, leaves and twigs being blown or washed away rather than accumulating and being incorporated into the soil to change its character and thereby its ability to support a different vegetation as a successor to the existing type.

These views of Shreve (1942) had more far-reaching implications than seem to have been recognized at the time. The point was made clear that ecological concepts did not possess the general validity and universality of application usually attributed to them. The idea of organism, with birth, maturity, and death, disappeared completely, along with less explicit assumptions of uniformity and orderliness in nature. In some respects it might be better to modify the definitions of ecology by omitting all reference to environment and saying only that ecology deals with organisms as they are found living together in space and time, and with the processes that create and maintain the element of organization among them.

In animal ecology there were conclusions that provided the opportunity for misuse of science for social theory. Hesse, in the Allee-Schmidt translation, emphasized the flocking habit as more conspicuous in the grasslands than in the forests. This view was widely held by zoologists and others outside of the ranks of science. Visher (1916) took a different view, however, and declared that birds displayed less tendency toward social flocking and that few mammals were gregarious. Carpenter (1935) took similar ground in listing only five of twenty-six mammals of the grassland as having the herd or pack habit. The views of Visher and Carpenter seemed to be in the minority, however, because in the period between world wars the drift was in the direction of collectivism and science was usually made to appear to support that trend of social theory.

Allee (1943) pointed out that on one hand the peck and nip orders in chickens and fishes, and on the opposite the cooperative order in other animals, were both soundly based in biology. In human affairs, he argued, the emphasis should be placed upon the cooperative drive to establish a new international system, and that such a program was soundly based upon biology.

On the other side of the social argument only one instance has been found. Among the students of birds; mostly since 1920, led by Howard, there developed the idea of territoriality among birds; that pairs are spaced through the pugnacity of the males in defense of their territory. Burt (1943) declared it was a display of property ownership which "reached its highest development in the human species." Thus the idea of property ownership was declared to be not a peculiarity of man, but a basic characteristic of animals. Heape (1931) had taken a similar view.

Soil determinism

The Russian theory of soil formation was based primarily upon climatic determinism which was modified under the influence of American soil scientists, but the idea of soil as a natural body became the central concept. Soils were treated as immature and mature and in some cases as degraded. All this fitted into the organismic concept made familiar by the ecologists. Jenny (1941) was the exception here, challenging the idea of natural body, and presenting the soil as a physical system and an open system to which substances could be added or subtracted. In every soil, he said, there was an element of organization.

In the Yearbook of the United States Department of Agriculture for 1938, Soils and Men, C. E. Kellogg made a social interpretation of soils. He related the great cultural systems of history to soils; Egyptian to alluvial soils, Arabian to semi-desert soils, Classical Greece and Rome to red soils, and Western Europe to leached forest soils. He said also that "race is rooted to the soil that gave it birth," and that migration produced individual and social conflicts. Furthermore, he maintained that "the traditions of individualism, developed among the people of the forested soils of Western Europe, came into sharp conflict with the demands of the grassland soils. The tendency appears to be toward the development of strong sentiment for cooperation." In rejecting explicitly the climate change theory of culture, he doubted the theory that soil exhaustion had caused the decline of ancient civilization, taking the ground that more probably, in a declining civilization, exhausted men caused the exhaustion of soils: "The rise and fall of cultural systems is no simple matter. It is possible that there are no causes and that the whole of human history may be explained as some sort of mystic rhythm of birth, maturity, and death." Nevertheless, he believed that men had "some responsibility for the direction of their destiny."

In his book, Soils that Support Us (1941), Kellogg again promulgated the idea that the forest soils (the Gray-Brown Podzolic) of the eastern United States developed the spirit of independence in farmers and business men, laissez faire of the political scientists. In contrast, he declared that the chernozem soils of the grassland gave birth to cooperation. Of course, cooperation occurred on occasion elsewhere, as in the case of the Tennessee Valley Authority, but in part he credited this to inspiration from the chernozem soils. Obviously this last reference was to Senator George W. Norris of Nebraska, in particular. If there was any truth to such a theory, it should be subject to some kind of proof. Norris was born in Ohio, and lived there during his formative years, moving to Nebraska at the age of twenty-four. The next eighteen years were spent in Nebraska, and the last forty, as he was in congress, a large part, possibly a half of his time was spent along the humid Potomac river. On the basis of such a calculation thirty-eight years could be credited to the chernozem country and forty-four to the humid regions of Ohio and Virginia, twenty-four of the latter being his formative years. Under these circumstances it seems a little puzzling how either a climate or a soil interpretation could be placed upon Norris's career. The present author challenges all such interpretations of social institutions as going beyond the facts of either science or history. They involve questions of cause and effect which cannot be properly tested by methods of scientific proof. As will be pointed out in another connection, other theorists have selected different factors as causes and have spun out equally "meaningless" webs of speculation.

Geography

In spite of the repudiation by representative geographers of geographical determinism in its environmental sense, Taylor's review (1937) of American textbooks showed how generally it persisted in the newest works and how unscientific it was, justifying fully Hartshorne's (1939) comments on the lack of responsibility of many geographers when indulging in social speculation. Thus the plains as defined by the geographer were made responsible for solidarity, or folk wandering, or imperialism, depending upon the author. Other writers tied democracy and Protestantism to an isothermal map, and others made Mohammedanism a desert religion. One geographer said that rich soils produce aristocracies. If that is true, then one must conclude that soil conservation is a mistake if democracy is to survive. Another attributed to oasis man pacifism, collectivism, slavery, polygamy, a highly developed sense of real property, and early development of the science of astronomy and geometry; mountaineers were individualists, and hill folk were liberty loving. It is one thing to find as a matter of history that a certain thing occurred at a particular place on the earth's surface, but it is quite another to insist, or imply, that it happened at that place because of climate or topography. The climate theory of civilization was associated primarily with Ellsworth Huntington, upon whom Taylor placed most of the responsibility for inspiring young geographers to go 'beyond the facts' in their social speculation. He insisted also that only plain speaking among geographers could stamp out such abuses and protect the good name of geography as a science.

As the low rainfall grasslands and deserts have been so persistently represented as producing cooperative or collective institutions, that particular aspect of abuse of social theory calls for comment. Large scale irrigation and some other kinds of activities do require cooperation or governmental support, but not because of geographical determinism. In any geographical area an enterprise of such scale or character as to exceed the resources or skills of the individual would make the same requirements; the trading corporation in England in the sixteenth century; deep-sea fishing, whaling; canal construction and operation; reclamation of swamps; harbor improvements, and flood control. None of these could be classified as determined by arid or desert climate. Nature is neither democratic nor republican; nor fascist, nazi, communist, monarchial, or democratic; nor individualist, or collectivist; nor Catholic, Protestant, Jewish, Mohammedan, or Buddhist. Man has constructed these concepts and must assume the full responsibility for the use, or misuse, he has made of them.

In his study of plant geography, Raup (1942) brought together a synthesis of the revisionist views of the plant ecologists and the geographers on the concepts of regionalism, succession, and organism. These revisionists concluded that natural regions were largely arbitrary in their limits, and were useful primarily as a convenient basis for organizing knowledge rather than as constituting an independent entity. Hartshorne (1939) was interpreted as doing for organism in geography much the same thing that Gleason (1917, 1926, 1927, 1939) had done for ecology. Both Raup and Hartshorne emphasized the use of the "circle of facts" methodology in preference to cause and effect; and in this sense Raup argued that "plant geography becomes simply a study of the areal differentiation of the world of plants," keeping dynamics or succession in a secondary position: "Our present ideas of the development of vegetation are based to a surprising extent upon inference ... rather than observed facts.... We would be better advised, therefore, to base our study of plant geography, and any system of classification for botanical regions that we may devise, upon the existing plant life rather than upon partially hypothetical trends of development." Any idea of region as an organism was wholly eliminated from such a body of concepts.

Although ecologists and the geographers challenged the organismic concepts in their fields, in that newest interest of American geographers, political geography, the organismic concept of nations appeared in virile form in such a popular text as Van Valkenburg, *Principles of Political Geography* (1939). This book was based on the cycle theory, as it was called, that nations passed through four stages, youth, adolescence, maturity, and old age. This was a direct importation from Europe, but was a characteristic feature of much of the thought there, as well as of the between-war political philosophy both in Europe and America.

Direct issue is taken here with geographical determinism in its general form and in all of its several special applications such as climate, soil, and soil erosion theories of history called into the service of political and social propagandists. For many years, geographers have repudiated in principle such theories of determinism, although many have surrendered in detail. In 1927, Sauer declared that the necessity of repeated repudiation of environmental determinism was becoming a "trifle wearisome." Twenty years after, however, public indoctrination in this determinism seems more complete than ever. In dealing with the grassland area, or the lesser area usually called the great plains, historians and social scientists, as well as geographers and biological scientists have committed themselves largely to some form of determinism in justification of social policies. At every turn, one meets the dictum that the low-rainfall climate makes necessary a collective form of society. In general American policy, a different but as prevalent a form of geographical determinism is derived from the open frontier formula of F. J. Turner, and in world policies from the closed-world-space doctrines of Mackinder and Haushofer. All these philosophical assumptions, along with the doctrine of organism, go beyond the facts, and must be rejected. Social policy must be grounded on facts, and tested independently upon its merits.

Genetics and environment

The problem of environment reached its most acute stage, probably, in connection with the conflict over genetics and environment. Lamarck (1744-1829) believed that acquired characteristics might be inherited, and in the twentieth century there was a Neo-Lamarckian revival (Conklin, 1944). Many ecologists, sociologists, and psychologists seemed to take some form of environmental adaptation for granted, but often exact statement of critical issues was evaded. Among the more outspoken advocates of a planned social order, Neo-Lamarckianism became one of the cornerstones upon which was erected their particular social myth.

The point of view of geneticists underwent a revolution, mostly after the opening of the twentieth century. Cytology based the transmission of characteristics on chromosomes and genes, and the Mendelian revival afforded a significant demonstration of that concept. The mutation theory, first presented by DeVries and substantially revised by others, was widely accepted by the second quarter of the twentieth century in the sense that all individual variation was in the nature of mutations and that speciation was a selective process operating on such variations. This statement of the genetic principle opened the way for a possible reconciliation of the geneticist and environmentalist, providing both were willing to agree that environment might operate as a stimulant to mutation (Mayr, 1942; Dobzhansky, 1937, Second edition 1941). By studying large numbers of specimens arranged in series, the sub-species could be established and the true basic characteristics of the species itself revealed; its distinctive features as well as all the variations which might intergrade into related species. It was not sufficient to show that two individuals differed, it must be shown that the differences justified the assignment to distinct species or sub-species. A name was not applied to the individual, but to the population, which included all the individual variants. In some instances, however, individual differences were so explicit that there could be no question that they were of different species. One of the most important facts to be observed relative to the problem of species was, not the uncertainty or indefiniteness of the species, but the incompleteness and inaccuracy of scientific knowledge about the material, and often, the intrusion of preconceived ideas about what the conclusions ought to be.

Under the influence of environment, plants or animals might undergo certain physical changes that could be measured and described. Two specimens of different species might develop very similar outward characteristics as a result of a particular environment, but there would be no genetic change. Two specimens of the same genetic value might be subjected to different environment, taking on substantial outward differences, but likewise there would be no genetic change. The changes were superficial, not intrinsic; a genotype was determined by chromosomes and genes, while a phenotype was "the resultant of the interactions between the genotype and the environment" (Dobzhansky, 1941).

A new concept of species developed out of the twentieth century genetics. The Old Systematics had assumed that species were definite and were differentiated by structure primarily. The New Systematics emphasized the importance of large numbers of individual specimens, and as no two were alike these were arranged in series according to variations. The approach was biological, recognizing morphology, but also geographical-ecological relationships, genetics, and cytology. Much museum material was not subject to genetic and cytological experiment, but the most complete results were possible only when the four criterions were applicable.

There appeared to be some differences in the details of the genetics of plants and animals, but the broader principles were the same. The most comprehensive long-term range of experimentation with plants was that carried on under the auspices of the Carnegie Institution,. reported by Clausen and others, (Yearbook, see especially Nos. 34, 1935; 35, 1936; 37, 1938; 39, 1940; 40, 1941; and Pub., No. 520. 1940):

Every transplant retained its individuality irrespective of the conditions of altitude, light and moisture.... A few transplants exhibited quite spectacular changes in vegetative characters when grown at different altitudes, but these modifications never obscured the individuality of the plant.... These modifications vary with different species, races, and even individuals of the same race.... All indications point to these modifications as being temporary, reversible and quickly induced.... No evidence is on hand which suggests yearly cumulative effects on perenial transplants (Yearbook 34, 1935).

Furthermore, regional races were found to be different in genetic composition and not mere environmental modifications (Clausen and others, No. 520, 1940). Clements (Yearbooks 1918-1938) had reported on his claims of having transformed plants by changed environment, but his full exposition announced in four volumes has not been published. Clausen indicated doubts about the validity of Clements's experiments on the grounds of errors in method and contaminations in his cultures similar to those of Bonnier (Clausen and others, No. 520, 1940).

The environmental thesis as it applied to origin of species suggested the question of which regions of the earth's surface stimulated the most active evolutionary processes. The most widely held assumption was that deserts and adjacent arid regions were such centers, and that was the conclusion of Clements and Shelford (1939). Dobzhansky (1941) was more cautious in his conclusions, limiting his generalizations to land forms of temperate and subtropical zones, where he stated that complex topography frequently proved to be cradles of rich variations in species, in particular, mountains and deserts. Implicitly such theories are grounded in the hypothesis that in such regions the struggle for existence was more severe than elsewhere with the result of greater variation being stimulated. The limiting factors were different, but that they were more severe is open to question. It would be necessary to demonstrate that these particular factors would stimulate greater frequency or quality of mutations, or both.

The clear weight of scientific evidence was on the side of heredity, rather than environment, as the basis of species determination, environment entering as a factor only on the hypothesis of influencing genetic changes in the process of mutations. When the historian and social scientist have occasion to use scientific material, they are under obligation to limit themselves so far as possible to areas of agreement in each field of science concerned, and when areas of disagreement must of necessity be utilized, the fact of differences should be clearly recognized and conclusions as bearing on social implications limited accordingly.

The most controversial aspect of genetics was the application to the human situation. Differences in details of genetic behavior of plants and animals seemed to be admitted generally. Among plants and animals differences between species and subspecies and their peculiar relations to environment, succession, and climax did not raise social questions. In ecology each plant and animal fitted into a particular niche in the biological equilibrium, and different species occupied a fairly definite sequence in succession series, or position in the total "circle of facts." The idea of equality of species and subspecies, and individuals, could scarcely arise. The breeder of livestock or field crops found it essential to insist upon genetic purity of blood strains, and upon selected hybridization.

Did biological principles as established for plants and animals apply to man? In human relations, the indiscriminate mixture of blood lines was often held up as the ideal condition of racial improvement. Race conscious extremists agitated race superiority or race equality without explicit determination of what either term meant, or whether either term possessed any scientific validity, or whether both terms possessed meaning only in relation to social policy. Differences among human racial groups were indisputable, but what interpretation or significance was attached to them? Dice's proposal (Science, June 9, 1944) for a program of comprehensive cooperative studies of the biology of man aroused a controversial exchange that demonstrated conspicuously how impossible it was to secure a meeting of minds among scientists on such a question. One is led to the conclusion that because of the abuse of certain advocates of the idea of race superiority, an emotional reaction has carried sentiment to the opposite and equally indefensible extreme of equality, and that incidental to the intolerance engendered by irresponsible extremists of both camps, science as the mediator in the brawl, had found itself the major casualty. The problem of verbalism should be stressed. No one has framed an acceptable definition of what is meant by equality and until the question is framed in terms that are subject to proof, the question is "meaningless."

Science and human relations

The practice of applying the ideas of ecology to human relations gained a strong foothold among sociologists, lead by R. E. Park, and E. W. Burgess (1915, 1921, 1925, Alihan, 1938), under the name of human ecology. They drew upon the work of F. E. Clements, botanist, W. B. McDougal, botanist, and C. M. Child, physiologist, including the organismic concept. In South Africa, J. W. Bews derived his philosophy from Jan Smuts's Holism, but used it in much the same way in his Human Ecology. R. D. McKenzie was associated with Park and Burgess, but in his book The Metropolitan Community (1933), he omitted the organismic concept, a fact which emphasized that such a hypothesis was unnecessary. In adopting the concept of organism the biologists appeared not to be aware that the idea was almost as old as philosophy, and that the historian of political theory had usually found it identified with the authoritarian state. In fact its popularity in the twentieth century had risen in association with the rise of collectivism and particularly with its totalitarian versions. The repudiation of organism by a large proportion of the biological scientists should be emphasized, however, in maintaining perspective.

The appreciation by the biologists of the social implication and responsibilities of science is evidenced conspicuously by the "Symposium on the relation of ecology to human welfare — the human situation" conducted by the Ecological Society of America, December, 1939 (Ecological Monographs, 1940). Along the same line, C. C. Adams published a list of "Selected references on the relation of science to modern life" (1940). He admitted the charge that scientists were largely "economic and social illiterates, "but countered with the accusation that "financial and administrative agencies are generally equally illiterate of pertinent science." He admitted also that "the scientist needs first to orient himself and put his own house in order, before advising all the world what to do," and it was in that spirit that he had compiled and published his excellent reading lists.

Emphasis upon the sciences in relation to history makes it necessary to take issue explicitly with the misuses of science for social purposes and to explore some of the possibilities and implications of employing science in the service of history and the social sciences. Science does not impose upon the world any version of determinism in social theory. Furthermore, science is not definitive, and its findings, or interpretations of its findings, cannot properly be imposed upon society under an "illusion of finality" and infallibility. As science, no reputable scientist would claim finality, but when the conclusions of science are carried over into social applications, there is a tendency to assume finality — science has proven, etc., — and therefore, anyone who challenges policies so buttressed is charged with attacking science and scientists, as well as being charged with engaging in partisan politics. Some conclusions of science are sound anywhere. Sometimes the scientist misinterprets his experiments or misapplies them. This is true particularly in interpreting and applying to problems of natural environment the results achieved under the controlled conditions of the laboratory or experimental plot. To be valid in nature, the science of the controlled laboratory must harmonize with life under natural conditions. But possibly of even greater importance, the scientists need to place more emphasis upon research under conditions of natural environment.

In the Christian Century, January 29, 1947, C. C. Morrison proposed the term "scientism" to describe the "prevalent mode of thinking" which raised "scientific procedure to the level of a metaphysic or ... a theology." Scientism, along with statism, have become major social myths that threaten freedom. Intellectual freedom and integrity of knowledge must not be shackled by any of these social theories allegedly derived from science. Misinterpreted and misapplied science may become as dangerous, or more so, than ignorance as customarily defined. To be valid socially, science must harmonize with social experience.

Chapter Twelve

THE COMMUNICATIONS REVOLUTION

There is a tendency to look upon the term regionalism as synonymous essentially with provincialism or localism in an isolationist sense. That is not the use made of it here. On the contrary, a machine civilization broke down local self-sufficiency and introduced a system of world economy which placed an emphasis on regional interdependence. The technological element which made this global economy possible, and bound it together, was communications. It was so important to the last third of the nineteenth century that the process should be designated as the communications revolution to distinguish it from the industrial revolution. The central unique factor in the industrial revolution was the application of mechanical power to the operation of machines in stationary installations. The communications revolution means the application of mechanical power to movement in space of persons, materials, and intelligence. Communications should be treated as an independent variable, as a thing in itself, and studied as such in the historical process. In the particular present instance, the communications factor is related to regional specialization, and inter-dependence in an emerging global economy. It was only through the communications revolution that industrial evolution was enabled to reach newer levels of volume and intensity.

David A. Wells, (1825-1898), Recent Economic Changes, published in book form in 1889, is the landmark in interpretative thought on the meaning of the latter part of the nineteenth century to the history of civilization, global as well as American. Although contemporaries and historians since were largely in agreement that the period after the wars of national unification of the 1860s was an era of degradation and demoralization, and applied to it various abusive epithets expressive of their disapproval or contempt, Wells declared that

> When the historian of the future writes the history of the nineteenth century he will doubtless assign to the period embraced by the life of the generation terminating in 1885, a place of importance, considered in its relations to the interests of humanity, second to but very few, and perhaps to none, of the many epochs of time in any of the centuries that have preceded it;....

In the preface Wells stated the argument of the book in a more effective manner than elsewhere:

The economic changes that have occurred during the last quarter of a century— or during the present generation of living men— have unquestionably been more important and varied than during any former corresponding period of the world's history. It would seem, indeed, as if the world, during all these years since the inception of civilization, has been working up on the line of equipment for industrial effort— inventing and perfecting tools and machinery, building workshops and factories, and devising instrumentalities for the easy intercommunications of persons and thoughts, and the cheap exchange of products and services; that this equipment having been made ready, the work of using it has, for the first time in our day and generation, fairly begun; and also that every community under prior or existing conditions of use and consumption, is becoming saturated, as it were, with its results. As an immediate consequence the world has never seen anything comparable to the results of the recent system of transportation by land and water; never experienced in so short a time such an expansion of all that pertains to what is called "business"; and has never been able to accomplish so much in the way of production with a given amount of labor in a given time.

Concurrently, or as the necessary sequence of these changes, has come a series of wide-spread and complex disturbances; manifesting themselves in great reductions of the cost of production and distribution and a consequent remarkable decline in the prices of nearly all staple commodities, in a radical change in the relative values of the previous metals, in the absolute destruction of large amounts of capital through new inventions and discoveries and in the impairment of even greater amounts through extensive reductions in the rates of interest and profits, in the discontent of labor and in an increasing antagonism of nations, incident to a greatly intensified industrial and commercial competition. Out of these changes will probably come further disturbances, which to many thoughtful and conservative minds seem full of menace of a mustering of the barbarians from within rather than as of old from without, for an attack on the whole present organization of society, and even the permanency of civilization itself.

The problems which our advancing civilization is forcing upon the attention of society are, accordingly, of the utmost urgency and importance, and are already occupying the thoughts, in a greater or less degree, of every intelligent person in all civilized countries. But, in order that there may be intelligent and comprehensive discussion of the situation, and more especially that there may be wise remedial legislation for any economic or social evils that may exist, it is requisite that there should be a clear and full recognition of what has happened. And to simply and comprehensively tell this— ...— has been the main purpose of the author.

In respect to the communications revolution described in the book, it is essential that the historian reorient somewhat his usual view of technology. It is not the date of first invention that is historically important, but the date, the period, at which the innovation became sufficiently perfected as to be applied widely and effectively in changing the manner of life of a region or the world — in this case the world. The time lag between inventions and use placed the realization of the communications revolution mostly after 1875; in ocean transportation of bulk commodities it awaited the installation of the compound marine engine and construction of ships of steel, in railroads the consolidation into through systems using steel equipment, and in electrical communications on consolidation and electrical efficiency in the ocean cable as well as land telegraph. The effect was global and touched intimately the daily life of every person on the earth even remotely within reach of modern communications. The grassland farmer found his products sold on newly organized or reorganized world markets, his local prices were based on world prices registered on those markets on the basis of world competition.

Few men indeed have had the gift at any time in history to interpret as significantly their own generation as did Wells. His book had its defects. At points when he should have rounded out the basic theme by a telling interpretative conclusion, the thread broke and he was diverted to statistical or theoretical economic argument with his contemporaries. By most any standard of measurement, however, it was an important book, but it just fell short of being a truly great classic. Even so, it is the most significant single book of that generation written as a historical interpretation of the author's own time.

Other factors entered into the communications revolution, however, that must be given recognition, factors that Wells did not include: Petroleum lubricants for heavy and high speed machinery; wood pulp paper for cheap newspapers, magazines, and books; the linotype for setting type mechanically, providing the first newspaper issued by that method on July 3, 1886; the camera, wet plate, dry plate, and film, in rapid succession for reproduction. Freedom of movement in space for commodities, persons, and intelligence achieved through these facilities their fullest measure of opportunity.

Albion (1932) proposed that the term "Communications Revolution" be used to designate phenomena related to movement in space as distinguished from the "Industrial Revolution", and that they be treated not as by-products, but as an entity in themselves. This was an excellent suggestion but fell short of the possibilities of the term because he considered it as beginning with turnpikes and canals rather than identifying it exclusively with mechanically powered communications. Power was the unique factor in both the industrial and the communications revolutions. First importance must be given to the technological innovations in mechanical power. They must be isolated for explicit study and then their applications become clearly significant. The compound marine engine in a freight ship, in contrast with a wind driven sail ship, was revolutionary; likewise the coal-burning steam locomotive on steel rails, compared with the horse-drawn freight wagon on a turnpike. These things became realities within the decade of the 1870s and the first half of the 1880s. The first historian to apply this data of the communications revolution explicitly to the writing of formal history was J. H. Clapham in his Economic development of France and Germany, 1815-1914 (1921), and also in his monumental work An economic history of Modern Britain (3 volumes, 1926-1938). It was in the midst of these complexities that the modern idea of regionalism emerged, the relation of natural environment to man's specialized utilization of regional resources, and the recognition of regional adequacy, and regional interdependence.

Chapter Thirteen

PIONEERING TOWARD GRASSLAND REGIONALISM

William Gilpin
Albert D. Richardson
Cyrus Thomas
Josiah Strong
Nathaniel Southgate Shaler
John Wesley Powell
Willard D. Johnson

Eugene Waldemar Hilgard
Frank Hiram King
William Ellsworth Smythe
Hardy W. Campbell
Ellery Channing Chilcott
H. L. Shantz
Isaiah Bowman

Introduction

    The Turner approach to American history was to trace the advance of the frontier across the continent. This idea was qualified, however, by the fact that this march westward was interrupted. Turner (1892) divided American history into four periods; the settlement of the Atlantic seaboard, the expansion westward to the Mississippi river, then the settlement of the Pacific coast and the Great Basin, and finally the filling in between the Mississippi river and the Rocky Mountains. The settlement represented in the third of these periods was primarily by water routes around Cape Horn or across the Isthmus of Panama, and secondarily across the Great American Desert by overland wagon routes. In this respect the major portion of the grassland and the desert constituted a great barrier which had to be avoided or crossed. It was only in Turner's fourth period that the barrier was broken down and the area settled. Turner's organizing principle of the march of the frontier, together with his supplementary ideas of physiographic regions and the section, did not provide a meaningful explanation of the sequence of periods three and four, and he did not provide a methodology for making an attack upon the problems inherent in those aspects of American history.
    Turner did not provide an adequate methodology for describing the process of maturing or stabilizing a society. Here is met a problem of verbalism. By maturing society is not meant the organismic idea of birth, maturity, and death, and the word stabilized as applied to society does not carry fully the meaning needed. What is meant is a continuing process of unstable equilibrium which reaches a relative stability somewhat comparable to ecological climax within the Gleason usage. Turner's

method assigned no particular reason why this area was left vacant; he did not even provide a substantial basis for assigning boundaries. Historically the use of the Mississippi river and the Rocky Mountains was inaccurate, and served its purpose at all only for the sake of brevity in generalization. The regional study that was most significant for historical synthesis, formulating in part a new approach to the problem, was Walter Prescott Webb, The Great Plains: A study in institutions and environment (1931), and in that sense it constituted a major landmark in historiography of the West. Before taking up a critical study of Webb's book, however, it is important to trace some aspects of the literature that preceded him; the work of some of the pioneers in regionalism contributed the essential accumulation of backlog of basic thinking.

The period of earliest contacts of forest man with the prairies does not fall within the primary scope of this book, but, of course, that aspect should be worked out carefully for the contacts east of the Mississippi river in the early nineteenth century. Within the mid-rainfall and substantially forested portion of the prairie only a part of the problems of the typical grassland area were met. Not until crossing to the west side of the Mississippi river did the larger implication begin to materialize, and even then the problem did not crystallize until the second tier of states was reached. Beyond was the Great American Desert of early nineteenth century tradition, covering a region of indefinite and varying boundaries. This desert tradition was so firmly embedded in the public mind that it became an obstacle to thinking about the area, and has been eradicated only in part.

During the 1850s this Trans-Mississippi West struck the public imagination. The Oregon migration in volume began in 1843 and the Oregon treaty of 1846 assured the United States of the mouth of the Columbia river and the Puget sound area. Texas annexation and the China trade treaty matured in 1844, and along with clipper ships, whaling, and the Mexican war, California, with its San Francisco Bay, came into American possession. The United States was a two front nation, with all that implied to a state facing both Europe and Asia, and the Isthmian canal link brought the treaty of 1846 with New Granada and the Clayton-Bulwer treaty of 1850. The California gold rush of 1849 spread during the next decade to include other notorious regions, the Comstock Lode and Colorado, along with hundreds of minor strikes and rumors of gold. The story of exploration and adventure was told in many forms, and the fur trade had produced ar adventurous breed of mountain men and legend. The wagon trade to Santa Fé had inspired Josiah Gregg's Commerce of the Plains. A definite Pacific railroad agitation began in 1843, culminating in the authorization, in 1853, of four surveys. The rivalry of routes was linked with the metropolitan rivalries of the Mississippi river and Great Lakes cities as eastern termini of the railroad, and as dominants in the western and Asiatic trade. The latter was an unrealized hope. The heart of North America was still looked upon primarily as a barrier to be crossed on the road to wealth. The Kansas-Nebraska agitation changed the situation, in part, bringing to a focus for the first time some direct consideration of the problem of climate, recognition of regional differences, conflict of opinion with respect to the boundaries of the desert, and the limits of the Anglo-American's capacity to adapt his way of life to the central grasslands.

    The American Civil War served both as an interruption and a promotion of the Pacific railroad, completed in 1869 by the North Central route connecting with Chicago, and of the westward spread of population. These hastened the killing of the buffalo, incited Indian wars and the liquidation of the Indian as a barrier, created a new crop of hero legends, revived the mining excitement, and expanded the livestock industry into the great range boom of the 1880s. Agricultural settlements gained a foothold at several points. Among the most significant development of this post Civil War-United States was the turning of the current of adventure into channels largely scientific, or allied with science, under the auspices of the federal government: F. V. Hayden, Clarence King, J. W. Powell, G. N. Wheeler. These activities culminated in the Geological Survey and contributed to the creation or enlargement of other agencies interested in science. The most of this western half of the United States was a scientific terra incognita, to which the incidental scientific contribution of the explorations prior to the Civil War had only whetted the appetite. American science was cataloging and describing a new region. All of these factors together broke down the Great American Desert tradition. Forest man's ideas were in process of a reorientation that would have as its objective the changing of the barrier into a most valuable asset. The first phase was the mistaken idea that settlement would change the climate, later that irrigation would neutralize the climate; and finally, that man must adapt his way of life to regional differences and complexities. This did not happen in formal stages, the threads ran through the period together with fluctuating emphasis, but the last mentioned phase at long last became the emerging dominant point of view among informed people.

    The impact of these western developments upon the East was carried through many channels. Among the most important was the variable stream of individuals who carried back their experiences to disseminate by the spoken word. High on a list of carriers must be the newspaper.

More conspicuous in the historian's mind, possibly because more tangible, was the immigration propaganda emanating from the western railroad boom and the erection of new states and territories, all of which carried on the most glowing advertising in the East and in Europe. Men with money to invest, both Americans and Europeans, were attracted to this new land and its resources, grass, forest, and minerals. Playboys with money and time on their hands amused themselves at ranching and hunting, and some of them wrote of their adventures. Much less appreciated by historians were Erastus Beadle's dime novels, the first of which was issued in 1860, the series reaching its climax in the two decades following the Civil War, with many imitators in the field (Goodykoontz, 1946). Western stories and heroes, irrespective of fact or fiction, were sensationally popular in the East. Among other things, they served as a psychological escape mechanism for the dwellers in an urbanized industrial society. But in all the wealth of incident and printed materials, the historian searches in vain for one or a combination of publications that properly could be said to provide a synthesis or interpretation of the Trans-Mississippi West in essentially accurate and striking form.

The processes of filling in the Trans-Mississippi West were varied; isolated islands of population, columns of settlement along railroads, river towns, military posts, and irrigation projects clustering around limited water supplies. There were vacant areas, some isolated by lack of communications, others by Indian or military occupation under the reservation system, and deserts or near deserts.

In historiography, the Great American Desert tradition gave way partly to the idea that all or most of the country was fit only for a pastoral civilization or a range livestock industry, along with mining and lumbering. The literature of the range livestock industry is so voluminous in elaboration of that particular form of utilization that it is given only incidental consideration in this chapter, but aspects of livestock adaptation to the low-rainfall environment will receive additional attention in the next chapter. Regionalism in any true sense for the Trans-Mississippi West possesses a broader foundation, and must be developed in terms of general adequacy and equilibrium in nature, and in addition, regional interdependence under the machine-age utilization employing a far wider range of its various resources. As a process this involved adaptation in conformity with principles derived from physical conditions and from the native biological system. The regional system included, besides livestock, mining, and lumbering, the development of agriculture on tilled land, both by irrigation and by dry-land farming. The least adequately treated in historical literature is the problem of agriculture, and on it, therefore, attention is focused.

William Gilpin

The views of William Gilpin (1813 ?-1894) on the political geography of North America have been called extraordinary or visionary, but however individual opinion may differ in evaluation, they occupy a significant place in ideas of western regionalism. Originally, Gilpin considered a military career but resigned from West Point and turned to law, settling at Independence, Missouri. His restless disposition led him to join Frémont's expedition in 1843, and the Mexican war found him with Doniphan in the southwest. Independence, as the terminal town of the Santa Fé, and Oregon and California trails, his expeditions into the western region and his participation in the Pacific railroad movement, from its early stages in the 1840s, form the background for his meditations on geography and history. His was a global view, with his "pivot of history" the North American continent, and more particularly the western United States.

Gilpin's views on the "physical Geography of our continent" were first presented formally in four letters to the National Intelligencer, written in August and published in October, and December, 1857. They dealt with the great plains, the mountain formation, the basin of the Mississippi, and the hemp-growing region. The subjects of the Pacific railroad through the South Pass, and of the mountain and gold region of the present state of Colorado were added to the first four papers and the whole published in 1860 as a book, The Central gold region, the grain, pastoral, and gold regions of North America, illustrated with maps. The emphasis on the gold region was largely a means of capitalizing on the gold fever to sell the book, because the heart of the work was his geographical study contained in the National Intelligencer letters which were printed without change, in spite of a certain contradiction introduced by the afterthought of the Colorado gold region. The book was reprinted in 1873, and again in 1874, in substantially the same form, but with the title Mission of the North American people, geographical, social and political. Still another book was published in 1890, The cosmopolitan railway compacting and fusing together all the world's continents.

In the preface of the book, The Central Gold Region, Gilpin argued that the fact that the continent of North America occupied an "intermediate position between Asia and Europe and their population, invests her with the powers and duties of arbiter between them. Our continent is at once a barrier which separates the other two, yet fuses and harmonizes their intercourse in all the relations from which force is absent." The second proposition enumerated in the preface was that two factors "the indefinite multiplication of gold coin; and international public works.... promise to enthrone industrial organization as the ruling principle of nations. America leads the host of nations as they ascend to this new order of civilization." Three major features of the continent of North America were singled out as the foci of the discussion; the Mississippi basin, the mountain formation, and South Pass. Before the American public could appreciate its opportunities and its destiny, he insisted that the people must be emancipated from the influence of Europe and maritime man's outlook, the many prevailing misconceptions about the West, "and develop an indigenous dignity — to appreciate Asiatic science, civilization, commerce, and population — these are the essential preparatory steps" (p. 14). He paid particular tribute to Humboldt for breaking the hold of European thinking in geology and geography because he saw the globe as a whole:

He only has spoken worthily of America to her own people. In him we recognize the intrepid pioneer who invites us to understand the gigantic proportions of our own great country, its order, its symmetry, and its grand simplicity of configuration. As Columbus led forth navigation and commerce from its lengthened tutelage in the Mediterranean Sea ..., this venerable pioneer of physical science and arts marshals us on ... to fit society to the broad foundation of the continents, and rear a country of civilization coequal with the globe.... It is for America that they [Columbus and Humboldt] have lived; to us they belong; apostolic citizens of our destiny (p. 93).

In contrasting the three continents, Gilpin emphasized the mountain masses which occupied the interiors of Europe and Asia, but

North America opens toward heaven in an expanding bowl to receive and fuse harmoniously whatever enters within its rim; so each of the other continents presenting a bowl reversed, scatters everything from a central apex into radiant distraction.... The American Republic is then <u>predestined</u> to expand and fit itself to the continent

As this was written on the eve of the Civil War, his views were more significant in declaring that thus "the holy question of our Union lies in the bosom of nature ..." (p. 20). The contrast extended to climate, which was maritime or moist in Europe and on the Atlantic slope, but continental or dry in the interior, especially on the plains and the interior plateaux, and was so novel that maritime man did not understand its superiority. The Syrian plateau had been the cradle of civilization, which moved into the Mediterranean basin, then into the Atlantic and he said in North America it was entering a continental plateau environment similar to that of its origin To Gilpin, as to many of the generation of 1857, science was the factor which should serve as the key to harmonious adjustment of man to environment:

Science develops how this harmony may be known and practiced. As we recede from it, turbulent force dominates, numbers are dwarfed, civilization withers, liberty is lost: as we approach it, civilization expands, charity smiles, order and empire rise (p. 108).

The mountain formation of North America, as described by Gilpin, consisted of five subdivisions or "paralled formations." From east to west, the five formations were the Black Hills (or outliers), about three hundred miles wide lying east of the Rocky Mountains, the Cordillera of the Sierra Madre (the Rocky Mountains), the Plateau of the Table-lands (the western basins), the Cordillera of the Snowy Andes (Sierra Nevada), and the Maratime Piedmont of the Pacific slope. The region as a whole was introduced by commentary upon the "immensity of the space they occupy, the grandeur of their bulk and altitude and the sublime order and symmetry which pervades them as a system and in the details" (p. 46). The breadth at 39° was given as 1600 miles and the length 4500 miles.

The Black Hills (the outlying hills east of the whole length of the Rocky mountain front) were described as an area of timber, and of little rain, but were "an important part of the pastoral region, are clothed in perennial grass, and abound in aboriginal cattle.... and infinite animal life...." This Black Hill country was pictured as interrupted in front of Long's and Pike's peaks, which were thus permitted to overlook directly the Kansas river basin (p. 69); "this delicious country, surrounding the very navel of our continent, and embracing its geographical centers, has from that fact a perpetual and permanent interest." The Kansas, the Platte, and the Arkansas rivers had their extreme sources "beneath the roots of Pike's Peak;—"

This Sierra Madre has its own characteristics, which are all of the grandest order. I am unable to illustrate it by comparison, because it stands supreme and alone, the standard to which all other mountain masses must be submitted.

Of the third formation, the Plateau of the Tablelands, Gilpin declared that,

... this appears to me the most interesting, the most crowded with various and attractive features, and the most certainly destined to contain the most enlightened and powerful empire of the world. At present it is no more known or comprehended, as it is, by the American people than was America itself to the poet Homer, and is to them as much a myth as the continent of Atlanta. Nevertheless, it is of such great area as to contain within itself three great rivers which rank with the Ganges and Danube in size, and five great ranges of primary mountains (p. 49, 74).

He emphasized "its variety and vastness," and also that it was a unit:

It is universally a rainless region, and nowhere is arable agriculture possible without artificial irrigation. Pastoral culture is the prominent feature, wherein it rivals the Great Plains. The air is tonic and exhilerating— the atmosphere resplendent with perpetual sunshine by day and with stars by night. The climate is intensely dry, and the temperature variant and delicious.... Timber grows upon the rivers and upon the irrigated mountain flanks. To arrange the arable lands for irrigation is not more costly than our system of fencing, which it supersedes. No portion of the globe can maintain a denser population.

He pointed out in several places that the ancient civilization of the Aztecs had flourished in the southern part of this area of the continent, and the Incas in a similar region in South America. In emphasizing the North American Plateau in comparison with those of other continents he hoped to "demonstrate its area, its climate, its capacity, and its geographical power in the world.(p. 72).

The Pacific Maritime Front also inspired Gilpin to superlatives,

In all these natural favors our western seaboard front is supremely more gifted than the classic shores of the Mediterranean and the Asian Seas, for fifty centuries the favorite theme of history, poetry and song.

After giving the overview of the whole mountain formation in Chapter 4 (a reprint of the letter of 1857), he developed in four other chapters the principal different sub-divisions in detail, especially important being the treatment of the interior basins and of the Rocky Mountains. The final chapter on the gold region was mostly an after thought. He had great visions of the future possibilities of the mountain formation emphasizing the calcareous nature of the soils, their wonderful fertility, the variety and importance of plant and animal life, the healthfulness of the dry continental climate, and the great possibilities of agriculture under an irrigation system. In the Rocky Mountain region: "The prominent agricultural feature of the Cordillera [Rocky Mountains] is fertility— pastoral fertility." And further he insisted that,

If the inquiring spirit ... will revive.... Counterfeit geography, promulgated with official dogmatism, will cease to be fashionable, or to defeat the divine instinct of the people. Patriotism, pioneered by truth and genuine science, will reveal and comprehend our continental geography as it is, huge in dimensions, sublime in order and symmetry, a unity of plan. Our political and social empire, expanded to the same dimensions harmonized to the same chequered variety, will assume a similar order, a like symmetry, and crown hope with a similar solid and enduring perpetuity (p. 71).

The chapter on the Sierra San Juan, the mountain area extending from southwestern Colorado southward into Mexico and lying west of the Rio Grande river included a section on the Great American Desert. This commentary revealed that Gilpin possessed a clear understanding, not only of the desert, but also of the Atlantic seaboard:

> The scientific writers of our country adhere with unanimity to the dogmatic location somewhere of 'a great North American desert'. Travellers under their promptings, especially search for it. It has been located <u>seriatim</u> in advance of the settlements, in Kentucky, in the North-west, in Missouri, upon the Plains, in California. No explorer or witness who has failed to find a desert, is allowed credence or fame. Yet there is none, either in North or South America; nor is the existence of one possible. On the contrary, the least fertile portion of our continent is the <u>silicious</u> maratime slope of the Atlantic States, whose climate is also the most inhospitable. Yet here is no desert, and none anywhere else exists. This dogmatic <u>mirage</u> has lately receded from the basin of the Salt Lake; it is about to be expelled from its last resting-place, the basin of the Colorado (p. 92).

He compared the land formations of this mountain-dominated area with the mountainous areas of Europe and equilibriums in nature and under cultivation:

> In readiness to receive and ability to sustain in perpetuity a dense population, it is more favored than Europe. Fertility of soil of the highest order, is the dominant and uniform characteristic of this immense region, [mountains, mesas, and gigantic terraces] ... It is the decay of lava, selenite, and carboniferous limestone that forms the soil. The pastoral fertility is developed by nature, which sustains its aboriginal herds as fish in the rivers and in the sea. The arable fertility needs the care of man, and awaits the economical development of artificial irrigation. For the reception of this system, the whole structure and contour of the surface is fitted, and the natural waters abundant.

Concerning the interior basins, Gilpin described "This Plateau, enclosed within the Cordilleras of the Mountain Formation, [as possessing]... characteristics new to mankind, and about to arrest the attention and sway the mental energies of America," (p. 72)....

> The soils of the Plateau are of the highest order of fertility, alike upon the mountains, the valleys, and the mesas or extensive plains. The dry and serene atmosphere converts the grasses into hay, and, preserving them without decay, perpetuates the food of

grazing animals around the year. This gives to pastoral agriculture an infinite capacity for production and superlative excellence. Meat food, leather, wool, fowels, fish, and dairy food are of spontaneous production.

The soils, accumulated from the attrition and decay of lava and of carboniferous and sulphurous limestones, possess an exuberent fertility. Spots of arid sands are few and insignificant; such as exist are from the auriferous granite, and contain placers of gold! These soils, then, composed of the essential elements of fertility and production, and warmed by an unclouded sun, need only irrigation to ferment their activity. For this, nature has provided in the configuration of the surface and the infinite abundance of snowy mountains, of streams and of rivers descending from their glaciers or bursting from their flanks.... [Irrigation] is understood and practiced by the aboriginal people. The laborious systems of culture to provoke germination, the uncertain yield common to our people of the maratime region of timber and uncertain seasons, are here unknown, and unnecessary. A perpetual sun and systematic irrigation (as in Egypt) dispenses with laborious manual tillage; The use of the plow is not indispensable; the waters for irrigation descend from a higher level and are constant. The laborious extermination of the primeval forest; fuel and refuge from the inclement seasons of heat and cold; periodical and uncertain inflictions of drought and saturation; dependence upon an atmosphere ever changing and forever fickle and treacherous; none of these vicissitudes are seen or known upon the Plateau. The adobe brick, of unburned clay, constructs fences and houses, inhabited more for domestic seclusion and convenience than from necessity.... The Plateau presents itself, therefore, prepared and equipped by nature in all departments at every point, and throughout its whole length, for the immediate entrance and occupation of organized society, and the densest population (pp. 77-79).

* * *

For agriculture, both pastoral and arable, no region in the world is more propitious, not even the Basin of the Mississippi, which is by its side. One remarkable characteristic pervades all of the rivers; their waters are supplied (as are those of the Nile) from the high mountains, whence they descend. Such rivulets as abound in maratime countries are not known.... (p. 81).

Such is the Plateau of America, transcendent in position, immense in area, superlative in climate, fertility, and variety of configuration. Here are blended all the elements which distinguish the other plateaux of the world. Its longitudinal form; the rainless character and perennial brilliancy of atmosphere; its perpetual vernal temperature; its alternate basins, parcs, and snowy sierras; its great rivers, its indefinite and propitious capacity to produce and to sustain population; its gold, metals, and gems; finally, its dominant position, butting over the Asiatic ocean on the one hand, over the calcereous plain on the other hand, continuously from the Polar Sea to the equatorial belt, all these arise successively and together to announce to the American people their accession to the most attractive, the most wonderful, and the most powerful department of their continent and country (p. 83).

Gilpin's second great division of the North American continent was "The Great Mississippi Basin" which he introduced with the declaration that

> The most remarkable feature of America is the Basin of the Mississippi. As yet the popular mind does not clearly comprehend its dimensions, and the understanding of its physical characteristics is indistinct and vague.

The basin was bisected by the Missouri-Lower Mississippi river artery, which was his unusual manner of approach, and in length 5000 miles, and in width 2500 miles: "Into this central artery, as into a common trough, descend innumerable rivers, coming from the great mountain chains of the continent." He delineated three main divisions distinguished by vegetation and climate. Timber occupied the area east of "an irregular line from the head of Lake Erie, running towards the south and west into Texas...." The prairie region occupied a belt west of this line for about 450 miles, "and within it artificial irrigation is not practiced, nor necessary, it being everywhere soft, arable, and fertile." This was a grass region, "though narrow lines of forest continue upon the saturated bottoms of the rivers and in the islands.... over which the fires annually sweep after the decay of vegetation." Beyond was the plains, extending to the mountains, an "immense rainless region" as it possessed "a compact soil, coated with the dwarf buffalo grass, without trees, and the abode of aboriginal cattle" — an "exclusively pastoral region" (p. 114). He estimated the relative proportions of the basin occupied by each form of vegetation: three-tenths forest, three-tenths prairie, and four-tenths plains. The valley's population as of 1850 was given at twelve million and the potential as ten times that number, and that of the North American continent two billion, or twice the world's population of that time. The confluence of the Republican and Smoky Hill rivers, or the vicinity of Fort Riley, was indicated as the geographical center of all three; North America, the United States, and the Mississippi Basin. Compared with the continents of Europe and Asia, inverted bowls, this basin was a bowl, right side up:

In geography the antithesis of the old world, in society we are and will be the reverse. Our North America will rapidly accumulate a population equalling that of the rest of the world combined: a people one and indivisible, identical in manners, language, customs, and impulses: preserving the same civilization, the same religion: imbued with the same opinions, and having the same political liberties....

> Thus, the perpetuity and destiny of our sacred Union find their conclusive proof and illustration in the bosom of nature. The political storms that periodically rage are but the clouds and

sunshine that give variety to the atmosphere and checker our history as we march. The possession of the Basin of the Mississippi, thus held in unity by the American people, is a supreme, a crowning mercy Viewed alone in its wonderful position and capacity among the continents and the nations; viewed, also, as the dominating part of the great calcareous plain formed by the coterminous Basins of the Mississippi, St. Lawrence, Hudson's Bay, and Mackenzie, the ampitheatre of the world— here is supremely, indeed, the most magnificent dwelling-place marked out by God for man's abode.

Gilpin's conception of the Great Plains was so significant that, in spite of some of its errors, it is worth reprinting in this study practically in full. It was one of the letters to the National Intelligencer, printed October 13, 1857, becoming chapter ten of the book of 1860:

There is a radical misapprehension in the popular mind as to the true character of the 'Great Plains of America,' as complete as that which pervaded Europe respecting the Atlantic Ocean during the whole historic period prior to Columbus. These Plains are not deserts, but the opposite, and are the cardinal basis of the future empire of commerce and industry now erecting itself upon the North American Continent. They are calcareous, and form the Pastoral Garden of the world.... [Between the meridian of the west boundary of Missouri and the Rocky Mountains] they occupy a longitudinal parallelogram of less than 1000 miles in width, extending from the Texan to the Arctic coast.

There is no timber upon them, and single trees are scarce. They have a gentle slope from the west to the east, and abound in rivers. They are clad thick with nutritious grasses, and swarm with animal life. The soil is not salicious or sandy, but is a fine calcareous mould. They run smoothly out to the navigable rivers. The Missouri, Mississippi, and St. Lawrence, and to the Texan coast. The mountain masses towards the Pacific form no serious barrier between them and the ocean. No portion of their whole sweep of surface is more than 1000 miles from the best navigation. The prospect is everywhere gently undulating and graceful, being bounded, as on the ocean, by the horizon. Storms are rare, except during the melting of the snows upon the crest of the Rocky Mountains. The climate is comparatively rainless; the rivers serve, like the Nile, to irrigate rather than drain the neighboring surface, and have few affluents. They all run from west to east, having beds shallow and broad, and the basins through which they flow are flat, long, and narrow. The area of the 'Great Plain' is equivalent to the surface of the twenty-four States between the Mississippi and the Atlantic Sea, but they are one homogeneous formation, smooth, uniform, and continuous, without a single abrupt mountain, timbered space, desert or lake. From their ample dimensions and position they define themselves to be the pasture-fields of the world. Upon them pastoral agriculture will become a separate grand department of national industry.

The pastoral characteristic, being novel to our people, needs a minute explanation. In traversing the continent from the Atlantic Beach to the South Pass, the point of greatest altitude and remoteness from the sea, we cross successively the timbered region, the prairie region of soft soil and long grasses, and finally the Great Plains. The two first are irrigated by the rains coming from the sea, and are <u>arable</u>. The last is rainless, of a compact soil, resisting the plow, and is, therefore, <u>pastoral</u>. The herbage is peculiarly adapted to the climate and the dryness of the soil and atmosphere, and is perennial. It is edible and nutritious throughout the year. This is the 'gramma' or 'buffalo grass'. It covers the ground one inch in height, has the appearance of a delicate moss, and its leaf has the fineness and spiral texture of a negro's hair.

During the melting of the snows in the immense mountain masses at the back of the Great Plains, the rivers swell like the Nile, and yield a copious evaporation in their long sinuous courses across the Plains: storm clouds gather on the summits, roll down the mountain flanks, and discharge themselves in vernal showers. During this temporary prevalence of moist atmosphere these delicate grasses grow, seed in the root, and <u>are cured into hay upon the ground</u> by the gradually returning drouth. It is in this longitudinal belt of perennial pasture upon which the buffalo finds his <u>winter food</u>, dwelling upon it without regard to latitude, and here are the infinite herds of aboriginal cattle peculiar to North America— buffalo, wild horses, elk, entelope, white and black-tailed deer, mountain sheep, the grisly bear, wolves, the hare, badger, porcupine, and smaller animals innumerable. The aggregate number of this cattle, by calculation from sound data, exceeds one hundred million. No annual fires sweep over the Great Plains; these are confined to the Prairie region.

The Great Plains also swarm with poultry— the turkey, the mountain cock, the prairie cock, the sandhill crane, the curlew; water-fowl of every variety, the swan, goose, brant, ducks, marmots, the armadillo, the pecary, reptiles, the horned frog; birds of prey, eagles, vultures, the raven, and small birds of game and song. The streams abound in fish. Dogs and demiwolves abound. The immense population of nomadic Indians, lately a million in number, have, from immemorial antiquity, subsisted exclusively upon these aboriginal herds, being unacquainted with any kind of agriculture or the habitual use of vegetable food or fruits. From this source the Indian draws exclusively his food, his lodge, his fuel, harness, clothing, bed, his ornaments, weapons, and utensils. Here is his sole dependence from the beginning to the end of his existence. The innumerable carnivorous animals also subsist upon them. The buffalo alone have appeared to me as numerous as the American people, and to inhabit as uniformly as large a space of country. The buffalo robe at once suggests his adaptability to a winter climate.

The Great Plains embrace a very ample proportion of <u>arable</u> soil for farms. The 'bottoms' of the rivers are very broad and level, having only a few inches of elevation above the waters, which descend by a rapid and even current. They may be easily and cheaply saturated by all the various systems of artificial irrigation, azequieas, artesian wells, or flooding by machinery. [Acequia, dike, canal, or ditch]. Under this treatment the soils, being alluvial and calcareous, both from the sulphate and carbonate formations, return a prodigious yield, and are independent of the seasons. Every variety of grain, vegetable, the grape and fruits, flax, hemp, cotton and the flora, under the perpetual sun, and irrigated at the root, attain extraordinary vigor, flavor, and beauty.

The Great Plains abound in fuel, and the materials for dwellings and fencing. Bituminous coal is everywhere interstratified with the calcareous and sandstone formation; it is also abundant in the flanks of the mountains, and is everywhere conveniently accessible. The dung of the buffalo is scattered everywhere. The order of vegetable growth being reversed by the aridity of the atmosphere, what show above as the merest bushes, radiate themselves deep into the earth, and from below an immense arborescent growth. Fuel of wood is found by digging. Plaster and lime, limestone, freestone, clay, and sand, exist beneath almost every acre. The large and economical adobe brick, hardened in the sun and without fire, supersedes other materials for walls and fences in this dry atmosphere, and, as in Syria and Egypt, resists decay for centuries. The dwellings thus constructed are most healthy, being impervious to heat, cold, damp, and wind.

The climate of the Great Plains is favorable to health, longevity, intellectual and physical development, and stimulative of an exalted tone of social civilization and refinement. The American people and their ancestral European people having dwelt for many thousand years exclusively in countries of timber and within the region of the maritime atmosphere; where winter annihilates all vegetation annually for half the year; where all animal food must be sustained, fed, and fattened by tillage with the plow; where the essential necessities of existence, food, clothing, fuel, and dwellings, are secured only by constant and intense manual toil; why, to this people heretofore, the immense empire of pastoral agriculture, at the threshold of which we have arrived, has been a complete blank, as was the present condition of social development on the Atlantic Ocean and the American continent to the ordinary thoughts of the antique Greeks and Romans! Hence this immense world of plains and mountains, occupying three-fifths of our continent, so novel to them and so exactly contradictory in every feature to the existing prejudices, routine, and economy of society, is unanimously pronounced an uninhabitable desert. To any reversal of such a judgment, the unanimous public opinion, the rich and poor, the wise and ignorant, the famous and obscure, agree to oppose unanimously a dogmatic and universal deafness. To them, the delineations of travellers, elsewhere intelligent, are here tinged with lunacy; the science of geography befogged; the sublime order of Creation no longer holds, and the supreme engineering of God is at fault and a chaos of blunders.

The Pastoral Region is longitudinal. The bulk of it is under the Temperate Zone, out of which it runs into the Arctic Zone on the north, and into the Tropical Zone on the south. The parallel Atlantic arable and commercial region flanks it on the east; and the Pacific on the west. The Great Plains, then, at once separate and bind together these flanks, rounding out both the variety and compactness of arrangement in the elementary details of society, which enables a continent to govern itself with the same ease as a single city.

Assuming, then, that the advancing column of progress having reached and established itself in force all along the eastern front of the Great Plains, from Louisiana to Minnesota; having, also, jumped over and flanked them to occupy California and Oregon; assuming that this column is about to <u>debouch</u> upon them to the front and occupy them with the embodied impulse of our thirty millions of population, heretofore scattered upon the flanks, but now converging into phalanx upon the centre: some reflections, legitimately made, may cheer the timid, and confirm those who hesitate from old opinion and the prejudices of adverse education.

It is well established that six-tenths of the food of the human family is, or ought to be, animal food, the result of pastoral agriculture. The cattle of the world consume eight times the food per head, as compared with the human family. Meat, milk, butter, cheese, poultry, eggs, wool, leather, honey, are the products of pastoral agriculture. Fish is the spontaneous production of the water. Nine-tenths of the labor of arable culture is expended to produce the grain and grasses that sustain the present supplies to the world of the above enumerated articles of the pastoral order. If, then, a country can be found where pastoral produce is spontaneously sustained by nature, as fish in the ocean, it is manifest that arable labor, being reduced to the production of bread food only, may condense itself to a very small per centage of its present volume, and the cultivated ground be greatly reduced in acres. [Statistical table of numbers omitted].

* * *

It is probable that the aggregate aboriginal stock of the Great Plains still exceeds in amount the above table. It is all spontaneously supported by nature, as is the fish of the sea. Every kind of our domestic animals flourishes upon the Great Plains equally well with the wild ones. Three tame animals may be substituted for every wild one, and vast territories reoccupied, from which the wild stock has been exterminated by indiscriminate slaughter and the increase of wolves.

The American people are about, then, to inaugurate a new and immense order of industrial production: Pastoral Agriculture.— Its fields will be the Great Plains intermediate between the oceans. Once commenced, it will develop very rapidly. We trace in their history the successive inauguration and systematic growth of several of these distinct orders: The tobacco culture, the rice culture, the cotton culture, the immense provision culture of cereals and meats, leather and wool, the gold culture, navigation external and internal, commerce external and internal, transportation by land and water, the hemp culture, the fisheries, manufactures.

Each of these has arisen as time has ripened the necessity for each, and noiselessly taken and filled its appropriate place in the general economy of our industrial empire.

This pastoral property transports itself on the hoof, and finds its food ready furnished by nature. In these elevated countries fresh meats become the preferable food for man, to the exclusion of bread, vegetables, and salted articles. The atmosphere of the Great Plains is perpetually brilliant with sunshine, tonic, healthy and inspiring to the temper. It corresponds with and surpasses the historic climate of Syria and Arabia, from whence we inherit all that is etherial and refined in our system of civilization, our religion, our science, our alphabet, our numerals, our written languages, our articles of food, our learning, and our system of social manners.

As the site for the great central city of the 'Basin of the Mississippi' to arise prospectively upon the developments now maturing, this city— (Kansas City, at the mouth of the Kansas) has the start, the geographical position, and the existing elements with which any rival will contend in vain. It is the focal point where three developments, now near ripeness, will find their river port. 1. The pastoral development. 2. The gold, silver, and salt production of the Sierra San Juan. 3. The continental railroad from the Pacific. These great fields of interprise will all be recognized and understood by the popular mind within the coming six years, and will be under vigorous headway in ten. There must be a great city here, such as antiquity built at the head of the Mediterranean and named Jerusalem, Tyre, Alexandria, and Constantinople; such as our own people name New York, New Orleans, San Francisco, St. Louis.

The relation of North America to the course of civilization was developed originally in the letter to the National Intelligencer, published December 3, 1857, reprinted in The Central Gold Region (1860) as chapter eleven, but in this second presentation another chapter, eight, was written on "The South Pass of North America" giving further elaboration and application of the idea to the pivot of history in North America. With the aid of a world map (opposite p. 81) he described the "Isothermal Zodiac or belt of equal temperature around the northern hemisphere of the earth. Within this are included all the civilized nations of Asia, Europe, and North America, about 850,000,000 white people in the aggregate nine-tenths of the human race." In another form, he stated his theory, that "This system of civilized society, of which we Americans form a part, is very ancient and is inherited. History is the journal of its geographical progress...." (p. 136). This represented a form of climatic determinism in history, and he argued that the great cities and empires of China, India, the Syrian Plateau, the Mediterranean, Europe, and America lay in the central current of this isothermal axis. Across the Atlantic and the Pacific oceans it curved north on approximate great circle routes, but across North America it made a southward curve from New York through St. Louis, Independence, Missouri, and then north of west following closely the Oregon trail through the South Pass to the Pacific Northwest and across the ocean. Within the central portion of North America the central current lay near the 40 parallel. By the hemp region, Gilpin meant Missouri, the area capable of varied production; hemp, tobacco, grain, blooded cattle, wool, and occupying the commercial cross-roads position in the Mississippi Basin, both by river, and by overland routes southwest and northwest. In this area the central current "traverses a line of greatest variety of production and largest distribution of groceries, dry goods, and manufactured metals" (p. 136).

The Mississippi Basin was contrasted with the Mediterranean Basin, the new center and the old center of civilization, but superior to the old beacuse the new valley was rich land surrounded by mountainous country, while the old center was an equal expanse of sea, similarly surrounded; the dry continental climate was compared with the maritime climate to the advantage of the former; and the easy river and rail communication was compared with the uncertainties of Mediterranean storms. Furthermore, the Mississippi Basin connected by imperceptible divides with the Arctic and St. Lawrence basins, only somewhat less important. The South Pass was fitted into this geographical setting as serving the same function to the Mississippi basin that the Pillars of Hercules did to the Mediterranean. It was only by the South Pass that the American people could find a gateway for their commerce across the one thousand mile barrier of the Cordillera plateau to Asia:

This Continental Railway is an essential domestic institution, more powerful and more permanent than law, or popular consent, or political constitutions to thoroughly complete the great systems of fluvial arteries which fraternize us into one people; ... and establish its structure so solid, that no possible force or stratagem can shake its permanence, ...[or its]... prosperity ... be impaired ... for want of room.

His was the third generation under the federal constitution:

The *first* gave us this sacred Union, ... the *second* has filled up the Atlantic half of the continent with States, secured the maritime connections with that ocean and Europe, and has blazed for us the way across the continent to the Pacific and to Asia. We, the *third* generation, receive from them the pious task to plant states toward that ocean; to complete the zodiac of fraternal nations round the globe, and to set deep and firm to their outward dimensions the foundations they have laid (p. 111).

Gilpin thought of American history in terms of westward movement of population, but on a world scale in the circuit of the globe. The westward march of the American people was interrupted at the meridian of the west boundary of Missouri, and he placed the blame on governmental policy:

In 1820, this middle column of the centre had reached the western frontier of Missouri, and opened trails along to the Pacific Sea; the flanks were then behind in New York, Lower Canada, and in Georgia. In the overwhelming revulsion of all previous political precedents, which pervaded our Federal councils from 1816 to 1828, central progress was forcibly interdicted. Abruptly stopped by an Indian barrier and Draconic code, and forced to recoil for forty years, the flanks have come up to an even front upon the right and upon the left (p. 82).

This explains in part his blast, already noted, at the "counterfeit geography, promulgated by official dogmatism" (p. 71), but he pounted out

Thus in 1842, commenced to agitate itself throughout America, the energetic geographical movement, to reorganize the column of central progress artificially stagnated in Missouri since 1820. Exploration, conquest, the conversion of the wilderness, have since advanced with celerity. As is the case with all normal instincts, war, peace, domestic and foreign schemes of opposition have each contributed to precipitate its advance and fire its activity (p. 27).

Another extract completes his thought for the development of the West:

Here, then, is an immense foundation wherefrom to grasp and control the expanding developments in front, consequent upon the obliteration of the Indian barrier, and the bursting forth of the pent-up flood of <u>central</u> progress, out over the prairies which undulate to Texas, Mexico, and the mountains. The front wave of this flood-tide is already in motion; its spray sprinkles the plains almost to the mountain foot. The achievements of the coming <u>decade</u> of years will differ from its predecessor. It will exhibit a greater mass of energy, concentrated in one direction, occupied by a single object, and moving with immense means over a very short line, which is perfectly straight and open. Heretofore the active force of progress has been operating round the rim of our territory, on Lake Superior, in California, in Texas, in Florida, in detached squadrons separated from the base of old society, by the diameter of the continent, or keeping up its communication round the circumference by <u>sea</u>. The opening <u>decade</u> beholds a concentric movement, flooding into the centre, and reducing all movements to the shortest radii !

Its career opens with a general force of 30,000,000 of population, having gold in hand, railroads, steamers, rivers and prairies on their banks. The difficulties of the wilderness are overcome, the temptation every way increased, the means of motion enormously accumulated.

Such is the prosperous future which shines over the <u>central west</u>, and fills the atmosphere to the remotest horizon.... Sound health, complete preparation, fresh and mature vigor, judgment, and a defined and finite object, all blend themselves with the immense and successful movement which closes in to occupy the centre of our country, to reunite its flanks, and to adjust its true and geographical balances for ever.

The manner of Gilpin's presentation invites contrast and comparison with the ideas of Turner, Mahan, and Mackinder. Turner found the key to American history in the expansion of the frontier across the continent. Gilpin viewed the history of civilization in terms of a circuit of the globe, in which American history was a phase, and he stressed the interruption of the westward march in the 1820s. This would terminate the frontier at that point — the frontier in the Turnerian sense of a forest process — and after the 1820s the filling-in of the Trans-Mississippi West proceeded from circumference to center by the shortest radii, and by means of "railroads, steamers, rivers, and prairies." His was a continental rather than a maritime view of historical destiny, or in the terminology of the twentieth century, he was anticipating the geographical land-mass pivot of history (Mackinder) rather than the sea-power (Mahan) interpretation of history. His pivot of history was the continent of North America, centering in the western Mississippi basin. To Gilpin, the new ruling principle of nations was not space, but industrial organization, and science held the key to the harmonious adjustment of man to the earth.

In the biographical sketch of Gilpin, Willard (D. A. B.) dismissed his maps as "quaint" and his argument as "more quaint", but such a verdict did not recognize them in their relations to other thinking of the nineteenth and twentieth centuries. A half century later, Ellsworth Huntington became conspicuously identified with a similar climate interpretation of history; Turner became the leader of a frontier school of history; and Mackinder promulgated his "pivot of history" theory, finding his pivot in the Eurasian land-mass instead of Gilpin's North American land-mass. If Gilpin's ideas were merely quaint, then the same dictum applied to Huntington, Turner, and Mackinder, only circumstances decreed that their ideas wielded more influence: All such thinking is subject to the limitations placed upon geographical determinism in any form, and upon single factor interpretation of history. On climate, in relation to soils and agriculture, Gilpin's views were exaggerated, but experience

of farmers in handling the land, and the soil science of
Hilgard were to demonstrate the substantial validity of
his point of view. Gilpin was constructive and positive,
not negative, in his estimate of the grassland. He did
not use deficiency terminology; his mind was remarkably
free from the perverted estimates derived from the mental
limitations of man conditioned by maritime climates — the
grasses and animals were products of natural equilibrium
in a continental climate, and he emphasized that such environment
had been the homes of great civilizations in
Asia, Africa, and in pre-Columbus North and South America.
In his land-mass interpretation of history Gilpin was
recognizing the advent into modern civilization of the
machine age of modern industrialism driven by mechanical
power. With the advent of the air age, the ideas of Gilpin
take on a new interest. The present author has presented
in another place his own views on the significance
of North America in this new age of communications (Malin,
Space and history, 1944). Gilpin possessed only an amateur's
knowledge of the science of his day, but his science
was supported by first-hand knowledge derived from
his life in the west and his contact with the men engaged
in the expansion movement of the mid-nineteenth century.
His biographers need yet to provide an adequate treatment
of his intellectual environment and the development of his
thought. And in any case, whether or not the twentieth
century reader approves or disapproves of his conclusions,
the Gilpin National Intelligencer letters of 1857, and
their elaborations in the book of 1860, constitute the
most provocative and significant estimate of the North
American grassland interior written during the middle
period of the nineteenth century.

Albert D. Richardson

One of the most popular, and certainly one of the
first, books to treat of the Trans-Mississippi West, in
any sense of wholeness, was Albert D. Richardson's Beyond
the Mississippi (1867). This was a newspaper man's
colorful account of his travels and adventures over the
decade 1857-1867. It was almost altogether descriptive,
containing very little that was significantly analytical
or interpretive of the region. That little, however, was
on the positive side, the "prefatory" being the most significant
portion in this sense:

> Twenty years ago, half our continent was an unknown land,
> and the Rocky Mountains was our Pillar of Hercules. Five years
> hence, the Orient will be our next door neighbor. We shall hold
> the world's granary, the world's treasury, the world's highways.
> But we shall have no Far West, no border, no Civilization, in line
> of battle, pressing back hostile savages, and conquering hostile
> Nature.

In exhaustlessness of resources, no other country on the globe equals ours beyond the Mississippi....

"Its mines, forests and prairies await the capitalists," but he suggested other attractions of interest to the antiquarian, the painter, the naturalist, the invalid, the immigrant, and as respects the last:

Its society welcomes the immigrant, offering high interest upon his investment of money, brains or skills; and if need be, generous obliviousness of errors past— a clean page to begin anew the record of his life.

The themes are fruitful. The Pacific Railroad hastens toward completion. We seem on the threshold of a destiny higher and better than any nation has yet fulfilled and the Great West is to rule us.

Scattered through the book were a number of significant comments, the muddiness of the Missouri and the Kansas rivers, and the scantiness of timber:

Here and there were scattered trees along far-apart streams; but they were like angel visits. This lack of timber was the most serious drawback of pioneers; yet the farmer would far better settle where he must go twenty-five miles for a house and fence lumber and firewood, than where he must clear away forests to make room for his corn and grass fields. The latter is the work of two generations....

In respect to the soil, he said that "where the streams have cut into it for thirty feet, the ravines displayed rich alluvium, black as jet, down to the bottom," which compared with that along the Muskingum and Miami rivers in Ohio. Obviously this statement was too sweeping as it could be true only of some localities. He subscribed to the theory that rainfall followed civilization, and that was in accord with the spirit of the time and place. In the desert, the soil was made productive by water as in Arabia, he assured his readers; in Nevada irrigation was from streams, but in California some water was pumped from wells by windmills, and "in time, simple and cheap machinery for irrigation will doubtless be introduced. Then the Great American Desert will become a thing of the past...." Only when "The Great Pacific Railway" was completed could anyone "comprehend the vastness and variety of our mineral resources." Furthermore, "the scarcity of lumber is a blessed thing for Kansas. It secures buildings of brick and stone, instead of log shanties and frame shells."

These pioneer explorers [Lewis and Clark] reported that the best land along the great river [Missouri] as between the mouth of the Platte. This undoubtedly embraces and borders upon the largest and best unbroken farming tract on the globe. Kansas has had only two injurious droughts in thirty years. With early planting and sowing, and deep plowing, she suffers no more from dry weather than New York or Massachusetts. Her soil is the very richest. There is not a swamp in the State. It is difficult to find ten acres of untillable ground. Coal underlies almost every county. Limestone and sandstone make excellent building material, and Osage orange admirable fences. Cottonwood, black walnut and maple grow large enough for sawing in five years from seed.

The average yield of corn is from forty to sixty bushels to the acre. With the best machinery, one man will plant, cultivate and gather fifty acres in a season. The hoe is never used. Weeds are kept down by plowing.

In addition to these over-optimistic comments, he summarized oat and wheat yields, the growing of Hungarian, and Chinese sugar cane (sorghum); for fruits Kansas was said to have no equal except California; "hay is a natural crop," and "stock raising is the most lucrative pursuit."

The idea of "out West" was a conspicuous part of Richardson's thinking and occurred in several forms in this book:

The early settlers upon Massachusetts bay, after exploring the country for twenty miles "out west," reported the fact with triumphant surprise, and boasted that the soil was tillable for that entire distance. Most adults remember when Buffalo [New York] was spoken of as "out West." How rapidly the application of that phrase has since moved toward the setting sun! Now on this remotest frontier [El Paso], I heard a merchant speak of sending goods "out west," about a hundred miles over into Mexico.

One of the later chapters opened with the theme: "There is a permanent westerly current in our social and political atmosphere like that which carries westward all material atoms after they rise to a certain height." The closing paragraph of the book dramatized the westward march by using a railroad metaphor, the calling of the appropriate end of line changes: at Chicago, at the Missouri river, at the Rocky Mountains, at Virginia City, and finally, at San Francisco — "Passengers for New Zealand, Honolulu, Melbourne, Yokohama, Hong Kong, and all points in Asia" change to the Pacific Mail Steamship Company.

Richardson's was not a systematic treatment of the Trans-Mississippi West, and these passages removed from their setting may tend to exaggerate by thus fitting them together consecutively from a book of 572 pages. Possibly the readers of 1867 would not gain the same impression, yet all these things are explicit within the framework of his descriptive narrative of adventures.

Cyrus Thomas

Cyrus Thomas (1825-1910) contributed substantially to the thinking about the Trans-Mississippi West as a region. His approach was quite different, however, from Gilpin's, and described another facet of this little understood country. As entomologist and botanist on the Hayden geological survey of the territories, 1869-1872, he was called upon to investigate the agricultural capabilities of the country under examination. In this connection he made four reports. A resident of Tennessee and Illinois, trained informally in medicine and law, he practiced law (1851-1865) in Illinois, where he turned to the ministry for a time before accepting service with Hayden. He had followed natural science as an avocation along with his professional activities, but in his later life (1882-   ) he was an archeologist with the United States Bureau of Ethnology.

The fundamental fact of regional differences between the East and the West (the forest, the grassland, and the desert) was recognized by Thomas in part in his first annual report (1867) and more completely later. He pointed out that eastern ideas of agriculture must be laid aside and the west must be studied from its own standpoint. For the drier portion, most of the country west of 100°, water and the method of its distribution for irrigation purposes were made the key to his idea of policy. To implement this policy he advocated that government must intervene and take over "absolute control of the system of irrigation or keep a watchful eye over it and guard it well by laws, regulations, restrictions, etc." Thomas rejected the theory so widely held that soils of low rainfall areas were sterile, and insisted that water, and water alone, was lacking to make them abundantly productive. He went a step further, and challenged the assumption that the necessity of irrigation was a regional handicap. He suggested that it was a regional advantage, because it would remove the hazards from agriculture, and insisted that in calculating costs of production, crop failures of rain-moistened regions should be balanced against irrigation costs.

The boundary line between the rain-moistened country and the irrigation-watered country he set tentatively at 100°, but in his last report of 1872 he was revising that generalization on the basis of more detailed information and was engaged in collecting weather data from which he hoped to arrive at fairly definite conclusions. He left the service of the Hayden survey, however, before this project was completed. Forests, and the supply of building materials, and fuel, and the absence of hard wood, were matters of particular concern and he proposed a policy of conservation and reforestation under governmental authority. The plains and the irrigated lands were peculiarly adapted, in Thomas's opinion, to the production of cereals, especially wheat, and to livestock. These two groups of crops were considered the key to the agricultural economy of the Trans-Mississippi West. (For a more detailed account of the development of Cyrus Thomas's ideas, see Agricultural History, 1947).

Josiah Strong

Another book of some importance was Josiah Strong's, Our Country, its possible future and its present crisis, published in 1885 for the American Home Missionary Society. Although his title indicated a national outlook, the major argument turned on the future supremacy of the Trans-Mississippi West: "the West is to dominate the East.... The West will direct the policy of the Government..., and therefore, destiny." Strong was deeply impressed with the space theory of history. He argued that heretofore during the whole course of history there had been unoccupied lands to which surplus populations could migrate, but within some twenty years the unoccupied public lands of the United States would be taken, and likewise of the earth: "Then will the world enter upon a new stage of its history — the final competition of races, for which the Anglo-Saxon is being schooled." By this line of argument Strong concluded that the West in that generation "commands the world's future."

But what of the 'Great American Desert' which occupied so much space on the map a generation ago? It is nomadic and elusive; it recedes before advancing civilization like the Indian and buffalo which once roamed it.... The vast region east of the Rocky Mountains, though once the home of the 'Great American Desert', really contains very little useless land.

There was worthless land in the East as well as in the West, and he indicated the location of this class of western land as within a triangle whose base ran from West Texas to Southern California, and whose apex lay in Idaho. Reclamation of a large portion could be accomplished, he argued, by irrigation, by deep plowing, by treatment of some soils with minerals, and by increased rainfall accompanying cultivation.

Nathaniel Southgate Shaler

Although his major interest lay in geology, the historian of the West is interested in N. S. Shaler (1841-1906) more largely because of his writings on geography and history. Necessarily this notice of his work must be brief, reserving a fuller treatment for another occasion. His outlook was that of a forest man to whom forests were the cradle of the race and the principal dependence in the evolution of civilization— "without them he can hardly maintain the structure of his civilization" (1884). This meant that Shaler was unable fully to appreciate the grassland although his evaluation of the prairie, as he called it, was one of the most significant aspects of his thinking on the theme of geography and American history.

In advancing westward the American had met the prairie in Ohio, Kentucky, and Tennessee, but more fully in Indiana, and Illinois. It was here that for the first time in history "a highly skilled people have suddenly come into possession of a vast and fertile area which stands ready for tillage without the labor that is necessary to prepare forest lands for the plough" (1884). As geologist and historian of Kentucky, he was familiar with the reactions of the frontiersmen upon first contacts with the "barrens", the forest man's conviction that land was sterile that did not grow trees. He explained the prairie as primarily a consequence of two factors, rainfall and annual fires. In the eastern grassland, fires were thought to be the controlling element, but farther west diminishing rainfall and fires (1889).

Acquaintance with prairie lands demonstrated eventually their great fertility and open country together with cheap steam transportation resulted in the phenomenal development of the West. Incidentally, the attraction of the prairies of one-third of the agricultural population had made unnecessary the clearing of much of the forest country, even permitting some reforestation (1884, 1889, 1894). In his book of 1889 he pointed out explicitly that already the first quality free lands were occupied and the tide of population had been turned back upon the second quality lands and those of lesser utility, and that the forest lands that had formerly been passed over were being threatened. The country was faced with the problem of reclamation of excessively wet and dry lands and others not available without special preparation. In 1894 he used the term "frontier," and pointed out that there was no longer an American frontier. It is important to emphasize, however, that he did not view this closed frontier, or closed space, as a closed world or as a calamity. It was an end of an era, but, as he had been emphasizing in 1889, in 1891, as well as in 1894, the new era meant a shift from an extensive and wasteful exploitation of nature's resources to a policy of conservation and intensive and economical utilization of man's inheritance. Science was assigned its rôle in this new era.

In interpreting the rôle of the frontier in American history, his treatment of 1894 was original and significant. The year 1830, so far as any year date could be fixed as a milestone, was the turning point. The frontier emerged from the forest into the prairie, and steamboat and rail transportation soon entered as a major factor. No longer did a pioneer spend a lifetime clearing a small farm. The rate of advance of the frontier during the first hundred years was less than one mile a year; it was speeded up during the second century to between two and three miles per year; and Shaler estimated that "after 1830 the frontier ceased to have anything like a definite line, for the immigrants swarmed to the westward along all the communications." During the latter part of the nineteenth century, he estimated the occupation of the interior grassland at a rate ten times as rapid as in any preceding period.

It is clear that Shaler thought of the frontier process, and of the frontier, as an advancing line in terms of the forest, a forest process which ended about 1830. The second phase of the occupation of first quality free lands, beginning about 1830, was a prairie process and was not associated with any frontier line. The end of this phase of the occupation of the grassland was completed in the 1880s. After 1890, the utilization of natural resources was to him a progressive process of developing second and lesser quality lands. Another important difference Shaler pointed out between the periods pivoting on 1830 was that of interdependence as between the pioneer and his background civilization. So long as the frontier was a slow-moving forest process performed by the pioneer in relative isolation, he was dependent primarily upon his own resourcefulness. After 1830 the balance had shifted and prairie and transportation had made possible the phenomenal occupation of the grassland through an increasing dependence upon the services of a mechanized background civilization which were paid for by mass production of foods and raw materials.

As a geographer Shaler was concerned with the concept of geographical regions and his interest in history emphasized his views on the relations of different geographic regions to man's occupancy. The preceding paragraph has developed his ideas on the forest and its relations with the eastern portion of the grassland. The next section deals with his reactions to the low-rainfall western grasslands. At the outset it should be stated that he never formulated a satisfactory definition of a physiographic region, but his several descriptions of the areal divisions contained significant observations.

In an article, "Improvement of native pasture lands...." (1883) he committed himself to the assumption that "the great part of the United States west of the meridian of Omaha is unfit for tillage," and that as a whole "its sole use is for the pasturage of cattle and sheep." Its value even for that purpose he argued was handicapped by the shortage of water and herbage. Wells and storage of rainfall could meet the first difficulty, but he urged a program of organized search for new plants to increase the yield of pasturage. Although the general idea was good, he revealed his inadequate knowledge of the ecology of the low-rainfall region by arguing that the wide spacing between plants should be filled in with new plants, thereby increasing the yield.

In 1884 Shaler thought that as a whole the country east of $100°$ was tillable, but that the Cordilleras reduced approximately one-third of the continent to sterility, except as it could be irrigated. Its resources were mineral, which would be exhausted within one hundred years and after that the country would "probably be to a great extent abandoned by man." In 1891 he again expressed himself explicitly on the Trans-Mississippi West, again making a similar division, but condemned the western portion to a "limited future." In 1894, he offered what might be termed a forest man's apology for misunderstanding the low-rainfall grassland, designating it, not as desert, but as "the country of scanty rain"; it was a grassland, but it was not sterile. He now went so far as to declare that "the Cordilleran district is on many accounts the most interesting portion of North America." Its vast array of minerals, including coal, oil, and gas, were well distributed, and were being combined with irrigated lands of "rare fertility in producing commercial values" that in twenty years would be "as large as that obtained in any equal area of the continent."

Scientist though he was, Shaler was keenly appreciative of the value of the accumulated folk experience of older civilization, a factor which was conspicuously lacking in America, especially in the Mississippi valley. He discussed it at some length, in his soil monograph (1891), urging the merits of science in helping to make good the deficiency. He returned to the subject in a broader setting in 1894 when he pointed out that on each new frontier a period of adjustment was necessary, but "after a time the culture becomes more or less reconciled to the environment...." As this was written during the depression of the 1890s, and Populist agitation, there was a particular interest in his further comment that "throughout the Mississippi Valley it seems to the present writer quite clear that this accord between the physical state of the country and the subjects of tillage is still quite imperfect." This concept of reconciliation with environment was fundamental to Shaler's thinking on the relations of man and the earth, and quite properly was fundamental to his approach to conservation of natural resources of soil, forests, and minerals. To him the passing of frontier, of new lands, of free land, was not a calamity, but the foundation for this stabilization of farming population and consequent reconciliation with environment.

Shaler's soil theory was not particularly original, being for the most part the prevailing physical-chemical explanation of soil origin and fertility, together with the rôle of animals, and of vegetation, and of the newest contribution, the as yet little-understood factor of bacterial action (1888, 1889, 1891). He emphasized (1891) that there was no satisfactory popular treatise on soils in any language and his writing on the subject was important primarily in that connection. He protested the popular conception of soil as unclean, arguing the importance of appreciation of its significance, "a clear recognition of the marvel and beauty of the mechanism", and proposing even drastic legislation, if necessary, to effect measures of conservation (1889, 1891). As a forest man, he considered forest soils the most important and enduring, especially those of limestone regions like central Kentucky where leaching and renewal were essentially in balance. He recognized the unusual fertility of grassland soil, but thought it only temporary and largely exhaustible in about thirty years. Grass roots were not sufficiently large and deep in penetration and low rainfall did not provide necessary leaching, according to his analysis, to effect fully the cycle of soil renewal (1889, 1891). Evidently, Shaler did not understand adequately the rôle of rodents, etc., the process of formation of lime accumulation zone, expounded by Hilgard, or the root ecology of grasses and deep rooted forbs in the grassland.

Forest conservation was ever foremost in Shaler's thinking, as well as soil conservation, and he looked forward to the exhaustion of the irreplaceable minerals. On this last point he emphasized the significance of light-weight metal (1889), aluminum— "an aluminum age would carry us almost as far beyond that of iron as we advanced when that metal replaced bronze in the mechanic arts" (1905). In connection with the exhaustion of coal for energy, he pointed to solar energy and faith in science (1889), which later seemed justified by the unexplored possibilities of radioactivity, even suggesting what might be termed an energy interpretation of history (1905). Furthermore, he emphasized the ocean as a source of both minerals and food, and in developing this subject revealed in the global sense his appreciation of interdependence of geographical areas of the surface of the earth in the economy of nature.

The use made by Shaler of the rôle of free lands (1889) and of the explicit term frontier make necessary some comparison and contrast of his thinking with that of Frederick Jackson Turner. Shaler used the concept of the passing of first quality free land in his book of 1889, which antedated anything Turner said on the subject, the latter's first use of the idea dating from 1892. Turner's use of the concept of the passing of the frontier, or the closed frontier, as marking the end of an era in American history, dated from his essay of 1893; first read at Chicago at the Columbian Exposition in the summer, and printed in 1894. Shaler's use of the same concept was evidently composed during 1893, because it was printed in 1894 as a part of the monumental three-volume collaborative work under his direction. There is no explicit or implied evidence that Shaler knew of Turner's essay, but it is evident that Shaler was familiar with the report of the director of the United States Census of 1890 from which Turner derived his explicit text for his essay of 1893. It would seem that both men derived their suggestion from the same source. It is evident also that, except for the term and the particular form of the frontier concept as a line, Shaler had used the essential idea in 1889.

The popular concept of Turner's frontier hypothesis was that he traced the westward advance of the frontier line across the continent. Strictly speaking, that was not true, because Turner had divided American history into four periods, the third of which referred to settlement of the Pacific coast and the advance of occupancy eastward, while the fourth dealt with the filling-in of the great plains area between the two advancing frontiers. Shaler used the free-land concept of the frontier but terminated the westward marching frontier line with the breaking out of the pioneer into the grassland about 1830. The passing of the frontier as a line occurred at that point. The next period, 1830 to the 1880s, marked the passing of first quality free land with the occupancy of the grassland. Shaler's distinctions bring out the fact that Turner's frontier as well as his own was a forest process, and was applicable only to the forest area. Shaler's interpretation of the period of American history after 1830 was a grassland process supplemented by interdependence with a mechanized civilization for transportation, machinery, and markets — the subjugation of a continent by a combination of processes. In Turner's writing on formal history he did not get his frontier out of the forest and the success of the Rise of the New West (1906) might be explained on that basis. Similarly the relative failure of his incompleted book, The United States 1830-1850, may be explained on the ground that the forest-process frontier formula was no longer adequate to explain the rise of a mechanized industrial civilization within the United States, and the relations of a two-front nation with the outside world, — an anticipation of a global economy based upon mechanized communications. Shaler differed with Turner on the originality of the frontier, holding that American originality lay primarily in its industrial innovation, otherwise American culture was primarily an inheritance. In conclusion, it should be pointed out that Shaler's philosophy might be termed a scientific supplement to Josiah Royce's, The World and the Individual, and that he maintained that the final test of any social system was the development of the individualities of its people (1905). (For a more detailed account of Shaler's ideas, see Malin, Essays on Historiography, 1946).

John Wesley Powell

A major work of scientific character, significant to the study of the Trans-Mississippi West as geographical regionalism in a sense pertinent to history, was John Wesley Powell's Report on the lands of the arid region of the United States, with a more detailed account of the lands of Utah, published in 1878. He was reporting as geologist in charge of the United States Geographical and Geological Survey of the Rocky Mountain Region, but it had been his intention to present a comprehensive description of all lands in the hands of the federal government, the two classes of most concern being those excessively wet, and those excessively dry. The former were swamp lands of the south Atlantic and Gulf coasts, Florida everglade lands, flood plains of rivers, and lake swamps of the Upper Mississippi river, and the Upper Great Lakes. Eventually the reclamation of wet land would increase largely the agricultural capacity of the United States, and he claimed that "their fertility is almost inexhaustible." In respect to the dry lands, he was more reserved in his expectations, "so far as they can be redeemed by irrigation [they] will perennially yield bountiful crops, as the means for their redemption involves their constant fertilization."

Powell was explicit in his statement of the requirements for the accomplishment of the desired object for both wet and dry lands:

> To a great extent the redemption of all these lands will require extensive and comprehensive plans, for the execution of which aggregate capital or cooperative labor will be necessary. Here, individual farmers, being poor men, cannot undertake the task. For its accomplishment a wise prevision, embodied in carefully considered legislation, is necessary.

The change in Powell's plan, restricting his work to the arid lands, was the result of pressure for immediate action on the arid lands and the meagerness of his data on the wet lands. This perspective is of some importance for evaluating the report. Powell was not particularly interested in the arid lands as such, or in them as constituting a problem in geographical regionalism. The whole projected study was conceived in the framework of policy for administering the whole Public Domain yet remaining in the hands of the federal government. The United States was divided into three parts; a humid eastern half, a sub-humid one-tenth; and an arid four-tenths. Of course, these were only rough approximations. The boundary between the humid and the sub-humid portions was indicated at the line of 28 inches average annual rainfall. On the accompanying rain chart this line was irregularly near 99° through Texas; thence northeast to near 96° at the Kansas line; thence north to near the mouth of the Platte river; and thence irregularly northeast to the west end of Lake Superior.

The boundary between the sub-humid and the arid regions was the line representing 20 inches of rainfall, following near the 101 meridian in Texas; thence northeast across Kansas to 100° in central Nebraska; and thence northeast irregularly to the Canadian line at about 93. The arid portion extended westward to the Pacific coast, except certain limited areas near the coast, especially in the Pacific Northwest.

Of the sub-humid area receiving between 20 and 28 inches of rainfall, Powell explained that in the eastern border drouths were infrequent, but near the western border they were frequently disastrous. Irrigation, he thought would be adapted early to the western part and eventually it would be practiced throughout the sub-humid region. The country was particularly attractive to settlers because it was ready for the plow.

But because of the lack of forests the country is more dependent upon railroads for the transportation of building and fencing materials and for fuel. To a large extent it is a region where timber may be successfully cultivated. As the rainfall is on the general average nearly sufficient for continuous successful agriculture, the amount of water to be supplied by irrigating canals will be comparatively small, so that its streams can serve proportionally larger areas than the streams of the arid region.

Taking a twenty inch rainfall as the lower limit of successful agriculture, Powell's introductory remarks indicated a tentative attitude:

Many droughts will occur; many seasons in a long series will be fruitless; and it may be doubted whether, on the whole, agriculture will prove remunerative. On this point it is impossible to speak with certainty. A larger experience than the history of agriculture in the western portion of the United States affords is necessary to a final determination of the question.

In respect to the arid area: "In all this region the mean annual rainfall is insufficient for agriculture, but in certain seasons some localities, now here, now there, receive more than their average supply. Under such conditions crops will mature without irrigation," but such favored seasons were infrequent and unpredictable. Three types of land were described; pasturage, timber, and irrigable, the last two being only a small fraction of the arid region. In the pasturage portion the grass growth was scant, in scattered bunches, and in low altitudes so scant as to be of no value — true deserts, the southern parts of the states and territories of California, Nevada, Arizona, and New Mexico. In the higher altitudes, and to the northward, Powell reported that the grasses improved.

Powell's recommendations of land policy were a substantial departure from the traditions of the humid East. The governing principles included a survey system conforming with topography; irrigable lands in small farms, pasturage lands in large tracts supplemented by irrigable lands; irrigation based on large streams, not small ones; reservoir sites set aside and protected; likewise, timber land constituting 20 per cent to 25 per cent of the area; residence should be grouped in conformity with topography and farm divisions; and pasture land managed in common, without fences, under community regulations for each topographical unit. To put this body of policies into effect, he submitted two draft bills proposing homestead systems for irrigable lands, and for pasturage lands. In either case nine or more persons might form a homestead settlement district; for irrigable districts, the farm unit not to exceed eighty acres, and for the pasturage districts, the farm unit not to exceed four sections or 2560 acres including not more than twenty acres of irrigable land.

The most important monograph, in terms of its actual influence in laying the physiographic foundations of American historiography of the twentieth century, was J. W. Powell's, Physiographic Regions of the United States, published with a colored map, in 1895. He repudiated the stream-basin concept of physiographic units, adopting the drainage slope concept under which the United States was divided into four slopes; the Atlantic, the Great Lakes, the Gulf, and the Pacific. Regions within the slope were treated as mountain, plateau, and plains, the three colors on the map representing these physiographic characteristics, and further sub-division was made into districts:

The regions here delineated are held to be natural divisions, because in every case the several parts are involved in a common history by which the present physiographic features have been developed....

In dividing the United States into a few great physiographic regions, it is not found possible always to draw the lines with exactness. Often one region blends with another, the transformation in general characteristics being marked by a general change. There are some lines of division clearly drawn by nature within narrow limits; other divisions are imperfectly marked by slow graduations from one to the other.

Powell did not recognize vegetation as having any significance in his scheme of classification. He indicated briefly three kinds; desert, prairie, and forest, produced chiefly by "inequalities of rainfall," less than 20 inches being deserts, 20 to 40 inches being grass with increasing forests, and above 50 inches insuring a heavy forest growth. Prairie and forest vegetation appeared on mountains, plateaux, and plains regions, these designations having a purely physiographic connotation. Thus the Lake plains were completely forested. According to his understanding, presence or absence of forests where rainfall exceeded 20 inches was attributed primarily to fire, which was partially controlled by moisture. Where rainfall exceeded 50 inches fires did not burn and consequently forests were dense. He argued that where moisture permitted, prior to the Columbian discovery, the Indians annually burned the whole country, destroying trees in some parts, and destroying undergrowth in others, creating open forests. Grass grew where forests were wholly or partially destroyed in this way. With settlement by white men and protection from fire the forests spread, undergrowth became dense, and thus on the Atlantic, Lake, and Gulf slopes, although there was less area in timber, he argued that the number of trees was as great or greater than in the primeval forests. His explanation of the grassland was contained essentially in the sentence: "The great prairie region of the United States is found mainly where rainfall is from twenty to forty inches, because the forests are destroyed by fire." Toward the humid East, forests were more frequent and vigorous and this he explained as resulting from broken or stony ground, especially near streams, "so that luxuriant grasses are not abundant to furnish food for fire." Rainfall, he insisted, was neither increased nor decreased by trees.

The prairie plains as he designated them on his map extended farther east than the grassland including in the physiographic concept much forest country, including west central Ohio, nearly all of Indiana and Illinois, one half of Missouri, central Oklahoma, eastern Texas, except a belt of Gulf plains; and in the north, Iowa, southwestern Wisconsin and Minnesota, and eastern North and South Dakota, Nebraska, and Kansas. This prairie plains was divided into two districts according to glaciation, the north part was the ice plains, and the south part was the water plains, the dividing line lying just south of the Kansas and Missouri rivers.

The term Great Plains was used by Powell with reluctance, his preference being the Great Plateaux, the zone beginning in British America at the north and extending into Mexico at the south. In respect to the boundary between the prairie plains and the great plateaux, he said that "the prairies merge imperceptibly" into the latter region. For purposes of mapping it was necessary to lay down an explicit line, beginning at the south the boundary was indicated by an arc of a circle intersecting the Rio Grande at about $100°$, curving slightly westward, and thence northeastward to the Kansas line where the Arkansas river intersected it; thence north by east to the mouth of the Blue (Manhattan); thence north and northwest to the mouth of the Niobrara river; thence following about 30 to 50 miles east of the Missouri river to the northwest corner of North Dakota. The western boundary lay east of the physiographic region designated on the map as mountains. The line was quite irregular through Montana and Wyoming, but from about $43°$ to $36°$ north latitude it followed closely the $105°$ west longitude; thence it ran south along the divide between the Pecos and Rio Grande rivers, which was near $105°$, to the south boundary of New Mexico. The characteristics of the great plateaux were described not as a plain, but as "a great group of elevated plateaux." Within the United States the plateaux were divided into four sub-groups or districts. At the north were the Missouri plateaux, the plateau blocks being cut by the tributaries of the Missouri, and including the country southward to approximately the divide between the Cheyenne and the White Earth rivers in southwestern South Dakota. The Platte plateaux extended south to the divide between the Platte and the Republican rivers. The Arkansas plateaux extended south of the Red river to include its southern tributaries, the boundary line running northwest into northeastern New Mexico along the divide between the headwaters of the Canadian and Pecos rivers. The Pecos plateaux included the Pecos basin, the southern end of the staked plains, and the headwater region of the Brazos and Colorado rivers of Texas.

To the westward of the continental divide, the Colorado plateaux connected in two places with the great plateau east of the divide; through the South Pass and through the pass in the vicinity of Santa Fé. The Columbian plateaux were disconnected. The remainder of the Far West was designated as mountain, even including the Great Basin deserts and the below-sea-level desert of southeastern California.

Powell's approach to physiography was purely physical, and it is evident that in 1895 he possessed no appreciation of the developments in ecology which paralled so closely in time the gropings of his own field of physiography in the direction of scientific adequacy in regional definition and understanding. Biology seems to have left his physiographic thinking and his soil theory untouched. The latter was the more serious, because agricultural development, as related to land policy, motivated much of his investigation. Powell's evaluation of the West was one that stressed limitations imposed by low rainfall upon utilization of the area by man, and the policies he recommended were based upon that fundamental assumption. Unconsciously, Powell was dominated by the forest man outlook and standard of values according to which the West was a deficiency area. In other words, as a physiographer he had not arrived at a satisfactory concept of regional adequacy in nature, or the corollary of interdependence of regions as the basis of utilization by civilized man. On the positive side, however, it should be emphasized that in his insistence upon differences between regions, and in consequence upon differences in policies to be applied to them, he was contributing constructively to social adaptation to environmental requirements. In respect to the soundness of his physiographic work and his originality Powell has been much overrated. In respect to his influence, it was tremendous, and not the least important of the channels through which it was wielded was Turner's dependence upon him, and later the use Webb made of his work.

Willard D. Johnson

The report of Willard D. Johnson, The high plains and their utilization, published in two parts, 1901, 1902, was in part in the tradition of Powell. Powell's report had appeared prior to the boom of the eighties which had resulted in the first occupation of a large portion of the country west of the 100 meridian by a farm population and small stockmen. The Johnson report was prepared after the disastrous drouth decade, approximately 1886-1896, climaxed by the world-wide financial collapse and depression of the nineties. His attention was focused upon the first of these and to it was attributed the hardships suffered by the plains people. Johnson remarked that the arid region which had appeared on early maps as the Great American Desert had contracted by "recognition of relatively humid tracts" and this had gone so far that "the name Arid Lands, as implying rather a grouping of detached parts, is coming into use instead."

Originally the desert was assumed to have its beginning in the treeless expanse of the Great Plains, including the whole. We now witness the spectacle of the highest type of agricultural development extended over a belt of States wholly within that early boundary. So much of the Plains has been mistakenly regarded as arid.

Reclamation by irrigation on a small scale, in connection with mining in nearby mountains, had demonstrated early the natural fertility of the soil, but general settlement did not come until the threatened exhaustion of desirable lands in the humid region: "It remained undetermined how far encroachment might profitably be carried without artificial aid," but under these circumstances the entire intervening space between the humid region and the irrigated area was filled. As he emphasized its apparent advantages, the high plains "presented the maximum inducement of fertile soil and unbroken smoothness of surface." The disaster of drouth and depopulation followed, and the focus of his study was to formulate a basis for sound utilization of the area that would not repeat that experience.

Johnson's study was directed, therefore, at a particular area and its problems, and appropriately, he attempted to define his subject carefully. First was the matter of definitions of terms and delimitation of the area to be studied. He did not discuss the term Great Plains, and apparently was using it as defined by Powell in his paper, "Physiographic regions of the United States," (1895). The term High Plains he defined as a topographical unit and as a climatic unit. In the first sense, it was a sub-region of the great plains lying between the Platte and the Rio Grande rivers, as north and south limits, and between the humid region and the desert as east and west limits. The great plains as a whole had been built by rivers which spread an apron of debris over the rock of the tertiary period, and the subsequent process of erosion had degraded the surface, except that of the high plains. To the east, the erosion had occurred in spite of grass sod cover because of the extent of rainfall, forming the rolling prairies, and to the west erosion had occurred because the grass was too scant to hold the soil even with low rainfall conditions— "this is the desert zone." Thus the high plains stood out in relief as an area scarcely eroded, or not at all, protected by a grass cover, level surface, and low rainfall. Viewing the great plains in original condition as a sloping area extending from the Rocky Mountains eastward to the forest region, he said the high plains "lie in an irregular belt from about midway across the long eastward slope." At another place Johnson said

The High Plains constitute a natural subdivision of the Great Plains, forming a belt extending north and south midway of its slope. The belt is cut across by streams from the distant mountains on the west, in parallel system, wide apart. Its distinctive character is lost, to the northward, in southcentral Nebraska. At the southern end, the Staked Plains of Texas constitute the largest individual plateau fragment.

Again, he put it in different words, "Thus, by reason of degradation on either side, the High Plains are in relief.... The High Plains are conspicuously uplands of survival," and at another place, "differential erosion of an original vastly extended plane surface has left there a fragment, or a close assemblage of fragments, in relief. The relief is not considerable. It is, however, sufficient to be dominating." These points of emphasis were important, and historians have not been careful in noting their limitations on the area under discussion; "though but a fractional part of the whole slope, they are yet absolutely of great size, and the traveller upon them immediately recognizes that they constitute the Plains proper. They alone have strictly the character which that term implies." It is clear that Johnson was not applying the term to any country north of the Platte river; that the smaller and variously eroded fragments which were not wholly typical were those cut through by the rivers Platte, Arkansas, and Canadian, and partly through by those lying between, the Republican, the Smoky Hill, and the Cimarron. The major fragment, and the only one completely typical of the somewhat idealized concept of the original great plains, and of the high plains survival of that formation, was the staked plains. The northern escarpment of the staked plains as a conspicuous feature of topography bordered the Canadian river, the eastern boundary southward through Texas, such rivers as the Red, Brazos, and Colorado having their headwaters in this geological formation. The Pecos river basin formed the western boundary, and the Rio Grande basin and drainage systems flowing to the Gulf of Mexico imposed the southern limits.

In regard to a climatic unit, Johnson was not so explicit or successful in formulating his definition and delimitation:

In its [the great plains] westward rise of thousands of feet it passes through climatic gradations from humid to arid. Although, necessarily, along a uniformly rising slope, the passage is gradual, so that any subdivision must be arbitrary, it may at least be said that midway, across a considerable breadth, the climate is semiarid or subhumid.... The boundaries of the topographical belt, to a considerable extent, have been given sharp definition by marginal recession— a work of headstreams sapping and encroachment from the eroded area— and the topographical belt in consequence lies somewhat contracted within the limits of the climatic belt, but substantially there is agreement in position.... The High Plains are equally the sub-humid plains.

It should be noted that, so far as climate is concerned, no northern or southern limits were set. In fact, nothing was said in this definition about the north and south variations within the unit, although some detail appeared in the chapter on climate. He was dealing exclusively with the high plains and defined his subject so narrowly that he made few commitments on matters outside of that area. He did not indicate a climatic dividing line between sub-humid and humid beyond what is quoted above indicating that it lay somewhat east of the topographical eastern edge of the high plains. In recognizing evaporation in relation to effectiveness of rainfall, he contrasted the high plains with the northern great plains, accounting for successful wheat growing there as the result of less evaporation and a greater net moisture available. Of the fifteen inches of high plains rainfall, he estimated that not more than three or four inches net was available for soil moisture and that was not sufficient to build up excess ground water. He repudiated explicitly the myth that climate was undergoing change in either direction, especially that it was becoming more favorable under the influence of occupation.

In turning to Johnson's treatment of utilization of the high plains it is essential to bear in mind his climatic distinctions. The high plains constituted a sub-humid or semi-arid region lying in a general north-south direction, but slightly northeast-southwest in trend, mostly between $100°$ and $104°$. Between the high plains and the Rocky Mountains was the arid belt, and east of the high plains was the humid country. According to this classification apparently everything east of $100°$ or some nearby line was humid. Johnson's views on utilization were so explicit and tersely stated that, for the most part, they can be presented best by quotation. His opening declaration was that

> Despite the persistent attempt at settlement already made and its utter and disastrous failure, the High Plains continue to be the most alluring body of unoccupied land in the United States, and they will remain such until the problem of the best means for their utilization shall have been worked out.

As the volume of rainfall fluctuated from season to season, at the minimum it was that of an arid climate, and "at intervals of about a decade," the humidity increased and might even "so continue for two or three consecutive seasons."

It should be recognized at the outset, however that for agriculture to be profitable in the long run. the subhumid country, like the arid, requires the artificial application of water, though in less degree; and that if the required additional supply is not to be had, the subhumid country in effect is arid also.

It is the purpose of this paper to show that the High Plains, except in insignificant degree, are nonirrigable, either from streams, flowing or stored, or from underground sources, and that therefore, for general agriculture, they are irreclaimable; but that, on the other hand, water from underground is obtainable in sufficient amount for reclamation of the entire area to other uses; that such reclamation has in fact already begun, and is in process of gradual but sure development; and that it will be universally profitable.

The problem will be found to be essentially one of well making and of the proper location of wells.

This was a livestock interpretation of the high plains, cattlemen particularly, in terms of large-scale operation only; and the water supply sought was stock water, irrigation water for small scale operations was feasible in some valley locations, mostly in connection with ranch management. The body of the Johnson monograph was devoted therefore to investigations of the ground water supply and to the nature of its occurrence. This concentration upon ground water was the outcome of his analysis of the supply of river water available which was inadequate in volume and which he concluded could not be stored practicably for high plains utilization.

The subhumid precipitation is insufficient for agriculture without irrigation and the only possible agricultural lands of the High Plains belt lie within the valleys, where small patches here and there are irrigable. The most evenly tillable area of the Great Plains lies within the High Plains, yet it is as a body nonagricultural, because a supply of Water for irrigation is not obtainable. Upon the uplands proper irrigation on a large scale is impossible, and even where it is possible on a small scale it is economically impracticable. All apparent sources of supply considered, there is not water enough for appreciable effect in general and permanent reclamation, and what small supply there is in fact is not economically available.

* * *

As to artesian waters, there is no instance, among many experimental deep borings upon the High Plains, of rise to the upland surface.

The only important crop found available to the high plains, he concluded, was sorghum for fodder. In some places it could be grown without irrigation, but even in the irrigable valleys it was indicated as the important crop, especially for headquarters ranches. If

Johnson's monograph is read with care, and the limits of his definitions rigidly observed, there could be little disagreement with his statement. In common usage, however, the term High Plains has been given a generalized application referring vaguely to the more elevated portion of the great plains from the Rio Grande river northward into Canada. When used in that sense, Johnson's commentary is not applicable.

Eugene Waldemar Hilgard

The place of E. W. Hilgard (1833-1916) in the history of soil science has been discussed briefly in its proper place in chapter six, but here it is important to trace something of the development of his ideas as they relate to regionalism. His experience in Mississippi, beginning in 1855 as a geologist, convinced him that agriculture, not minerals, was the true interest of that state, and he turned his energies so far as circumstances permitted to soil studies. After two years in Michigan (1873-1875) his career after 1875 was identified with agricultural experimentation in California, and particularly with soils. Beginning in 1877 a continuous stream of reports flowed from the experiment station.

In a fairly broad interpretative vein, he presented, in 1878, his early impressions of California agriculture. Sharp contrasts with the high rainfall agriculture of the East were predominant in everything he had to say:

> Agriculture in California possesses many peculiarities arising partly from climate causes, and partly from the somewhat exceptional history of the industrial development of the state.

He pointed out that the frontier process did not apply to California:

> But unlike the great agricultural States of the Mississippi valley, California has not undergone the slow and regular process of settlement by pioneer farmers, who, fleeing from the too close approach of towns and neighbors, as well as from soil exhaustion, keep selling out and moving west as part of their normal existence.

California farmers "soon discovered that in a great many respects the rule-of-thumb experience and practice of the older countries would not avail them here and casting loose from precedent, they tried a 'new deal' in constructing for themselves a practice adapted to the new conditions. One of the controlling features was the scarcity and high price of labor, the introduction of labor-saving machinery was among the very first needs, instead of being a late fruit of long discussion and costl experience." Robbery of the soil increased in direct proportion to the increased efficiency of the new big machinery, without rotation of crops or other effort at renewal.

Hilgard was chosen as special agent for the Tenth Census of the United States (1880) in charge of the study of cotton production, and in part he made it a study of soils, assembling a large volume of data on soil analysis for the cotton states and California. In this he was building in the direction of his comparative evaluations of eastern and western soils, but drew no regional conclusions (1884). He was associated with the Northern Trans-continental Survey, 1881-1883, and in 1882, with T. C. Jones and R. W. Furnas, made a report on the arid regions of the Pacific slope for the United States Department of Agriculture. A summary of the early investigations of the California station was published in the Report of the United States Department of Agriculture, 1889, in which a characteristic statement was made that "People living east of the Mississippi river have very little conception of the nature, number, and importance of the problems which confront new settlers or older farmers in California and other States west of the one hundredth meridian." From the outset of the work of the station in 1876 study was directed to a large number of fundamental problems from entirely new standpoints: Nearly all soils were found to be calcareous, and lime fertilizers were useless; the accumulation of lime in arid soils was a consequence of climate; potash was present in excess compared with soils east of the Mississippi river; and phosphoric acid in small amounts had been found profitable in some lands.

Hilgard's basic regional monograph came in 1892, "A report on the relation of soil and climate." Using a physical and chemical approach to soil formation, he showed how climatic factors modified soils. Although pointing to absence of detail and exact experimental data, he concluded that high temperatures accelerated and low temperatures retarded soil formation; also, a wet climate accelerated and a dry climate retarded the process. Rainfall on permeable soils percolated through the soil mass carrying away dissolved mineral compounds to the sea; on impermeable soils it tended to run off without leaching; in low rainfall regions it penetrated the soil to a limited depth and then evaporation caused deposit of mineral compounds in the soil or on top as efflorescence — alkali.

Hilgard classified soils as light (sand or silt), heavy (clayey), and humus, and only the first kind was characteristic of the dry climate. If clay occurred there it was derived from clay shales of a prior wet geological era. This introduced the subject which the soil scientist in the twentieth century came to call soil structure. By clay, Hilgard meant "the gelatinous, plastic 'colloidal' substance that results from its physical disintegration, and alone plays the part of rendering the soils coherent." In soils of high rainfall areas the proportion of clay entering into its composition was high and such soil possessed coherence. In dry climates, the proportion of clay was small or wanting and the soil lacked coherence. In this manner he explained characteristics of the extremes of soil conditions in the dry region which eastern forest men seldom understand.

This coherence of the soil material in arid climates, resulting from the scarcity of clay, becomes obvious to the traveler in the sand and dust storms that sometimes annoy him while traversing, e.g., what has been conventionally known as the 'Great American Desert', a desert only so long as the life-giving influence of water is withheld from it. Droughts may render the surface of the country in the Atlantic States, or in Europe, as dry as the great plains themselves; yet away from highways and cultivated fields little or no dust will ordinarily be raised by the strongest wind, because of the coherence of the soil, which will generally be found covered by a hard-baked crust. In the arid region, under the same condition, a puff of wind may raise a cloud of dust, and a wind storm becomes, almost unavoidably, a sand or dust storm also. 'Dust soils', which during the dry season are even in their natural condition so loose as to rise in clouds and render travel very uncomfortable, are not uncommon in Washington and adjacent parts of Oregon, on the uplands bordering the Columbia and Snake rivers....
The same general facts are known of the other arid regions of the globe, whether in Asia, Africa, or Australia.

Without discussing structure in the transitional country, the other extreme in soils, the clays, were the text of his explanation of the ideological standards of the high rainfall areas:

But so generally has the idea of inherent fertility been associated, in humid regions, with soils of more or less clayey character, that the terms 'strong', 'substantial', 'durable', are habitually applied to them in contradistinction from 'light', 'unsubstantial' ones of the sandy or silty type. Hence the newcomer will frequently be suspicious of the productiveness and durability of soils in the arid regions that experience has proved to be of the highest type in both respects.

Physical differences between the soil and subsoil (Hilgard had not yet adopted the terms horizon A and B) were pointed out for humid climates, because the percolation of water carried colloidal clay downward and deposited it in the subsoil rendering it more compact and less penetrable to roots. In arid climates there was little or no difference between soil and subsoil:

Very commonly hardly a perceptible change of tint or texture is found for depths of several feet; and what is more important, material from such depths, when thrown on the surface, oftentimes subserves the agricultural uses of a soil nearly or quite as well as the original surface soil.

The soils of the arid region were deficient relatively in humus, a condition which was attributed to the slowness with which vegetable matter decomposed and to the manner of decomposition, without any essential fermentative action, and leaving little but mineral residue. Burning of stubble, rather than plowing it in, was practiced by farmers because it quickly returned these minerals to the soil which could be accomplished only slowly otherwise. Thus for full value of humus accumulation he recommended composting refuse before applying it to the soil and then plowing it under deeper.

In respect to the consequences of the chemical process of soil leaching, the greatest attention was given to the lime factor. In the humid climates, it was dissolved, along with other soluble compounds, and carried off by country drainage, but in arid climates, where water penetrated only to limited depths and then rose to the surface in evaporation, the lime was deposited in the soil or even on the surface as an alkali efflorescence. In humid climates therefore, the lime content of the soil must be renewed constantly, either naturally by decomposition in limestone regions like the Bluegrass region of Kentucky, or by artificial treatment, but in arid climates, as the lime carbonate solution was not carried off, but was deposited, there was a constant natural accumulation. Not only were the soils of arid regions thus richer in lime, but the same principle held for most of the other mineral compounds essential for healthy plant growth. Hilgard emphasized that he was placing on record, for the first time, the exact and detailed data in support of this explanation of the unusual fertility of the soils of arid climates. Where alkalis existed in excess quantity, or in injurious forms, remedial measures were available and were described for neutralizing them.

The nitrate question came in for special consideration, emphasis being placed upon the chemical condition of the soil thought to be favorable for the well being of the "nutrifying organism:" "We have thus in the necessary calcareous nature of the soils of the arid region the fulfillment of one necessary condition of ready nitrification."

In conclusion, Hilgard reminded his readers of the scant and incomplete data available on climate and soils, and the difficulties encountered because of the looseness and unsystematic nature of so much of the observations upon which the student of the problem must depend: "It should not, therefore, be surprising to anyone if in this first attempt at a systematic exposition of the subject that there are many important gaps and omissions that might have been avoided had a more complete library and a larger share of leisure been at command." He hoped, however, that his monograph would stimulate a study and lead the way to investigation of the lands of various regions "and of the means of utilizing them to the best advantage."

By 1892 Hilgard anticipated Marbut and associates in distinguishing between the contrasting lime characteristics of western and eastern soils, but did not attempt to fix an exact dividing line except his reference to the 100 meridian. Marbut's solum groups were only a refinement of Hilgard's basic thinking. Hilgard did not have first hand experience with central grassland soils and in consequence did not deal with the transition area, low to mid-rainfall climate, and its soil characteristics. His insistence on the absence of essential differences between soil and subsoil in arid regions, as being a normal condition, was a challenge to the later concept of maturity of soils, emphasized by the Russian-American school, and one that was never met in a satisfactory manner. Shreve (1942) pointed out that the idea of succession in plant ecology would never have originated in a desert. The present author would point out that the concept of soil maturity was the product of a mid-rainfall climate, the Russian chernozem country, and likewise would never have originated in a dry climate. Neither concept possessed validity as a universal rule, and the attempts to apply them on that basis are typical of the tendency to set up a striking peculiarity of one environment as a standard of measurement for all environments. Robinson (1932, 1936) warned that all soil classification thus far was only provisional, and Jenny (1941) questioned the assumptions of maturity as not being proven, but that there was an element of organization in all things. Hilgard's physical-chemical explanation of the soil forming process was already being modified to admit the entrance of the biochemical idea in relation to bacteria and the nitrogen problem. In comparison with his eastern contemporary, Shaler, Hilgard was far in advance in all aspects of his soil theory. Of course, the next half century was to modify and add to what he said in 1892, but his claim was valid that he was presenting the first systematic study of soils in the low rainfall region. He did it on a sound and meaningful comparative basis. The present author has not found anything else of his period that can approach it in originality and understanding. Hilgard was doing basic thinking.

In 1894, Hilgard, with M. E. Jaffa, presented the landmark study of humus and nitrogen in arid soils. Soil of the arid region contained less humus than did soils of the the humid region, yet it was found that the land was not nitrogen hungry. According to the methods of soil analysis then in use, more nitrogen was found in arid soils than in humid soils and the conclusion was reached that "the humus must be richer in nitrogen than usual," and in this manner equalized the total nitrogen supply as between arid and humid soils. This work was basic thinking for approximately twenty years before any advance was made to a fuller understanding of the problem.

Hilgard's interest was not confined to California soils, but extended to a comparative study of the qualities of fruit grown on arid soil under irrigation, and European fruits. During 1893 and 1894 he reported his findings; that in all fruits except the orange, the nutritive value of those grown on arid soil was greater, in some instances by as much as 50 per cent, and he called attention to the rise of ancient civilization in arid regions.

The alkali soils constituted one of the major problems in the arid regions and especially when it was accentuated by irrigation. Hilgard's first report on the problem came in 1880, but the major study was that of 1886 and later. In 1896 he made a popular presentation of a decade and a half of study under the title, "Steppes, deserts, and alkali lands."

Continued investigations made since have given additional confirmation, and have developed new facts having important bearings upon the possible utilization and productive value of vast land areas thus far considered either irreclaimable or adapted only to scanty pasturage.

From his previous comparison of eastern and western United States he passed to generalization of world scope, asking and answering his own question, why the great ancient civilizations had developed in or near deserts. The former material was reviewed to show the cumulative effect of arid climates on plant nutrients.

It logically follows that, inasmuch as actual examination shows practically _all_ arid soils to be calcareous, 'arid countries are rich countries' whenever irrigated; and the actual and concordant experience of mankind corroborates the conclusion.

In other words, the ancient civilizations have, consciously or unconsciously, chosen countries having naturally rich and durable soils, capable of supporting for a long time a denser population than the forested regions, without resort to artificial fertilization beyond irrigation....

But if these things are true, then the steppes and the alkali lands deserve the most earnest attention, both of agriculturists and of students of natural economy; for in them lie possibilities for the abundant sustenance and prosperity of the human race that have thus far been almost left out of account. While it is true that irrigation water may not be practically available for the whole of the arid regions of the globe, so much remains to be done in the study of the most economical use of water, of appropriate crops and methods of culture, that even an approximate estimate of actual possibilities in this direction can not yet be made. At all events, it is of the highest interest to study the problem of the reclamation of these intrinsically rich lands in all its phases.

* * *

When once the high productive value of alkali lands is generally realized, enormous areas will be added to the producing lands, not only in the arid region of the United States, but in the Old World.

Irrigation experiments led to an important conclusion (1898) relative to the remarkable drouth resistance of plants in arid soils. He attributed it to root depth, and therefore formulated the principle that irrigation water should be used so as to induce deep rooting of crops, also, he adopted the soil mulch idea designed to protect soil against evaporation.

Hilgard's final formulation of a lifetime of thought on soils was presented in 1906 in his only book, Soils, and is summarized in Chapter 6. Fate had launched young Hilgard on his career as a geologist in Mississippi, in 1855. His removal to California seems equally a matter of chance. Possessing a creative mind, had he remained in Michigan, to which state he moved from Mississippi, in 1873, in all probability he would have distinguished himself in some manner, but that is not history. The dry climate of California provided the necessary stimulus, in contrast with both Mississippi and Michigan, to set his creative mind to work upon this unique contribution. Sharply contrasting environments were to an original mind the nemesis of provincialism. His understanding of all soils was sharpened by the experience in contrasts. Hilgard's treatment of the two areas made regionalism significant, whether from the standpoint of soils, geography, or history.

Hilgard had limitations, but they need not be emphasized in comparison with his other qualities. As an irrigationist, he did not appreciate sufficiently the possibilities of dry-land farming. As a physical scientist, he did not appreciate sufficiently the possibilities of the biological factor in soil farming processes, but he did recognize almost from the beginning in Mississippi the fact that certain kinds of vegetation were associated with particular soils, and he developed the idea of native plant indicators as guides to agricultural crop policies — a procedure so conspicuously identified with the career of Shantz many years later. One of the most significant aspects of his work was a recognition that the new soils of the United States presented a problem quite different from long-used European soils.

It is vital to the study of Hilgard, and to the study of American soil science, to call attention to his foreign relations and the recognition by others as well of soil work being done abroad. Born in Germany, although reared in America, Hilgard's natural interest in Germany was developed by his graduate education there. In December, 1892, he attended the fifth convention of the association of agricultural experiment stations of the German Empire at Berlin, and before his return was awarded, by the Royal Bavarian Academy of Science at Munich, the Liebig silver medal in recognition of his soil work. At the same time the Liebig medal was awarded, also, to Lawes and Gilbert, of Rothamsted, England. In 1903, the University of Heidelberg reissued his degree of Doctor of Philosophy as "a golden degree," one half century after it was first conferred. Many of Hilgard's papers were published in Germany, and the monograph on arid and humid soils was translated and published, in enlarged form, in both Germany and France.

Although the soil scientists of the United States Department of Agriculture, in the early 1890s, and of its Bureau of Soils, created later under the direction of Milton Whitney, were otherwise preoccupied, the Office of Experiment Stations sent a representative, W. O. Atwater, with Hilgard to Berlin in 1892. Furthermore, the publication of that Office, The Experiment Station Record, listed European soil literature and published abstracts, including Russian soil work. The views of the great Russian triumvirate, Dokuchaev, Sibirtzev, and Glinka, and others are found there. For the Record, 1900-1901, Dr. Peter Freeman translated and condensed a series of studies by Sibirtzev, which was introduced by an unsigned editorial recommending the Russian investigations to Americans in general and declared further:

They should be of special interest to American investigators, since the soil conditions of Russia are to a considerable extent duplicated on this continent, a fact which has been recognized by Hilgard and others, particularly in the study of the virgin soils of America.

A summary of Sibirtzev's most recent report and soil map was promised in a later issue of the Record, but for some reason did not appear. The relations with Russian soil science were not a one-way proposition as was shown by Sacharov's study of Russian alkali soils in which comparisons were made with Hilgard's alkali soil work. (E.S.Q. 18, 1906-1907). In 1909 Hilgard attended, and participated in, the Budapest meeting of the first international agrogeological congress. In E. J. Russell's account of the congress (Nature, London, 1910), three men were singled out for particular comment; Hilgard, Ramann, and Glinka. In commentary on Glinka's paper on Russian soils and other plans of soil classification, Hilgard concluded that each region should adopt its own classification, because characteristics significant for one region were not necessarily significant for other regions. The second congress was held at Stockholm and was attended by Hilgard where he read a paper on soil classification. This meeting recommended the establishment of an international soil journal, which was accomplished the next year, with Hilgard as the American representative and contributor of one article to the initial volume. Possibly one thing that obscured these developments was the fact that Hilgard was at the end of his long career, and soil work in the United States came to center so largely around Whitney's Bureau of Soils.

Evaluations of Hilgard's career on the part of his contemporaries in soil science appear in review of his book, Soils (1906), particularly one by King, and one by the Editor of the Experiment Station Record, and one as a tribute to him, by the same Editor[W.E.Allen] on the occasion of his retirement in 1909. These were more carefully prepared than the obituary notices of 1916.

About the time of his death, in 1916, the question of humus and nitrogen came up for revision. The climax for Hilgard's California experiment station came in 1916 when Charles B. Lipman, his successor in soil science, announced a repudiation of the Hilgard-Jaffa conclusions that the percentage of nitrogen in the humus of arid soils was greater than in humid soils, thus tending to equalize the nitrogen content in the two regions. New methods of chemical analysis had revealed errors in the historic calculations, and besides the whole humus problem in soil science was about to undergo a revision, especially the theory followed by Americans. This was no reflection upon Hilgard, but was a part of the process of discarding and reconstructing by which science makes its advances and enlarges the adequacy of its understanding.

There remain a number of unanswered questions — Why did not California appreciate him and tell the world about Hilgard and California? Here was a man very much worth boasting about. Why did Marbut refuse to recognize his contribution? It would seem that Hilgard had formulated, independently, many of the same ideas developed by the Russians, and that Hilgard, not Marbut, was the first to recognize, and to a degree to utilize, Russian soil ideas in his final synthesis of 1906. The more one studies Hilgard, the more insistent becomes the need of an adequate biography and evaluation of his place in American history and particularly his contribution to a positive or constructive interpretation of the characteristics of the regions of low-rainfall and dry climates. He performed a valiant service toward breaking down the ideologies and provincial prejudices that dominated the eastern United States as the heritage from the forest man, and provided in his concept of the relations of climate and soil a principle of unity, both significant and important for the Trans-Mississippi United States as a region.

What had been thought of as a barrier of sterility blocking American advance, one which must be bridged to the Pacific by railroads, under the influence of his science, in confirmation of the accumulation of folk experience, became one of the most valuable assets of the nation. Furthermore, his influence was incalculable as it was transmitted through the work of many others who built upon his foundations both at home and abroad.

Frank Hiram King

F. H. King (1848-1911) was professor of agricultural physics at the University of Wisconsin and a man of outstanding distinction in his field. Hilgard wrote, in 1904, that upon the death of the German soil physicist Wollny, by common consent, his mantle fell upon King. Wisconsin was partly a prairie state and was related closely enough to the so-called arid region of the 1880s and the 1890s to appreciate something of the West. Although not interested directly in regionalism, King did recognize the regional contrast, basing that aspect of his work largely upon Hilgard. In 1888 he outlined his research program in soil physics (Baver, 1940). His three major books, The Soil (1895), Irrigation and drainage (1899), and Physics of Agriculture (1899), contained material important to the historical development of the regional problem. As a soil physicist, he approached the soil problem much as Hilgard did and recognized the biological aspect of soil formation largely in terms of the new discoveries in microbiology rather than vegetation in the general sense. He summarized or quoted Hilgard on the distinctive characteristics of soils of arid regions in respect to lime, magnesia, and granulation, and cited in detail Hilgard and Jaffa (1894) on humus and nitrogen. His chapters dealing with moisture in the soil explained capillarity, percolation, etc., but in the matter of conservation of soil water he was making some new contributions from his own experimental work which were of peculiar importance to the low-rainfall areas.

The use of soil mulch to conserve moisture was an accepted practice in agriculture but little was known quantitatively of the differences of soil mulches and the means of maintaining an efficient texture. Thick mulches were more effective experimentally than thin mulches, and a coarse mulch of quartz sand was about three times as effective as pulverized fine clay. In terms of tillage these matters were thought essential to moisture conservation in the arid regions or in drouth periods in humid regions, and a generation of tillage practices was built largely upon King's data. King pronounced the moldboard plow the most efficient mulch former as it removed the soil slice and turned it over completely, breaking the slice into "a layer of loose crumbs." As plowing was not always desirable, especially in dry climates where rapidity of operations was critical, the disc harrow was highly recommended. Ridging of soil, as with the lister, was condemned as wasteful of moisture. A light rain, which was not sufficient to result in percolation of moisture, caused a rapid rise in subsoil moisture, which must be minimized by cultivation to restore the loose crumblike mulch. This was emphasized as particularly important to the plains country, where the rains fell mostly during the spring and early summer heat, but was less significant to California where the rains fell in the cool winter season. Rolling the soil accelerated the rise of moisture and dissipation by evaporation, and therefore if a roller was used it must be followed by a harrow to restore the mulch. Destructive effects of winds on Wisconsin sand districts gave King something in common with the arid regions, and he recommended, for breaking the wind velocity, the planting of groves, hedgerows, alternate narrow strips of grass and crops, and crop rotation.

The crumb texture (the modern term is structure), which King emphasized, he associated with all virgin soils, but especially with grass, and was "always more or less completely restored whenever soils have been laid down to grass for a sufficient length of time." Cultivation tended to break down this granular or compound texture into incoherent particles and rains ran them together into close texture and they dried into hard masses. Cultivation as soon as possible after a rain tended to establish the granular condition, winter weathering and certain fertilizers contributed, but periodical return of grass was most effective.

Summer fallowing received substantial attention from King, in both books, for both its physical and chemical effects; moisture storage, aeration, nitrification of humus, development of microorganisms, contribution to granular texture by winter weathering, and cultivation. He reviewed historically the old system of intertillage inaugurated by Jethro Tull, in England, who advocated drilling small grains in rows wide enough apart to permit of cultivation (horse-hoeing) between the rows. Hunter modified the system into strip-farming using nine-foot alternated strips of planted ground and cultivated fallow. The argument was that the planted strips borrowed moisture and plant food from the fallow strips so that no excess was wasted. The Lois-Weeden system developed by the Reverend Mr. Smith called for the planting of one peck of wheat per acre in three rows spaced one foot apart with grains three inches apart in the row, and then a fallow strip three feet wide which was carefully cultivated. He maintained in this manner a yield of 18 to 20 bushels per acre without fertilizer. Although the practice of fallowing was generally discontinued in humid climates, King concluded that

These cases of old and now generally abandoned practices are called up here because they involve a principle which, when correctly applied, is of great importance in sub-humid climates, where water for irrigation is not available.

The problem of the sub-humid region, he argued, was "to devise a system of planting for the various crops" that would determine "how many plants can be matured upon the ground with the available water," and fix "the distance between the rows and the distance between the hills in the row." Furthermore he specified that "in the subhumid regions, the limiting factor is water alone, and the distance between plants must be made such, if necessary, that the roots of one will not encroach upon the feeding ground of another." King rejected broadfield summer fallowing because of the danger of soil blowing, and waste of soil fertility, especially nitrate surpluses. Instead, he recommended a plan similar to the Lois-Weeden system; that small grains be planted in strips of four to six drill rows spaced nine inches apart, with cultivated fallow strips of thirty inches between. This is summer fallowing, but, he admitted,

... in a very different way, and for quite a distinct purpose from that usually had in mind. Of course, it would not be urged, except on soil and in climates in which there is an insufficient supply of moisture to mature the crop under ordinary methods of farming.

He admitted that the yields would not be as great, nor the costs as low, but

All that can be asserted, or can be reasonably expected, is that better crops can be raised by it in sub-humid climates and on lighter soils in humid climates, than can be raised by the ordinary methods.

In one chapter (1899) King summarized "The extent to which tillage may take the place of rain or irrigation." As there was insufficient water in the low-rainfall regions for irrigation of more than a small part of it, tillage as a partial compensation for deficient rainfall was crucial. Soil could not derive moisture from the dry air through tillage; the rate of transpiration of plants could not be reduced by tillage to diminish the water requirement of crops; tillage could only increase the percolation of water into the soil and reduce evaporation loss; and to these ends "tillage to conserve soil moisture is chiefly effective in saving the winter and early spring rains." The methods were by fall plowing, subsoiling, earth mulches, early spring tillage to reduce evaporation, summer fallowing, inter-tillage, reducing frequency of cultivation as season advances, adjusting depth to seasons, rolling under certain conditions, and reducing the destructive effect of the winds. It was emphasized also that all of these methods of conserving moisture were equally applicable to either dry land or irrigated farming. It was pointed out also, that the crops best adapted were spring or early summer maturing crops, and that those whose growing season was the hot mid or late summer should not be planted.

The United States department of agriculture created the Bureau of Soils, in 1895, with Milton Whitney as chief. He published a general paper in the Yearbook, 1894, on "Soils in their relation to crop production," the last part of which dealt with the arid region. According to his division, humid country extended to $97°$, semi-arid to $100°$, and the arid region was west of that line where the annual rainfall was less than 20 inches. Twenty inches of well distributed rainfall, he said, was sufficient to make a crop:

Thorough preparation of the land, with subsoiling where this is necessary to break up a compact subsoil, followed by shallow but frequent cultivation of the surface, will undoubtedly make the crop much safer and surer in the arid regions of the West.

In the next Yearbook, 1895, Whitney discussed the "Reasons for cultivating the soil." The objectives of plowing were to increase receptiveness to rainfall, to maintain moisture near the roots, to admit air, and to facilitate root penetration. The objectives of cultivation after plowing were to kill the weeds which robbed the soil of moisture, and to provide a loose dry soil mulch to retard evaporation. He commented on the tendency of a plow to establish a plow sole — if plowed to the same depths, which was sometimes unfavorable, and described the spade as the ideal soil tool. In a paper, Bulletin No. 21, Means (1903) presented the problem of reclamation of alkali land by drainage, as though nothing had ever been done on the subject in the United States.

The bureau of soils was engaged mostly in making soil surveys and classifying soils according to standards in which physical characteristics of texture dominated, little attention being given to chemical analysis. This trend of soil theory came to its climax in the bureau's Bulletin No. 22, (1903) by Whitney and Cameron, Bulletin 23 (1904) by the same authors, and Farmers' Bulletin 257 (1906), Bulletin 55 (1908), and Bulletin 57 (1909). The argument of the Whitney theory was that "the ordinary methods of determining the plant foods which a soil contains have not given results apparently related in any definite way to the yield of crops," and a conclusion was reached that, with the presence of sufficient moisture all soil solutions contained the necessary mineral plant food, "and the soil [served]as a reservoir and distributing agent for this solution." The controlling factor in crop yields was the physical condition of the soil as determined by climate and cultural methods. Associated with this theory was a revival of the contention of de Candolle, that roots excreted a toxic substance which accumulated in the soil if repeatedly planted to the same kind of crops. Rotation of crops, according to this theory allowed the leaching out of the poisons and restoration of productivity for that crop. According to these assumptions soils would never wear out, and fertilizers were unnecessary except as they influenced the physical texture of soils. King was a thoroughgoing believer in soil chemistry and in fertilizers to maintain plant food in soils. His experiments carried out in 1903 for the bureau led to the preparation of six papers which King interpreted as confirming his views. Whitney disapproved of King's conclusions. He rejected three of the papers outright and published (1905) three of them with the explanation that they represented views disapproved by the bureau. King resigned from the bureau, and published the rejected papers. Hilgard (1904), and Cyril G. Hopkins (1910), of Illinois, and representatives of the 47 experiment stations, rallied to the defense of King and of the cause of conservation of soil fertility. King admitted fully that chemical analysis of soils had not yielded satisfactory results, but his conclusion was not to abandon the theory of soil fertility in terms of plant food, but to intensify experimental work to discover the adequate body of scientific principles. Under these conditions it is clear that there was little common ground as between the advocates of the official view, and the view of soil scientists and agronomists outside official circles. Grassland regionalism and its peculiar problems found the situation confusing when scientists were unable to agree.

William Ellsworth Smythe

Aside from the propagandist boasting about the wonders of the West, especially by land agents and by immigration boosters, Hilgard was the most important, if not the first, to present in a positive scientific sense the idea of the adequacy or the superiority of aridity as an environment for civilization in America. An extreme use was made of his ideas, for propaganda purposes, by W. E. Smythe, in support of an irrigation interpretation of the Trans-Mississippi West.

William Ellsworth Smythe was a journalist on the Omaha Bee in 1890. Becoming interested in irrigation, under the impact of the drouth then devastating the West, he made irrigation publicity his major interest. In 1899 he published a book, Conquest of Arid America, rewriting it in 1905.

As the revised version most completely represented his matured views, it is used here as setting forth his thought. His enthusiam included public control of water, with water rights based upon effective use; small farms associated with village-centered agricultural communities; a diversified farm program; and cooperative control of water and marketing, combined with private ownership of the land. The home-building instinct, he maintained, was the key to American history, which had already passed through three eras of migration; to the foothills of the Allegheny mountains by 1770, to the Mississippi river by 1860, and to the mid-plains by 1890.

The movement never paused until it encountered an obstacle beyond the power of the individual settler to overcome. This obstacle was aridity — the failure of rainfall to meet the demands of agriculture. The impetus of the movement carried its vanguard across the dangerline and into the territory where existence could not be maintained without recourse to methods then little understood, and indeed not fully developed. Upon this strange boundary of prosperity, which nature had marked with indelible lines, the hosts engaged in the third colonization era trembled and hesitated for several years, then fell back baffled and disappointed.

This discussion was the preamble, of course, to his main theme, the fourth era of American history, which was the conquest of the arid region through irrigation. In this approach he took a positive attitud , dividing the nation in half by the 97 meridian, the East the humid half, the West the arid half — "the better half of the United States." Contrary to popular belief, he insisted that it was the humid region that was subject to uncertainties of rainfall and weather conditions, while the western half possessed aridity as its chief asset because aridity made necessary and possible controlled moisture, scientific agriculture in its highest development, because under irrigation each crop would receive water when needed and in the amount requisite for the optimum yield in both quality and quantity. He quoted Hilgard on the relation of irrigation to cooperative effort and to social stability, but probably more important was the use he made of his scientific conclusions that the phenomenal fertility of the soils of the West were to be explained by the fact of aridity, soils from which the nutrients necessary to plant growth had not been leached out by the rainfall of the humid climate. Not only was the West the better half of the United States because of aridity, but he insisted that it held the greater body of natural resources of all descriptions, and to Smythe, opportunity lay in the development of these resources, agricultural and industrial. It is evident that this positive view of the superiority of the arid region was one that deciduous forest man, who dominated the culture of the United States, could not be expected to understand, even if it had been presented in a more moderate and a more defensible form. From the standpoint of historiography, the important contribution Smythe made was in repudiating the apologetic deficiency attitude toward his region, and in asserting aggressively that differences from the humid climate were advantageous.

Hardy W. Campbell

The first settlers who entered the low-rainfall areas of the West met the hazards of farming according to the traditional techniques of the high-rainfall areas. They worked out many adjustments, however, mostly as a folk process of experimentation and individual resourcefulness (Malin, Winter Wheat, 1944). After the passing of the prolonged drouth period of the 1890s, population again moved into the drier areas and forced the issue in a large way. Hardy W. Campbell formulated a soil culture system, and, with the enthusiasm of a crusader, publicized his views on dry-land agriculture. Born in Vermont, he settled in South Dakota in 1879, and to him the problem of farming in such an environment became a challenge. He experimented with different methods and invented new machines much as other farmers were doing, but he differed from them in the will to systemize his conclusions and publicize them. His first publication of a pamphlet on the subject appears to have occurred in 1893. He interested railroad companies in his ideas and with their support he carried on extensive cooperative experiments with individual farmers in several plains states during 1895, 1896, and 1897, and 1898. He concluded that the farmers were not following closely enough his instructions and turned to the idea of a model farm directly under his supervision. He persuaded James P. Pomeroy, of Colorado Springs, to sponsor such an experiment on land the latter owned near Hill City, Graham county, Kansas. Operations began in March 1900. In 1902 Campbell published the first of a series of Soil Culture Manuals, the principal revisions appearing in 1907, 1909, 1914, and 1916. This does not seem to be the place to trace, in detail, the evolution of his ideas through the formative twenty years, but his thinking on dry-land farming evolved through several stages; first experimental period, 1884-1891; first formulation of principles, 1891; summer fallowing, or alternate-year cropping, advocated to 1893 inclusive; continuous cropping, 1894—; return to alternate-year cropping, or two crops in three years, under the name of summer culture sometime between 1900 and 1902; and lastly, a return to continuous cropping, with an emphasis on spring crops, but not to the exclusion of other systems, 1916.

The earliest body of material possessing continuity, from which the main aspects of the evolution of Campbell's more mature views can be traced, is his farm paper, The Western Soil Culture, launched from Sioux City, Iowa, in June, 1895, and continued with changes in title, location, and management until 1898. In the issue of December 1, 1896, Campbell stated that his method had been proved "not only on our own farm in South Dakota in '92, '93, and '94, but on fields in North Dakota, Nebraska, Kansas and Colorado in '95 and '96..." In the issue of August, 1895, he recounted in general terms the stages in the development of his methods. In 1891, for the first time, he said, he had discovered significant principles; that moisture could be controlled largely by mechanical work, and that the movement of moisture was governed largely by the looseness or the firmness of the soil. In the spring of 1891, a field had been sown with grain broadcast by a seeder mounted on the rear of a broad-tired wagon. Of course, seed sown thus was covered by harrowing. The field blew badly, ruining the crop, except in the wagon tracks:

> We discovered by this that packing the soil firmly at the bottom and leaving a loose mulch on top, which was the condition in this case, good results could be attained. From that time to this we have worked carefully and persistently upon this line until some phenomenal results have been attained. This has been accomplished only by a careful study of soil physics, agricultural chemistry, and botany, in connection with careful watching of the experiments we have carefully conducted year after year.

The principles to which he referred were summarized as deep plowing, subsurface packing, and frequent surface cultivation. He was so optimistic in his enthusiasm as to predict that within five years the northern and central plains states would be dotted with 80 and 160 acre farms, and this "in the face of the generally conceded idea that large portions of the level prairies of the so-called dry belt must sooner or later be abandoned to the stock raiser and ranchman."

Campbell's abandonment of summer fallowing was explained as follows:

> Upon this point, we shall only touch lightly; principally because we do not agree with the majority. Up to the season of 1894 we were advocates of summer fallowing, but we are now strongly of the opinion that it is a waste of time and money; for we are quite

inclined to the belief that thorough tillage of a growing crop (as explained in our June and July numbers) will leave the soil in a more fertile condition than the average summer fallowing. Another vital point especially in arid or semi-arid districts, is the amount of moisture that can be carried over in the subsoil for the coming season's crop. Many tons more moisture can be retained in the subsoil by thorough cultivation of a growing crop than by summer fallowing. This, coupled with the fact that a good profit may be gained from a well-tilled crop as against no revenue from summer fallowing, is the convincing argument; but if the farmer has too much land to properly till and must crop it in the old slipshod manner, then by all means summer fallow a part of it. In this case we are strongly in favor of at least one-third summer fallowing each year, but don't plow before the middle or last of July, when you can turn under a rank growth of weeds in their green state. Put them under at least eight inches of good soil, then follow with a subsurface packer.

In the opening statement with which he launched his paper, The Western Soil Culture, June, 1895, Campbell declared that

There is no business or science that the great American people have so sadly, yea, shamefully neglected, as that of soil culture. Its scientific principles are almost entirely unknown to the average tiller of the soil; in fact, it is not comprehended by the professors of our colleges yet, but is today being most earnestly studied all along the line.

Lamenting the backwardness of American agriculture compared with European in both quantity and quality of production, Campbell proposed not only to print his own views and the results of state and federal experimental work, but to demonstrate them. Because mere statement was not sufficient to convince farmers of error, he had selected five demonstration fields, in as many different states, which were being farmed during the season of 1895 under his supervision, and he promised to report upon them from time to time. The fields were located in North Dakota, South Dakota, Minnesota, Nebraska, and Colorado. Preparations were announced for additional fields, where work would be begun in the fall of 1895, Kansas being added to the list of states with a field near Oberlin, Decatur county. Operations were to continue over a period of three to five years. He was convinced that he could prove in these fields his contentions already stated, and, furthermore, that over a period of years, the supply of moisture could be built up through storage in the soil, and the supply of plant food or fertility could be increased. He went so far as to claim that hot winds would not injure crops materially, so long as there was adequate soil moisture.

Campbell's indebtedness to the literature of scientific agriculture was not only acknowledged, but he drew upon it freely, citing or quoting authorities in support of his contentions. In the first issue of *The Western Soil Culture,* he reviewed the classic experiments in England, of Jethro Tull, 1728-1741, with horse hoeing, emphasizing three propositions: Drilling small grain in rows or strips with wide spaces between to permit cultivation through the growing season, intertillage; soil cultivation as a substitute for manure as fertilizer; continuous cropping without manure and with increase in soil fertility. Campbell then reviewed the Lois-Weeden system of the Reverend Samuel Smith in Northhamptonshire, England, also the experiments with intertillage at Cornell University, New York, 1874-1877, where wheat operations with a one-horse drill and a one-horse cultivator resulted in broad stooling, until at harvest the wheat seemed to cover the whole of the ground, giving the appearance of a solid field. Campbell argued that the land between the rows was not waste, but that the roots penetrated laterally for moisture conserved by the soil mulch or dust blanket established by the continuous surface cultivation.

The virtues of mulch, any kind of covering over the soil, were discussed repeatedly, and publications of the agricultural experiment stations were cited or quoted, especially those of Missouri and Minnesota, and Wisconsin. Mulch retarded evaporation and thereby conserved moisture. Agreeing with these mulch theories, a Cornell University experiment station bulletin emphasized that while herbage mulch was nature's method both for conserving moisture, and for restoring fertility, this method was not practical under agricultural conditions. The necessity of plowing once a year destroyed the herbage mulch, and besides, except in unusual cases, it was impossible to secure sufficient herbage to afford an adequate mulch for fields.

Soil mulch was the method for retarding evaporation agreed upon by the agricultural scientists of that day, and articles were printed in *The Western Soil Culture* from the best of them. All used the standard term soil mulch, and some used the term dust blanket as a synonym. Campbell used the terms interchangeably. For the theory of the soil mulch, Campbell cited particularly the work of F. H. King, as published in the reports of the agricultural experiment station of Wisconsin, 1889, 1891, 1893, and quoted liberally from them. This first citation of King was in the first issue of Campbell's paper, an article by King was published in November, 1895. King's new book, *The Soil,* was praised in December, 1895, and in the last issues of the year 1896, articles by King were featured. W. M. Hays, of the Minnesota Agricultural Experiment Station, contributed an article, "Soil, air, and water" to the issue of October, 1895, explaining the theories of soil mulch or dust blanket. He used both terms

In connection with soil mulch, Campbell was faced with the question of what would prevent soil blowing? He argued that subsurface packing would accumulate moisture, stimulate root growth, and when the soil was turned up the following year by the plow "it would be like new soil, and cannot be easily pulverized into fine dirt" (October, 1895). Later he wrote on the question of whether or not soil mulch would blow.

If so it is because the soil turned up by the plow was so dry the year before as not to permit the fine roots to live in it. Soil kept moist by proper tillage will be filled so full of little hair-like roots, that the soil will be held together in little pellets so large that they will not blow" (December 1, 1896).

It was clear that Campbell had gone part way, but was not fully appreciative of the problem of soil structure in relation to the soil physics involved in its maintenance and the techniques of soil management

Campbell recognized two aspects of soil fertility or productivity, using those words either together or interchangeably. He cited European and American experimental work as warranting his conclusions "that fertility or productiveness of the western soils will, under proper tillage, increase in productiveness by being constantly cropped, instead of decreasing, as is generally believed, and all without adding fertilizers except what is added by nature through natural channels." He urged the plowing under of all vegetable matter to contribute to the natural process, subsurface packing to accumulate moisture in aid of decomposition and nitrification, and surface cultivation to establish a soil mulch to conserve moisture. In another place Campbell argued further that

There is not a man in the least conversant with soil physics but will admit that there is no soil so filled with fertility or plant food as the prairie soil in this so-called dry belt, lying for the most part in the two Dakotas, Minnesota, Kansas, Nebraska, Oklahoma and Colorado, and the only thing necessary for putting this food in a valuable condition that the plant can readily assimilate it is moisture. Now with the fact clearly demonstrated that the natural rainfall may be utilized what more is wanted than intelligent soil tillage.

Campbell cited Professor Snyder of the Minnesota agricultural experiment station as saying that these soils contained enough plant food for 125 successive crops of wheat, but that little was available at one time; it was released gradually by the decomposition of vegetable matter, by air, by moisture and by microorganisms. Campbell quoted other studies to the effect that cultivation had a decisive influence on nitrification, and greatly accelerated that process (July, 1895).

Campbell's independence of mind in selecting from scientific literature is illustrated repeatedly, and in the matter of climate change he went against the current of the day, using, July, 1895, the writings of J. B. Sage of the Iowa Weather Bureau to support the view that climate was not changing. Sage pointed out that the idea of climate was a "delusion in all ages and in nearly all countries." In the late eighteenth century the conviction was prevalent, and Jefferson concurred, that the cutting of forests moderated the climate. Humboldt had investigated this myth and concluded that there was "no evidence of any appreciable change" of climate in America. Referring to the eighteenth century views, Sage remarked:

> In those days, it will be noted, the current theory was that clearing away forests had made the climate more equable, but in later days the opposite view has obtained. But, in fact, neither view is correct; that is, there has been no appreciable change in climate.

The concept of the permanence of climate was essential to Campbell's scheme of things. If the popular view was correct, that rainfall increased with settlement and development of the country, then his method of soil culture could possess only a minor and temporary interest. If permanence of climate was demonstrated, then soil culture, according to his method, was essential and also permanent, and led to his dictum: "Don't try to change nature's laws to fit your own methods and habits, but change your notions and habits to conform with nature's ways and thereby utilize both." In another place he said:

> The supposed semi-arid region has the best soil in the world and plenty of water. To utilize the rainfall is the problem. It is much simpler to adjust agriculture to a scant rainfall with the exhaustless soil of the northwest than to struggle against the floods and the impoverished soil of regions supposed to be more favored (December 1, 1896).

Soil was treated by Campbell as a storage reservoir for moisture; by deep plowing and subsoiling the reservoir was built, by subsurface packing it was filled, and by surface cultivation and weed-killing the leaks were stopped: "Don't pray for rain; save the water you have." In the high-rainfall regions of the East, estimates were cited that half of a forty-inch rainfall was run-off, the other twenty inches making the crop. In the West, if eighteen or twenty inches were available and all was saved, a crop was insured. This led to the plausible conclusion that moisture conservation, not more rainfall or irrigation, was all that was needed in the West.

By tillage according to his methods, Campbell insisted that not only would the hazards of climate be minimized and crops made certain, but yields would be increased— 100 per cent to 400 per cent, or doubled or trebled, but his predictions varied. Soil culture, not financial theories was the key to agricultural prosperity: "The farmer whose soil is properly cultivated has little time to discuss theories of finance over the barbwire fence with his neighbor of opposite political views."

Tillage tools were important to Campbell's soil culture. His principal dependence was upon the moldboard plow (a 16- or 18-inch sulky), a subsurface packer of his own invention, and a spring-tooth surface cultivator. He disapproved the disc harrow except in preparation of land for the plow, or the spike-tooth harrow, and the roller, if used should be followed, as specified by King and Shaw, by a tool that would loosen the top soil to restore soil mulch. Campbell resented the misconceptions that gained circulation that his soil culture was a matter of special machines, and replied that methods not machines were the important thing. At another time he reminded his readers that the only special tool recommended was the subsoil packer.

A list of demonstration farms supervised by Campbell is not easily reconstructed. Apparently plans for some of the projects announced were not executed and unannounced shifts occurred. The reports promised in The Western Soil Culture on the outcome of the experimental work were incomplete. In some cases, additional information, especially local reactions, can be compiled from the newspapers of the several localities. In February, 1900, in a statement to the Kansas City Star, he made corrections of a news story, saying that his unusual success in 1894 came to the attention of the Northern Pacific railroad. At their request, he conducted experimental fields at five points along that road in 1895; in 1896 he inaugurated similar demonstration fields at five points along the Burlington railroad in Nebraska, Kansas, and Colorado. By 1897 his program had expanded to include 43 fields in five different states under the auspices of four railroads. By the fall of 1898 he decided that this proxy farming could not bring satisfactory results and discontinued all such work. In November, 1899 he made the acquaintance of J. P. Pomeroy and made the arrangement to take over the full management of one of Pomeroy's farms near Hill City, Graham county, Kansas.

The Campbell system as it became widely publicized, represented primarily the practices presented formally in his series of manuals, 1902-1916. The dry-land-farming boom-literature began to appear in the general magazines in 1906, Campbell being the principal center of attention, and a series of annual Dry Farming Congresses was inaugurated in 1907.

The soil culture manual of 1902 was based particularly on the Pomeroy Model Farm operations in its first two years, plus, of course, Campbell's accumulation of ideas. This meant that it dealt primarily with winter wheat. In studying Campbell's manuals it is necessary to recognize a dualism as between caution and enthusiasm. He said that fundamental principles were important, not rules:

There cannot be laid down any rule by which to be guided in the cultivation of the soil under all conditions,... but in the great semi-arid area of our western country we believe a general rule may be applied, and if followed diligently the resulting storage and conservation of the natural rainfall in the soil will produce, in the average years, as good crops of cereals, and of all the vegetables that are commonly grown, as can be produced in the humid central portions of the United States.

Again, he said, the difference between the Campbell method and the traditional methods was the difference between forty bushels and eight bushels of wheat per acre.

The two prime objectives were storage in the soil of all moisture that falls and the conservation of that stored moisture against dissipation, especially by evaporation, and "the storage and conservation of the rainfall in the soil by our method of cultivation is the only means of saving that great section and making it bloom and prosper." He called his system summer culture, rather than summer fallow, in order to emphasize a particular plan of operations for the cultivation through one season to store and conserve moisture and raise a crop the second season, thus producing one crop on two years' accumulation of moisture.

His description started with early spring operations for the summer tillage year; double discing the ground so as not to leave an open furrow. This would increase the penetration of spring rain and snow, provide a loose soil mulch to retard evaporation, and to mix air with the soil. After each rain, as soon as the soil was dry enough, it should be cultivated, preferably with an Acme harrow, to maintain good physical condition. In early June it was to be plowed six or seven inches deep and packed the same day with a subsurface packer, followed by an Acme harrow, preserving the loose soil mulch on top. Cultivation with the Acme harrow was to continue during the summer as necessary to kill out absolutely all weeds and after rains to preserve good physical condition of the soil mulch. Wheat could then be planted in the fall, using only a small amount of seed, about half the usual amount to permit vigorous growth of individual plants without too intensive competition for moisture. The next year, immediately after the wheat was harvested, the land was to be double-disced to start again the new two-year cycle with moisture storage. As adapted to spring wheat or corn, discing or plowing during the early winter, after summer tillage, was recommended to improve the storage of moisture and put the soil in condition for spring tillage and planting.

The machines used in these operations were given a particular importance in the system. "To my mind," Campbell wrote, "there is scarcely an agricultural implement more important to the western farmer than the disc harrow." He used it to precede the plow, mixing organic matter into the top soil to facilitate decomposition, and to render the soil as receptive as possible to moisture. Small discs, fourteen inches in diameter, were recommended because they would pulverize the soil more completely than large, slow revolving discs. The Acme harrow was an implement with curved knives that he insisted contributed most effectively to maintaining good physical condition of the soil mulch. A weeder made of flexible arched rods was another important tool. The moldboard plow was used, but in conjunction with the subsurface packer, a series of narrow wedge-shaped wheels which were designed to firm the soil at root level, leaving the surface loose. This machine was of his own design. He preferred a shoe drill with a chain cover for planting wheat, and, although he thought the perfect drill had not been developed, insisted that the drill was less important than the soil culture operations that had preceded. For corn, he preferred the combination of plow and check-rower, but in the higher altitudes and the northern latitudes, he accepted the lister.

From the standpoint of soil science, Campbell was leaning heavily upon Hilgard for his understanding of the nature of soils of low-rainfall areas as contrasted with humid climates, and upon King in matters of soil physics. Interest centers in the latter phase of the problem as it related to physical condition of the soil and to movement of moisture in soil. On soil mulch, Campbell was following King and cited his experiments, which indicated that soil mulch retarded evaporation, and that a coarse granular mulch was more effective than a fine dust mulch. Movement of moisture in soil by a capillary action was also accepted from King, and firming the soil with a smooth roller tended to accelerate the upward movement to the surface where it evaporated. It was in this connection that Campbell had developed his idea of the sub-soil packers; assuming that if the soil was firmed at root level the rise of moisture to that point would aid in plant growth in a moist firm seed bed, and the establishment of the soil mulch above the firmed subsoil would prevent evaporation. In the advertisement of the subsoil packer in the back pages of the manual, Professor Collins of the Kansas State Agricultural College was quoted in an endorsement of the machine. Campbell followed King in assuming that a light rain, insufficient to result in percolation into the subsoil, had the effect of increasing evaporation. To retard this effect, cultivation after rains was recommended to maintain an effective soil mulch. In all this, Campbell thought of his system as improving upon King's practices, especially in details and in continuity and timeliness of cultivation. Stored moisture, he argued, was more important than rainfall. He challenged the idea that the region did not receive enough rainfall: "The real difficulty in the semi arid belt is not a lack of rainfall, but the loss of too much by evaporation...."

The nature of the soil mulch was important in Campbell's system; the mulch must be "composed of minute lumps," a condition that could be achieved only when the soil was cultivated at exactly the right time; not wet, not dry, but moist. A dust mulch composed of incoherent particles would blow, but "these minute lumps made from slightly moist soil when dry never blow." Light soils that did not readily assume the granular texture, or never did presented difficulties, but Campbell maintained that they could be overcome with the proper soil culture. If all the principles of his system were carried out with efficiency and timeliness Campbell maintained that "These facts, when fully comprehended, must and will make of this great semi-arid belt the best and most desirable farming country we have in the United States."

Campbell's second soil culture manual issued in 1905 was substantially like the first one, but there were modifications of language in places and four new, or substantially new, short sections which crowded out other material; The physical condition of the soil, the air in the soil, the crop rotation, and a list of the necessary farm tools for operation of a summer tilled farm. Of most significance was the section on crop rotation in which Campbell repudiated the practices except as a matter of convenience. For western Kansas and Nebraska he modified the summer tillage program for alternate years to allow for two successive crops of winter wheat and summer tillage every third year.

Campbell's 1907 Soil Culture Manual was a much more comprehensive presentation of the system than those of 1902 and 1905. He called it a complete guide to scientific agriculture as adapted to the semi-arid regions. He thought that the world was on the eve of an Agricultural Revolution in the twentieth century, including plant and animal breeding, diversification, planting, harvesting, and marketing. Dry-land farming was only one phase of this larger conception. He did not claim for the manual "a universal guide to success in farming; but the reader will find something in it of value for him;" he did not claim for it perfection, but it had gone a long way toward the application of the principles of nature; it was not "a code of imperative rules to govern every act of the farmer in his culture of the soil," but it did the important thing in explaining "the reasons for doing things — then leave it to the intelligent farmer to do the rest." He had gained such assurance, however, from his success that he asserted that originally the claim had been made to double production, now he could say to quadruple crop yields.

His system had been subjected to adverse criticism and a defensive attitude was evident. His reference to Chilcott is a case in point. Campbell and Chilcott had started farming in South Dakota about the same time, and both became interested in the problem of dry-land farming. Chilcott became identified with the agricultural college and experiment station and later with the federal bureau of plant industry. Campbell had become identified with the western railroads and private aid in developing his ideas. There was evident a rivalry between the two men. Chilcott emphasized rotation of crops, (South Dakota Experiment Station Bulletin 98, 1906). The The problem was complicated further by the conflict between King and Whitney over soil fertility, Campbell quoting both men, also Hopkins of Illinois. He consoled the farmers by admonishing them not to be discouraged, because the experts also were in a haze about soil management. He followed Hilgard's assurances of adequate fertility in the low-rainfall soils, and defined his own understanding of the elements of soil fertility under three heads:

Soil fertility is due to the proper combination of elements in the soil.

Soil fertility is developed in the soil by the proper tilling of the soil, so as to have available the right proportions of air, water, and other elements.

Soil fertility is possible to a high degree in almost every soil, and the addition of fertilizers is only one way of gaining this condition.

Again we repeat that all the processes of agriculture look to development and maintenance in the soil of this available fertility which is so essential to plant growth.

The soil mulch was one of the most prominent features in Campbell's system and was often miscalled dust mulch or dust blanket. Campbell himself had used the phrases interchangeably, 1895-1898. He had stated the matter correctly, however, in his manual of 1902, following King, but not with sufficient emphasis to avoid being misunderstood. In the manual of 1907 a short separate chapter was devoted to the subject, making the correct distinctions a clear cut issue:

Soil mulch is the true name for the loose soil on the surface intended to conserve moisture below, and this mulch should be composed of lumps of soil ranging from the size of a pin head to that of a walnut. To secure such a mulch may seem difficult, but it is not if the soil is cultivated or harrowed when moist, not wet, not dry.... The soil under this condition takes granular form. The sun dries these granules or lumps and no blowing will be noted whatever.

Two difficulties had to be met in dealing successfully with a soil mulch, blowing, and baking. Rains or repeated cultivation tended to dissolve and break down the granular structure. Hot sun after a rain baked the dissolved soil into a hard crust unless quickly cultivated. Unless the cultivation occurred at just the right moisture condition the granular structure did not result, but became instead a dust mulch. The lighter or sandy soils might not form into granules. Campbell's assurances about soil mulch in terms of granules and lumps was on the optimistic side.

The issue of barnyard manures and stubble in relation to dry-land farming was met directly in this Manual of 1907, Campbell taking definitely the side of King and Hopkins, against Whitney, on this particular matter. Much organic matter was thrown away or burned rather than plowed under in dry-land farming operations, because experience had shown that frequently it injured rather than aided crop production. The difficulty lay in the slowness of decomposition and nitrification in dry climates, the presence of such unincorporated material contributing to rapid evaporation. Campbell insisted on the value of organic matter: "In no section of the country is the soil of such a character as to respond more quickly and effectively to the use of barnyard manure and in no place will the effect of such manures last longer, or be of such permanent improvement." It was the method that was important in Campbell's view, and he prescribed discing the manures into the top three inches of the soil, then plowing about six inches deep, followed by subsurface packing, thus thoroughly shredding and mixing the material into the soil and firming it to promote moisture absorption and decomposition.

Tools became of particular interest to dry-land farming, the emphasis being on large tools to complete operations quickly while conditions were exactly right, — timeliness of operations. Campbell was operating under a horse-power age and was thinking in terms of the small farmer. The farm unit, therefore, was presented in terms of 80-100 acres under the plow, and his power in terms of four-horse teams. His list of tools for 1907, as for 1905, for such a farm included a gang plow (14 inch bottoms), disc harrow, improved harrow, combination weeder, sub-surface packer, a two row and a one-row cultivator, and such planting and harvesting machines as were necessary. He gave special emphasis to the sub-surface packer first marketed in 1895, and called attention to the expansion in sales during the last three years. The kind of plow was of less importance to him than the manner of its use. He disapproved listing wheat, and although he did not specify just what kind of machine he had in mind, he must have referred to the lister drill. He recommended again the shoe drill.

Campbell approved the search for new plants, but warned against over-confidence; a combination of such a program with soil culture would produce the best results. He was still partial to corn as a row crop for grain feed and forage, apparently not appreciating the significance of the sorghums. Campbell disapproved emphasis upon prices and programs for production and price control. A high price policy that depended upon some one else having a failure was wrong, he said, because crop failure reacted unfavorably upon everyone. The only sound price structure was one based on certain and stabilized production, ample for all needs, and furthermore he argued that if uncertainty in production was eliminated, the volume would adjust to demand: "Scientific soil culture gets at the root of the problem .... making certain of a good crop every year...."

Campbell took much the same positive view of the area of his interest as Smythe and in a chapter on "Advantages of the semi-arid region" appealed to farmers:

> Don't apologize for being a farmer of the semi-arid region. It is not advisable to be boastful beyond that which is easily demonstrated; but at least do not feel that in conducting the business of agriculture in a region where the rainfall is small you are defying nature. It is true that you may be defying the traditions of the past and doing violence to the old accepted theories on agriculture, but you need not concern yourself about these things.
>
> .... Perhaps it is better on the whole not to have so much water.... [In view of the virtues of the unleached soils of low rainfall regions], this is an advantage.

In connection with dry-land farming publicity of 1906, J. L. Cowan wrote in the *Northwestern Miller* (November, 1906):

> The National Department of Agriculture and the various state agricultural colleges have not endorsed or given official recognition to the Campbell system. They have recognized its results, if not in official documents, in the far more significant form of establishing numerous experiment stations on the high, dry plains that they have always hitherto regarded as hopelessly arid. There they are demonstrating along independent lines the very facts to the proving of which Hardy W. Campbell has devoted more than twenty years of his life. This belated government action, taken when the results of the Campbell system could no longer be denied or ignored, is, in fact, the strongest endorsement that the Campbell system of farming without irrigation in the semi-arid region could receive.

Campbell summed up his own reaction to the historical status of his system as follows:

Many of the things he had to learn by patient experiment would have been taught him by the schools, or could have been reasoned out had he been thoroughly grounded at the start in scientific methods. Perhaps, however, if his idea had been conventionalized by too much of the science of college curricula he might have accepted the dictum that the reclamation of the semi-arid lands was impossible, and the Campbell system would never have been born.

* * *

Some have gone so far as to assert that most of the methods taught by Mr. Campbell were advocated by Jethro Tull, a hundred and twenty-five years ago. Inasmuch as Jethro Tull never visited America and probably never heard of the American plains, it would be remarkable indeed if he had devised a system of agricultural procedure suited to conditions there.

It is, of course, true that many of the facts of the Campbell system were known long before Mr. Campbell's time. Some of the methods used are applicable to farming the whole world over, and have been practiced for generations. Some of the processes have been worked out under the pressure of necessity by hundreds, or perhaps, thousands, of farmers on the plains. If Campbell had done no more than collect, organize and classify these disconnected facts and methods into a coherent system of practice adapted to conditions in the semi-arid belt, he would have accomplished a work of the very highest utility.

In his book of 1916, <u>Progressive Agriculture</u>, Campbell announced changes in his views, as he said, on the basis of better knowledge and more practical field information. He was not looking at the future in terms of a closed frontier defeatism, but declared that "There are millions of acres of unoccupied land, waiting for the 'Home Maker'," and farming in the semi-arid region was still in its experimental stage, and like Edison and the electric light, experimental work must continue until a way was found. He now wrote of three classes of tillage: spring tillage for spring planting, which was now his principal emphasis; summer tillage for fall planting; and fall tillage for crops to follow immediately the one just harvested.

Of spring tillage he said that "it promises more than our plan of summer tilling outlined some years ago. ..." The tillage procedure was essentially the same as for summer tillage, but plowing was earlier and planting was timed for late spring or early summer: The matter of planting time was most clearly the element of innovation, and it was combined with emphasis on quick maturity of the crop. Crops and varieties should be chosen on the basis of quick maturity, and late planting had proven experimentally to mature the same plant as early or earlier than early planting, because growing conditions were better through more thorough tillage, and warmer temperatures. The special recommendation for spring tillage was a crop every year rather than every second year.

For all three of his systems of tillage, Campbell emphasized that the disc harrow must follow the harvester, and in the photographs used as illustrations pictured a horse drawn tandem disc following the harvester in the field, and a tractor drawn harvester with a tandem disc attached. This killed weeds and opened the soil to receive such rains as might fall.

Corn figured in Campbell's farm planning as one of the essential crops, even though wheat was the staple crop, because of the necessity of corn for livestock. The emphasis on spring tillage stressed corn more than formerly, and he still favored checkrowing on plowed ground rather than the listing:

However, if you wish to put the corn in the ground and let Providence take care of it, we would advise the listing plan, as Providence would have a better chance than if the field was check rowed.

This was an interesting admission, more significant than he realized, and was a typical corn-belt reaction to the lister. It was reminiscent of the 1880s in Kansas when listing was first introduced — and was denounced as the lazy man's method of raising corn. It was equally the cornbelt prejudice to refuse to recognize the sorghums as superior to corn in crop certainty and nearly, if not quite, equal in feeding value. So independent in much of his ideology, Campbell had not emancipated his mind in respect to the corn complex.

In shifting his emphasis from summer tillage to spring tillage, he insisted that he was not abandoning the summer tillage:

The difficulties militating against its success are two; first, a lack of full and thorough understanding of the principles and the observation of every detail; second, lack of adequate tools for handling the land more especially after the land has been plowed and sub-packed, to then completely keep out all the weeds without destroying the seed bed by too deep cultivation.

At another place in the book he returned to the subject:

Suitable tools for this kind of work are not available, therefore, the task of keeping the weeds out with such tools as we have is not an easy one. It is hoped, however, that some day the real merits of summer tillage as it is now understood will be sufficiently appreciated to demand proper tools, but so far there have been so many failures because of the many mistakes that the interest is waning.

In this book Campbell emphasized the soil auger, a tool which he had been using since the mid 1890s:

> There is no implement so little used on the farm and yet so capable of conveying a wider scope of valuable information and practical assistance as the soil auger. Its intelligent use will disclose much of the folly of our past efforts.

By means of boring holes in the field he would discover the depth of moisture penetration, and the variations in depth, and their relations to the effectiveness of tillage and the growth of crops. He had not yet arrived at a body of quantitative data and conclusions, but later, in other hands this approach was to yield important guiding principles in dry-land farming. (See Chapter 14).

Ellery Channing Chilcott

The various state experiment stations did something with the peculiar problem of dry-land agriculture, but there was no plains state during the last quarter of the nineteenth century, or in the twentieth century, that produced a man of anything like Hilgard stature. Kansas had E. M. Shelton, 1874-1890, who possessed much independence of thought, but nothing like the energy, originality, and prolific productivity of Hilgard. There were occasional agricultural articles and bulletins from the federal department that touched on western problems (Holmes, 1902; Chilcott, 1903; Failyer, 1906), but in no comprehensive manner. Failyer's bulletin on soil management was conspicuously influenced by Campbell's system. In South Dakota, E. C. Chilcott carried on experimental work which led to his appointment, in 1905, by the United States Bureau of Plant Industry, to take charge of its dry-land investigations. The great plains area which fell within his program of experiment was bounded on the east by the 98 meridian, on the south by the 32 parallel, on the north by the Canadian line, and on the west by a line connecting points with altitudes of 5000 feet. A paper of 1907 opened with this paragraph:

> There is perhaps no phase in the agricultural development of the United States which has attracted more attention during the past year nor any which may have greater importance to the Nation as a whole than the problem of the best utilization of the semi-arid lands of the western United States.

The reasons given were "that government lands suitable for ordinary agriculture are almost a thing of the past;" increasing population made utilization of land of low-rainfall areas necessary; and two or three years of more favorable moisture gave rise to the false assumption of a change in climate. The burden of the argument was mostly in the nature of warning:

> Actual additions to existing knowledge of the subject have been relatively few, but public attention has been directed to the work until the idea prevails that much is now possible in the way of utilizing a limited rainfall which was never possible before. There is some foundation for such a conclusion, but nothing to warrant the many exaggerated statements which are now current.
> The conquest of the semiarid West, to be successful and to be accomplished without large and costly failures, must be made slowly and by careful application of definitely ascertained scientific facts.

In respect to soil fertility, his views were essentially Hilgard's, and the improved methods and practices for dry-land farming including moisture conservation methods of handling the soil, drouth resistant plants, farm management on an experimental basis recognizing that it had only begun, rotation of crops to maintain organic matter rather than summer fallow which was of doubtful utility and depleted the organic matter causing deterioration of the physical condition of the soil and consequently its moisture-holding capacity. Tillage should be used to conserve moisture, plowing being done as soon as possible after harvest, followed by a sub-surface packer with V-shaped rims (Campbell's), or a disc harrow set straight, and less than the traditional quantity of seed planted per acre to avoid overcrowding of plants.

Chilcott's comprehensive experiments were begun in 1906 and were expanded from time to time, stations being maintained from Montana and North Dakota, to Texas. His first major report was made in 1910 (Bureau of Plant Industry Bulletin, 187) because "urgent demands are being made by settlers, actual and prospective, for information concerning the best methods of farming in the Great Plains.... It is hoped that the tentative conclusions drawn and the suggestions made may be useful." The major point was that

> ... our investigations lead strongly to the conclusion that the devising of systems of rotation adapted to local conditions is of greater importance than tillage methods..., [but] tillage methods must be carefully studied and intelligently practiced.... [the two to bring about] the best possible physical and biological condition of the soil.... No hard and fast rules can be established. Each farmer must study his soil, his climate, and his crop requirements and must adopt such systems of tillage as experience and observation have shown to bring the desired results under the peculiar combinations of conditions which prevail at the time and place and are most likely to exist during the growing period of the crop....

Chilcott recognized clearly the cash problem—
"the need of immediate cash returns"— and its bearing
on cropping methods, the accent being on wheat as meeting
that need more adequately than any other crop, although
running counter to a program of crop rotation in combination with livestock.

In 1912 Chilcott published his important paper on
"Some misconceptions concerning dry farming," (Yearbook,
1911), and in his opening declaration said that twenty-
five years of settlement on the great plains should have
shown some accumulation of "a store of experience that
could serve as a foundation for future effort, but such
does not appear to have been the case." He then condemned the accumulation of misleading systems, sensationalism
in the press, and declared that no new system had been
discovered, that agriculture was not an exact science,
that dry-land farming was particularly critical and could
not be reduced to a system because of the numerous factors of uncertainty.

> It is practically useless to attempt to lay down any hard
> and fast rules for the guidance of the dry-land farmer. All the
> help he can hope to obtain from outside his own experience is a better understanding of some of the general principles involved in the
> production of crops under semi-arid conditions. For the application
> of these principles he must rely almost entirely upon his own judgment, experience and powers of observation. Anything which tends to
> lessen his self-reliance and leads him to hope that someone else can
> lay down rules for him to follow is a step in the wrong direction.
> The work yet to be done in the agricultural conquest of the semi-
> arid West is pioneer work of the most strenuous kind.
>
> * * *
>
> This undertaking calls for men possessing the same sterling qualities of self reliance, initiative, and ability to meet and overcome
> new and unforseen difficulties which have always been characteristics of the successful pioneer. Any form of paternalism which fails
> to take into consideration these traits of character in the farmers
> of this region will be resented by those who possess these characteristics and will prove ineffectual to help those who do not. These
> farmers need all the assistance that the United States Department of
> Agriculture and the State Agricultural Colleges and experiment stations can give them in their agricultural conquest of the semi-arid
> lands, but this assistance should be in establishing principles
> rather than in teaching practices.

He indicated the extent to which he was willing
to go in formulating principles of tillage:

> It is fairly well agreed that three things are necessary:
> (1) A fine, moist, mellow, but somewhat compact seed bed of a depth
> sufficient to afford the best conditions for the germination and
> early growth of seed; (2) a receptive condition of the soil, so

that the rains falling upon the surface may be absorbed as rapidly and as completely as possible; and (3) a retentive condition of the soil, so that the water which is absorbed may be held within the soil for the use of the crops instead of being evaporated from the surface. Thus far we are on comparatively safe ground, but when we attempt to go one step further and ascertain from a consultation of the literature of the subject how these desirable conditions of the soil are to be produced and retained we encounter a maze of conflicting theories and generalizations, all based upon far too limited observation, experience, and experimentation.

His five conclusions were as follows:

> In conclusion, the following misconceptions concerning dry-land farming may be mentioned as the most serious: (1) That any definite 'system' of dry farming has been or is likely to be established that will be of general applicability to all or any considerable part of the Great Plains area; (2) that any hard and fast rules can be adopted to govern the methods of tillage or of time and depth of plowing; (3) that deep tillage invariably and necessarily increases the water-holding capacity of the soil or facilitates root development; (4) that alternate cropping and summer tillage can be relied upon as a safe basis for a permanent agriculture or that it will invariably overcome the effects of severe and long-continued droughts; and (5) that the farmer can be taught by given rules how to operate a dry-land farm.

Chilcott's Bulletin 268 (1915) marked a substantial change in his point of view on several aspects of dry-land agriculture. Six bulletins, each one on a separate crop or group of crops, preceded this and the winter wheat Bulletin 595 (1917) was in preparation, therefore this "crop production" study was designed to generalize upon the whole field. He emphasized the wide diversities in climate, soils, and adaptable crops between the 32 and the 49 parallels, as well as the different methods and tools that might be used, presented comparatively the experimentation at fourteen stations dealing with seven different soil handling methods; fall and spring plowing, discing corn stubble, subsoiling, green manuring, summer tilling, and listing: "Summer tillage is, with the exception of green manuring, the most expensive and the least profitable method under trial." He then set out to demolish the summer tillage theory on other grounds:

> The purpose of summer tillage is accomplished by the prevention of vegetative growth rather than by the maintenance of a mulch. Numerous experiments made in connection with this work have furnished an abundance of evidence that when vegetative growth is restrained the loss of water from a mulched surface is practically the same as from an unmulched one.

The cheapest and most efficient methods of weed destruction necessarily form a soil mulch. The results accruing from the prevention of weed growth have been very generally attributed to the mulch itself when the mulch is, in fact, only incidental.

The strangest doctrine in this Bulletin 268 was that relating to after-harvest tillage:

Tillage for the purpose of destroying weeds after harvest is warranted only in those exceptional cases when sufficient water remains in the soil to start weed growth after harvest or when heavy rains come soon after. In such cases early fall plowing is the most effective method of destroying the weeds and thus saving the moisture that would be used by them if allowed to grow. The same object may be accomplished by disking soon after the weeds have started. This method has the advantage of being more rapid than plowing, thus making it possible to cover more ground with the same number of teams and men. But as the land will have to be plowed before another crop is sown, the labor of disking is mostly lost, although the labor of plowing the disked land may be somewhat less than if it were not disked and a better job of plowing may sometimes be done on the disked land. The cost of early fall plowing or disking, when the weather is hot and the men and teams are needed for stacking, threshing, and hauling grain, is greater than later in the fall, when the weather is cooler and there is less other work for men and teams. All of these facts should be taken into consideration before going to the extra expense of tillage to kill weeds immediately after harvest.

One of the important conclusions was that use of the lister in the preparation of the ground did not result in markedly different yields, but at most of the stations where tried, and in most years, it was the most profitable. At the Hays, Kansas, station, listing had a marked advantage in yields of winter wheat. The key to this superiority of the lister, as a tool, lay in the rapidity and cheapness with which operations could be completed.

Conclusions respecting the regional adaptation of crops described the southern range of hard spring wheat as somewhere in Nebraska, winter wheat was grown from Montana southward, especially western Kansas, but it was said not to be profitable in southwestern Kansas and in northwest Texas. Corn was reported as a cash grain crop as far south as North Platte, Nebraska, corn and fodder as far south as the north line of Kansas, as a competitor of the sorghums in northwestern Kansas, and as less valuable than the sorghums in southwestern Kansas. Milo maize and Kafir corn were well adapted to western Kansas and southward.

Nine conclusions closed the bulletin: When climatic conditions were favorable, all stations produced profitable crops by all the several methods; under unfavorable conditions, no profitable crop could be produced by any method; under normal conditions over a ten-year term, some crops were profitable at the labor and grain price levels then existing; no single crop was profitable for the whole great plains region, by any method; forage crops could be grown profitably at all stations. In respect to farming methods, Chilcott emphasized that good farming was as essential on the great plains as elsewhere:

> Good farming may involve methods either intensive or extensive, either expensive or inexpensive; and it must be practical and economical as well as scientific and thorough in order to be good farming. It is just as poor farming to go to too much expense as it is to go to too little expense to accomplish a given result. These investigations show that the largest net profits have usually been obtained from crops raised by cultural methods involving a low cost of production rather than from high yields obtained under methods involving a high cost of production. Lessening the cost of production without proportionally lessening yields should therefore be given first consideration. In other words, extensive rather than intensive systems of farming should be followed.

Other conclusions dealt with different soils under different climatic conditions, the personal equation of the farmer and his family in relation to each farm situation, and finally that "Dry farming in the Great Plains, in common with all farming, to be successful must be systematized, and in order to accomplish this, some definite rotation of crops should be established," and the farmer himself could decide this better than any outsider.

H. L. Shantz

In an earlier chapter (2) reference was made to H. L. Shantz (1911) as the first botanist to make a comprehensive ecological study of the grassland. In "natural vegetation as an indicator of the capabilities of land for crop production in the Great Plains," he was doing that and more because of the bearing of that ecological approach upon the practical question of agriculture of a distinctly regional character. Hilgard had used native forest vegetation as an indicator in Mississippi and in other parts of the south during the 1870s, and in the census report on cotton (1884). Shantz cited Hilgard in some parts of his study but credited the indicator idea to German precedents. He emphasized that the entire plant cover was a better index than any single species, and repudiated any single factor interpretation of plant distribution, in particular challenging the Merriam temperature zones criterion. A single factor became a determinant, he said, only when others were essentially uniform, and "variations, either physical or chemical, are greatest in importance when they approach the limit of favorable conditions." Furthermore, any correlation for one region or set of conditions would break down when changed to another region or conditions. Most of the work for the study was done in eastern Colorado, and therefore, conclusions applied primarily to that area. The prevailing grass cover, in relation to depth of moisture penetration, showed that short grass (grama-buffalo) had not more than two feet of moisture under it; bunch grass (little bluestem, Andropogan scoparius) had moisture to a much greater depth; and between the two was wire grass (Aristida longiseta). Under the short grass the moisture was available only during the spring and early summer, but under the bluestem, being of greater depth, it lasted longer. On the same type of loam soil, with moisture as the variant, short grass would grow in eastern Colorado with 15 to 18 inches of rainfall, wire grass would grow in western central Kansas with 22 to 24 inches of rainfall, and bluestem would grow in eastern Kansas with 26 to 30 inches of rainfall. Again, on the same type of loam soil, with both moisture and temperature as variables, short grass grew in the panhandle of Texas with 21 inches of rainfall, in eastern Colorado with 17 inches, and in Montana with 14 inches. The moisture content of the soil was an inadequate index, because sand, with 18 per cent moisture, was nearly saturated, while clay, with 18 per cent of moisture, was nearly dry.

These conditions offered certain guidance for crop adaptation. On short grass land only early maturing, shallow-rooted crops could be expected to produce, and on wire grass and bunch grass land successively, only later- maturing and deeper-rooted crops matured. In favorable years the loam short-grass lands produced most heavily and sandy bluestem land least heavily. In unfavorable years bluestem land produced best, and short grass land produced least, or failed altogether. In either case the wiregrass land was intermediate between the extremes and proved most reliable over a term of years. Short grass land with an admixture of Psarolea tennuflora, a deep rooted legume with nitrogen fixing qualities, or with an admixture of wiregrass, was almost as reliable as a wiregrass association land.

Shantz's final paragraph revealed that settlers had arrived at his conclusions ahead of him and without the aid of science, although he did not elaborate upon that aspect of the matter:

> Many of the older settlers in eastern Colorado have moved from short-grass land onto wire-grass land, or even onto bunch-grass land, where they claim there is much less likelihood of crop failur

but the newcomer in the region or the speculator almost invariably chooses the hard or short-grass land because it is darker in color and looks more like the soil he has been accustomed to farm successfully in the East. On this account short-grass land brings the highest price and changes hands oftener, while wire-grass and bunch-grass lands are not so readily sold and bring much less per acre.

Isaiah Bowman

Certain aspects of the evaluation of the idea of regionalism have been indicated, without any pretence of complete coverage. In botany, and zoology, the ecologists were working in a similar direction as indicated in preceding chapters, and geographers were arriving at more adequate regional definitions. Joerg (1936) traced briefly and clearly the history of the problem as related to geography. It was in this connection that he remarked that Isaiah Bowman gave "the first adequate statement of the regional principle in an American work on physiography," in his Forest Geography (1911, pp. 108-110). Although this book centered on forest geography it included important chapters on the prairie plains, the great plains, the Columbia plateaux and other sub-divisions of the Great Basin, and significantly, the book was dedicated to Hilgard.

Conclusion

It is not desirable to fasten artificial division points upon historiography, and in this instance there seems to be a minimum of the artificial in designating the period interrupted by World War I as a highly significant point of culmination of all of the several lines of interest under review. The period between the two World Wars was essentially a new and different generation based upon this preparatory work. It is difficult to evaluate the dry land farming situation as it was embodied in the work of Campbell and Chilcott. Both men shifted positions as the experimental basis of their defense changed, and neither was free from serious weaknesses and inconsistencies. If the emphasis is placed upon the conservative and cautious side of the Campbell dualism, his approach was remarkably sound in relation to the surrounding concepts of agronomy, and particularly of soil physics. If the emphasis is placed upon the boastful egotism of his claims for his system, he cuts a sorry figure. His adverse critics made the most of his weaknesses rather than recognizing his constructive contributions. Among the weak points in his culture system were too frequent tillage, which tended to break down the soil structure; excessive costs, which were entailed by such frequency of tillage; the small scale of operations, which piled up costs in proportion to the total possible income; the corn tradition; the failure to recognize the value of the lister as a tool, and sorghum as a substitute for corn. His important contributions included his emphasis upon timeliness of tillage, which included the necessity of having the means of speedy completion of operations while conditions were absolutely right, his emphasis on weed killing to conserve moisture, his realization that the secret of achieving these objectives lay in more adequate machinery not yet available, although he had not placed his finger exactly on the right spot, which was the farm power question more than the tool itself, however important that was; his insistence upon stabilized production as the key to farm prosperity rather than an artificial price structure subject to attempts at controlling production and marketing.

Chilcott's publications revealed both a dualism and an inconsistency. He was campaigning against the systems of others, and at the same time was building a system of his own; he denied that the folk process had accumulated experiences that possessed much value, yet he pointed out that the application of scientific principles must be left to the acquired experience and good judgment of the individual operator. His first paper of 1907 was mostly negative and in the zeal of his attack on the systems, which meant principally the Campbell system, he committed himself to an untenable position from which it was difficult to retreat. The weak points in his work included the objectives on which he justified his emphasis on rotation; his inadequate recognition of timeliness. His important contribution included emphasis on low cost per unit of production in relation to the going price and yields; large scale operations as the means of bringing down unit costs in terms of overhead and capital investments in equipment; his recognition of the importance of the lister, although belated; a partial appreciation of the rôle of the sorghums; his warning against paternalism and any kind of outside control.

Both Campbell and Chilcott failed to recognize the disc plow as a factor in dry-land farming. About the time of World War I, the disc plow was converted into the one-way disc as a basic tractor tool. The soil mulch problem presents a special case. Both men subscribed to the soil mulch at first, and in this they were in general agreement with the soil physics of the time. Chilcott repudiated the theory in 1915, holding to weed eradication as the only virtue of the soil mulch operation. Some other research workers took the same ground later, but still others found conclusions in favor of the soil mulch in its own right. Baver (1940) seems to defend the soil mulch as a practical fact, irrespective of the differences in the methods of rationalizing the reasons for maintaining it in terms of good soil tilth. It became the fashion to repudiate Campbell, or even to ridicule him, but in order to keep the record straight, it should be pointed out that the theories of his rivals were also largely discarded. It was not a case of one being right and the other being wrong. In the light of subsequent conclusions both were wrong in much of their practice and in most of their theory; and both were right in important matters. Possibly it would be more accurate not to say that either was wrong, but to say that both were superseded by later conclusions, which in turn were again superseded, and all was a part of the continuous process in which their failures, as well as their successes, constituted the foundations upon which further advances were based. Neither man appreciated how much all his work had been anticipated by the folk process among the plains farmers, a fact which the present author has traced elsewhere in his several studies of adaptation in sample Kansas communities.

Chapter Fourteen

SOIL PHYSICS AND TILLAGE

The nineteenth century produced a significant succession of soil physicists, Schübler, Schumacher, Wollny, and King, and it was out of their theoretical work that tillage practices that prevailed in the opening year of the twentieth century were rationalized. By the second quarter of the twentieth century there was a revival of interest and a new cycle of soil physics theory was emerging, but was not yet formed. The capillarity theory of the nineteenth century succession gave way generally to the energy potential theory, although some distinguished soil physicists still defended the former theory, at least in part (Puri, 1939). The soil mulch theory of retarded evaporation was challenged, but not altogether disproved (Baver, 1940). The unsatisfactory state of the science was dramatized by the publication of E. H. Faulkner's Plowman's Folly, (1943) in which he challenged the soil scientists to present a scientific reason for plowing. He maintained that plowing was contrary to the laws of nature and compared so-called scientific agriculture to the sowing of wild oats, and called for a sobering up and a return to first principles. His idea was to work all organic material into the topsoil in imitation of the state of nature. The confusion caused by this book was ample evidence that there was very little in soil physics on which scientists were in agreement The most convincing replies to Faulkner came from those who admitted much of validity in his challenge, but argued that different soils and climates required their own kind of treatment, and that Faulkner's system had no greater general applicability than the one he denounced. In 1947 with some redefinition of terms and objectives in the interest of precision, Faulkner returned to the conflict with a book entitled, **A second look**. He insisted that the heart of both books was to be found in the stress on highly developed biological activity in the soil in relation to plant growth, and that his repudiation of the moldboard plow was based upon the contention that it retarded that essential process. He was still convinced that the primary object of tillage, as thus defined, was most effectively accomplished by mixing organic matter into the surface soil.

The dry-land farming literature of the between wars period emphasized that losses of water from the soil that were subject to control, were those resulting from weed growth. The prime objectives in tillage methods in low-rainfall areas, according to this formula, were storage of all of the moisture that fell, and weed eradication to prevent unnecessary loss of moisture. The maintenance of the soil in good tilth, however, was recognized as essential, irrespective of the theoretical explanations of the function of such soil structure. Possibly the soil scientists of the 1940s might profit by restudying Hilgard's conclusions and reinterpreting them in terms of modern soil knowledge. In the perspective of half a century, Hilgard's basic thinking had gained remarkably in significance, even though much of his technical explanation of reasons had been superseded by subsequent scientific investigations.

As far west as the 97 meridian in Kansas the moldboard plow was the principal tillage tool, with the lister extensively used on upland for corn. West of that line, the lister and its modifications were more extensively used for both soil preparation and planting of corn, sorghums, and wheat. Introduced into eastern Kansas and Nebraska in the late 1870s for corn, the lister became a part of the agricultural scene. The farmers of eastern Kansas and Nebraska, and western Iowa and Missouri invented and developed a new machine, the lister cultivator. Within a decade of the first mention of the lister in Kansas as a corn tool, it was being adapted to the planting of wheat and other small grains; the first Hollinger lister drill being built in 1887, and patented in 1888. The wheat was planted in rows eleven inches apart, the furrows being made by small lister-shaped shovels. The use of the new tool was limited but the important thing was the entry of the new method into the process of adaptation of the winter wheat culture. By the opening of the twentieth century, the regular corn lister was being used extensively, along with the lister-cultivator, in preparation of wheat ground, instead of the moldboard plow, these corn tools being taken over bodily into wheat culture. The listing principle was of advantage because it lowered costs by reducing the number of operations, it got over the ground faster, it ridged the soil against the prevailing winds, and it retarded or prevented blowing of the soil (Malin, Winter Wheat, 1944). During the second decade of the twentieth century the lister drill lost popularity for a time, but in the 1920s it was revived in much its original form, and also in a new form as a deep-furrow and a semi-deep furrow disc drill, with spacing between the rows of fourteen or ten inches. (See Kansas Experiment Station Bulletins 248, (1929); 250, (1930); 262, (1932); 293 (1941); Technical Bulletins 13, (1924); 18, (1925); USDA Farmers' Bulletin 1917, (1942, 1944). Photographs of the machines and of fields treated by these machines appear in most of these bulletins). Salmon reported, in 1924, that "beyond a reasonable doubt" planting small grain in furrows was an advantage. In 1929 the disc type furrow drill was preferred on account of lighter draft, and the ability to cut through trash, but the principles were the same (Bulletin 248). For sorghum as a row crop a modified lister tillage, in the extreme western part of Kansas, was recommended (1933); blank listing in the fall, and nosing out the furrows and planting in the spring; or fall plowing, with the spring planting done by shallow listing or by a planter equipped with a lister or furrow opener.

Disc machinery became conspicuous in American agriculture during the 1880s in the form of disc harrows and disc cultivators, and after 1895 included the disc plow — sometimes called rotary machinery. In the low rainfall areas the disc forms were particularly attractive. Disc plows, to replace moldboard plows, became common, and about the time of World War I the disc one-way tool was developed for tractor farming and became very popular. It did not cut as deeply as the disc plow, but cut much deeper than the disc harrow. Disc machinery did not turn the furrow slice over completely and covered the trash only partially leaving the surface relatively rough to resist blowing. The disc one-way, and the lister, occupied largely the tillage field in the country west of 98°, although many still preferred the plow. Much depended also upon the type of soil in choosing from the several possibilities of available machines. During the drouth decade of the 1930s the disc one-way incurred some unpopularity because there was a tendency in some quarters to make it the scapegoat for soil blowing. It survived, however, and with a seed box attachment was used to some extent for planting grain as well as for tillage purposes, making it a single all-purpose wheat culture tool.

One of the most important experiment station studies was reported in Kansas Bulletin 273 (1936) on the relation of the amount of soil moisture at planting time in the fall, and the probabilities of a wheat crop the next summer irrespective of winter and spring rainfall. A preliminary report was made in 1930, but the full report of 1936 provided for the first time a quantitative basis for approaching this problem. Moisture, at plowing time, to a depth of three feet or more virtually assured a crop, and a lesser depth of moisture meant an increase of hazard and below a determinable point it was useless to plant. Of course, differences in soil characteristics and topography had to be recognized.

The drouth decade of the 1930s, combined with the depression after 1932, put all the accumulated knowledge of the grassland to the test. The drouth was so extreme and prolonged that the farmers of the tall-grass country as well as those of the transition mixed-grass region, suffered severely and were forced to turn in a substantial degree to the unfamiliar methods of the low-rainfall plains, and in many respects they suffered worse than the farmers of the plains proper on account of unpreparedness. The custom of burning the wheat stubble, in the west, had deprived much of the land of necessary organic matter. When crop failures from drouth intervened, little or no root influences on soil structure were present as well as little or no addition of organic matter. The structure of the dry soil deteriorated and control of blowing was all but impossible. Drastic mechanical measures had to be resorted to, especially evident in development of the basin tools, the most conspicuous of which were various forms of damming listers designed to hold all moisture that fell. Lister drills, used in the most exaggerated form, were attempts to prevent blowing until the crop had provided a ground cover. When the drouth prevented growth sufficient to hold the soil, it was necessary to run lister furrows through the wheat fields, at intervals of 20-30 feet, to stop the drifting soil that otherwise would destroy the whole field. The emergency called out in a conspicuous manner the inventive genius of the plains farmer, and the Dodge City Fair, of 1936, made a feature of the exhibits of plains machinery developed by individuals, by farm equipment firms, and by the state experiment station. Of course, all such unusual mechanical devices for soil control were emergency makeshifts. The only remedy was the restoration of moisture, and of organic matter to the soil, when the weather changed to a wet phase.

Summer fallow studies growing out of long-term experimental work, climaxed by the drouth period, were reported in Kansas Experiment Station Bulletin 293 (1941). East of $98° 30'$, summer fallow was not recommended; between that line and $100°$, one fallow year followed by three crops was recommended; between $100°$ and $101°$, one fallow year and two crops; and between $101°$ and $102°$, the west boundary line of the state, the fallow and crops years alternated. The conclusion was reached also that fallowing did not reduce fertility. Fallowing was recommended, however, within the limits indicated, only on heavy soils, not on sand soils or on hill topography. Here is one of many conspicuous instances of where the Triple A restrictions on production and enforced fallowing ran contrary to proven agricultural practices.

With the turn of the weather, in the 1940s, to a wet phase, there was a tendency to revert to less drastic methods of soil control. This was encouraged by war demands for food and the relaxation of acreage allotments and penalities, all of which added up to a widespread abandonment of summer fallowing and a disuse of the basin tools and other more extreme methods, except in the critical soil control areas. In respect to the disc type of furrow openers for grain drills, the demand shifted to the semi-deep furrow type. The lister drill proper was not only in demand from manufacturers specializing in plains tools, but the demand was reported to the present writer to be several times greater than the supply available under war restrictions on materials. The amazing capacity of the plains country to produce, after its drouth experience of the 1930s, is demonstrated by the sixth successive large crop in 1947 throughout a large part of the area formerly called the dust bowl, a continuous crop record never before equalled in the history of the plains.

Out of the decade of the 1930s had come several new tools or a new emphasis on old ones. In the latter class are the springtooth cultivators and several variations of the sweep, or duckfoot-type shovels. Among the innovations, the most unusual was the rod weeder, a revolving rod that turned just beneath the surface. As a means of reducing machinery investment for tractor farming, most of the farm equipment companies developed a single all-purpose tool, with quick-change attachments, that made possible all year-round tillage and planting operations.

An apparent new departure in soil tillage was emphasized in the 1930s by the USDA and the agricultural experiment stations — that was called stubble-mulching. The method was not to plow but to use sweeps, either knives, or shovels similar to the duckfoot cultivator, which ran under the surface loosening the soil but not turning it over or disturbing materially the stubble on top. The arguments were that the mulch would prevent blowing, reduce evaporation, and return organic matter effectively to the soil. The emphasis on reduction of evaporation revived the question of soil mulch, but in a somewhat different form.

Bonner (1945) called attention to the work of Garland D. Harmon, in Georgia, in the 1850s, who held that plowing was at best a necessary evil and advocated surface cultivation by sweeps or scrapers. In his essay on soils, in 1891, Shaler condemned the plow. Whitney (1895) had made similar criticisms. Wheat raising possibilities in southwestern Kansas were discussed, in 1893, before the Kansas State Board of Agriculture, and the argument was made that "we should have some instrument by which we can loosen up the ground deep, but not turn the stubble under." Patents were issued for such tools about the same time, and at Chapman, Kansas, in 1897, an "underground cultivator" was placed on the market (Malin, Winter Wheat, 1944). Thus when Faulkner's controversial book, Plowman's Folly, was published in 1943 some of the ideas in it were not as new as he seemed to think. Tools for shredding organic matter and working it into the top soil were not available for Faulkner's purpose and he used the disc harrow, but new tools were described by Haystead, in Fortune (January, 1945). For stubble-mulching as it was being advocated in the central plains in 1942, and later by the USDA and the experiment stations, the farm equipment companies had provided the tools. USDA Farmers' Bulletin 1917, issued in December, 1942, and reissued in revised form in August, 1944, described the method and gave photographic illustrations of the tools and the work being done in the field. Drilling grain in the stubble-mulch presented a particular problem and placed a new emphasis on the lister drill in some form which would be able to deposit the seed beneath the mulch.

With another return of severe and prolonged drouth conditions, crop failures, and resort to extreme summer fallowing, stubble will not be available at all, or only in limited quantities, for stubble-mulching and again reliance will have to be placed upon mechanical means of soil control. Just what machines will be used must await the event, but according to the principle of probabilities, the lister and disc types of machines, now basic to so many of the soil control practices of the area, will provide the foundations. For instance, already at least two farm equipment companies were offering (1945) rotary listers; one without a point, for soft ground; the other a combination of a lister point with rotary (disc) moldboards. It is not a question of final rôle of any machine, but one of a continuous process of adaptation.

Several factors were involved in these different types of planting machines. The ridging of the soil crosswise to the wind was most obvious. A second factor of regional significance was the spacing of the rows in relation to moisture, man's approach to nature's method of wide spacing of vegetation in dry regions. For natural vegetation of the grassland, with sufficient moisture, the grass formed a sod; with less moisture, it assumed a partial bunch grass tendency; and with still less moisture, but within the range of the species, a pronounced bunch formation, leaving much of the space unoccupied. In nature, the dominance of a particular combination of plants was established by natural competition, the spaces between remaining vacant because the possible competitors were too weak to survive in the spaces. In man's cultivated fields the same principles operated, only the dominance of the desired crop over the weeds needed man's encouragement. In climates of greater rainfall, the plants which were closer together shaded the competitors and the planted crop maintained its dominance. In drier climates, wider spacing was required between the plants of the crop to insure their vigor, but not so wide as to give the weeds too great an opportunity to occupy the spaces. Where cultivation was practiced between the rows, it was primarily to aid the crop against this weed competition for moisture. In the high-rainfall areas, a six inch spacing between the rows was often used for small grains; on the low-rainfall prairie-plains the standard width was eight inches. The commercial drills of the semi-deep furrow disc-type used ten inch spacing, center to center, and the deep-furrow disc-type used twelve or fourteen inches. Both of these deep-furrow types were often called lister-type drills, but the true lister drills used a lister-shaped furrow opener and usually were spaced fourteen inches apart, or twenty inches from center to center. For row crops like sorghum, the normal forty inch spacing was sometimes increased to eighty or one hundred and twenty, or in other words, only every second or third row was planted. The width between the rows for either kind of crop was not arbitrary, but was a matter of adjustment, by the farmer, of the crop spacing to the environment in terms of rainfall, temperature, and soil characteristics. During periods of extreme and prolonged drouth, it was desirable for the wider spacing to be pushed east toward the tall grass prairie, and conversely, during wet periods the desirability of wide spacing was pushed well west of the 100 meridian.

In the harvesting phase of wheat production, the header, not the binder, was the plains tool of regional significance. It proved its place in central Kansas in the 1870s and became the basic machine upon which experimentation was practiced in attempting to create the header-thresher combine in the 1880s. The successful tool of this type was not available, except the California type, until after World War I and the advent of tractor power.

Chapter Fifteen

WEBB AND REGIONALISM

The publication, in 1931, of Walter Prescott Webb's (1888-   ) The Great Plains, was a landmark in the regional approach by historians to the Trans-Mississippi West, and it synchronized closely with the rising tide of criticism of the Turner frontier school of American history. Webb explained that he did not use the term Great Plains in the sense usually employed by geographers and historians. It was an area described in terms of topography (level) vegetation (treeless), and rainfall (sub-humid), but all three did not apply to the whole area embraced in the great plains environment. The eastern part was treeless and level; the western part (interior basin) was treeless and arid; the middle part was the great plains proper, because it possessed all three characteristics. He placed the dividing line between east and west at $98°$, and contended that early in the nineteenth century, upon coming out of the timber area in western Missouri, the frontier halted for about a generation. The line of this halt was near the dividing line, where it was necessary to await a preparatory period before a further advance took place. West of this line the environment differed so materially as to impose a new formula of living, the line constituting an institutional fault similar to a geological fault. The east failed to recognize the difference and to modify legislation accordingly or to permit western men to follow the dictates of the new region. The differences between the East and the West were defined further upon the basis of climatic characteristics, and geographical distribution of grasses and animals. The physiographical description was based primarily upon Johnson's monograph on the high plains, without a clear distinction of his differentiation between the high plains remnant and the idealized great plains.

In contrasting the pioneering process, Webb emphasized the dependence of forest man upon available wood, water, and natural water transportation, while in the treeless West the pioneer found it necessary to buy substitutes with money and transport them to the West. A change in weapons and warfare was given particular notice The availability of these substitutes depended upon the Industrial Revolution, with its mass production of standardized goods at prices within the reach of the pioneer. According to this line of analysis, most of the necessary items did not become available until the decade of the 1870s, the renewed settling-in movement gained initial momentum slowly and then closed in the 1880s with a rush. The cattle business was emphasized as the principal industry of the great plains and the leading example of regional influence on institutions. Irrigation, its limitations, and the reconstruction of institutions under its influence, was another major theme, while dry-land farming received minor attention. In closing the account, the geographical influence was presented in literature, in song, and in the intangible factors of regional psychology.

In 1937 the Social Science Research Council undertook a series of critiques of significant works in the social sciences. Webb's The Great Plains was chosen in the history field and was assigned to Fred Shannon. His appraisal was presented and a conference with a committee of nine was held at Sky Top, Pennsylvania, in September, 1939. It is not the purpose of this chapter to summarize in detail Shannon's appraisal or the report of the conference, as they must be read in full in order to be adequately understood. Shannon seemed to challenge Webb's basic assumption of definition, of regional uniqueness, of the importance of differences, and of his particular form of geographical determinism. Other subjects of adverse criticism were Webb's handling of reasons for delayed settlement of the plains, the revolver, the origins of the cattle business, barbed wire, windmills, artesian wells, irrigation, dry-land farming, etc. Webb refused to accept Shannon's work as a fair and competent appraisal, and declined to make a formal reply. He did attend the Sky Top conference and made a statement there of his position. Omitted from the membership in the conference committee were persons competent in plant or animal ecology, general physiology, soils, microbiology, and climatology. Colby, a geographer, participated, but his specialty lay elsewhere. Neither Webb nor Shannon, nor members of the committee, revealed any specialized knowledge of these fields. Colby made suggestions relative to geographical matters, especially the soil question, but did not arouse any significant response from the other members on these matters. Wirth, a sociologist, made some important points on general methodology and interpretation. As a sort of footnote to the Sky Top conference, attention should be directed to Shannon's book, The Farmers' Last Frontier (1945, Chapters 1-2), in which he continued in his denial of the regional distinctions of Webb's book.

In 1942 the Rockefeller Foundation sponsored two conferences on the great plains; one at New York, in April, and the other at Lincoln, Nebraska, in June. At these meetings the membership represented history, literature, the press; and at the Lincoln meeting, anthropology was added. Among those present there were no significant conflicts of opinion in evidence. All appeared to be essentially orthodox in accepting the Webb type of regionalism. Peculiarly enough the Lincoln session was held at the University of Nebraska, the institution that had given Bessey, Clements, Weaver, and Shantz their opportunity as ecologists, and who in turn had given the University of Nebraska world-wide fame, but at this conference there was no ecologist present, no physiologist, no soil scientist, no climatologist, no geographer.

At the opening of the session at the Sky Top conference, Schlesinger as chairman stated, and later restated, that "no difference of opinion exists between Mr. Webb and Mr. Shannon regarding the principal thesis of the book, namely, that human institutions and ways of life changed upon reaching the Great Plains...;" the differences lay in Webb's vagueness of definition of the great plains and his inconsistency in application (Pp. 141,194). As the present author reads the record, and the same impression was evident in the minds of some of the participants, the difference included the problem of regionalism. Shannon's own remarks seem to repudiate regionalism by his insistence upon his view that the pioneering process was one of continuous and gradual adjustment from the eastern coast to the desert. At any rate he made no indication of what he considered an adequate definition of great plains regionalism, and did not explain whether or not he thought any definition of regionalism in general, or this one in particular, was even possible. Wirth, the sociologist, thought the concept of regionalism should be explored, and asked Colby, the geographer, to discuss it. Colby concluded that the use of the term great plains "should not be an issue" (p. 147). Thus Webb became the "whipping boy" between the different views in which his use of regionalism, or the term great plains, was defended and condemned without reference to the basic issues of whether or not any regionalism was valid, and if so, whether or not an acceptable single definition was possible upon which all or even a majority could agree. Having failed upon that question, the further deliberations of the conference were largely futile.

As the present author understands Bridgman's position on the problems of methodology in physics, most of the controversies arise out of failure to define adequately the problem to be studied, and then to prescribe a set of procedures by which impersonal and objective conclusions could be reached. The Sky Top conference is a good example of the failure to do everything that Bridgman's dictum would demand before a conference could even begin. No two participants appear to have had the same idea of what Webb intended, or of what Shannon intended, or of the purpose of holding a conference at all.

An idealized approach to such a problem might start with a survey of the literature of physiography, of climatology, of physiology, of genetics, of plant ecology, of animal ecology, of soils, of soil microbiology, of agronomy, of the geographers' attempts at regional definition, and the question raised by some, whether or not regionalism possesses any validity.

Powell (1895) declared that he repudiated the basin concept of physiographic regions in favor of the slope concept. The basin concept, as applied to the Mississippi, would exclude the rivers of southwestern Georgia, southern Alabama, and southwestern Mississippi, because they drain directly into the Gulf of Mexico. They are, however, a part of the Gulf slope. Likewise, in the West the divide between the Red river and the rivers of Texas would, for the same reason, exclude most of Texas and New Mexico from the basin. They, also, were included within the Gulf slope. If the subject of study was mountain formations, as done by Gilpin and Shaler, then the divides between drainage basins would not separate the physiographic region being studied, but would become the central feature of the area. Slopes diverging from the crest of the mountain peaks and divides run to sea level and below. At what point between these extremes of elevation would the mountain formation become something else? Gilpin and Shaler both discussed basins, slopes, plateaux, and mountain formations without quibbling over the obvious fact that they overlapped and that the same area was discussed as a part of two or more different regions. It was not a question of one treatment being right and the other being wrong; they merely emphasized different aspects of the same body of facts which possessed a particular interest when related in different patterns, both of which were valid. Some purists, who insist upon carrying the logic of verbalism to bitter-end extremes, arrive at a destination of nihilistic futility by denying that there can be any valid physiographic regionalism.

In the field of climatology, the traditional approach was to define characteristics of climatic regions on the basis of mean averages of rainfall, temperature, and other properties. This approach emphasized uniformities. The more recent studies, which were directed at the problems of the relations of climate to living things, place the emphasis upon the extremes and frequency of fluctuations (cf. Chapter 4). Of particular interest to the present considerations are the systems of Russell (1934), and Kendall (1935), although those of Crowe (1933), Lackey (1937), and Thornthwaite (1941), may be fully as significant.

The general physiologist is concerned with the problem of protoplasm, irrespective of whether it is organized in the form of cells in plants, or in animals, and in its behavior as living matter. The phase of the protoplasm problem that bears most directly upon the historian's problem of regionalism, is to follow the physiologist's experiments under different conditions of moisture, heat, light, and photoperiodism, and the extent of his ability to determine the limits of tolerance for fluctuations as respects frequency and extremes. The literature of the physiologist demands attention so long as biologists are of the opinion that the secret of the relations of plants and animals to cold, drouth, or light is determined in part or largely by the properties of protoplasm.

The geneticist is concerned with problems of inheritance and the relations of speciation to environment. The theories have been discussed (Chapter 11) and make clear the bearing of such problems on the possible significance of geographical environment upon mutations, and the problem of isolation in the establishment of species and geographical races.

The literature of plant and animal ecology (Chs. 2, 3, 10) deals with communities and their relation to the occupied area, the problems of distribution, the relation of species and individuals to each other, and to all the factors of environment. If geographical regionalism is admitted to possess any validity these findings of ecology become essential elements in any valid determinations of boundaries.

The literature of soils is less satisfactory in giving aid at present, but that is not because of any lack of importance. When soil science is more developed, it should possess a more important rôle, especially if some current assumptions are correct that microorganisms exhibit a regional distribution as definite as the higher forms of plant and animal life.

The literature of the agronomist deals with the problem of agricultural crops. Whittlesey's (1936) maps of agricultural regions did not coincide altogether with the assumptions that might be derived from native plant regions, and the same is true of some type-of-farming area studies and mapping. There are some broad similarities, but with conspicuous differences and peculiarities of detail. The distribution of crops, geographically, under controlled conditions of cultivation necessarily presents a changing map of regionalism, especially under the influence of plant and animal breeding and introduction of species from other areas.

When the geographer and the historian have canvassed all of these fields of specialized literature, as they bear upon the problems of regionalism, they are in a position to attempt synthesis, employing the whole circle of facts or independent variables. It is only within such a comprehensive framework that they have a right to discuss definition of regionalism and the validity of employing the regional approach in their disciplines. Obviously, the particular regional definition that is framed will depend upon the purpose for which it is to be used, and more than one good definition is possible, each emphasizing valid aspects of the area of the earth's surface being studied as the habitat of man. And as the first organizing principle of history is time, the historian must recognize that factor in his treatment of man and region. If all parties to the Sky Top conference had possessed such a background for their consideration of Webb's book the outcome should have been different. A large part of this controversy turned on quibbles over verbalisms, much as had occurred in the Turner controversies, rather than upon constructive discussions of the obvious intent of the author.

On some points, Shannon was disconcertingly candid. He admitted (p. 5) that he had never read Webb's book through until the previous year, presumably about the time he accepted the assignment to write the Appraisal. Furthermore, he confessed that "I was not employed as an expert on the history of the West..." (p. 194). It is not in the least remarkable that Webb refused to recognize Shannon's critique as a competent Appraisal. In both the Sky Top conference and the Rockefeller Foundation Great Plains conference, the backlog of available scientific information and basic thinking which might be given a significant application to the study of history was not used. The members of the conferences did not appear to be acquainted with it, and the conferences were too narrowly organized to bring that kind of material into the discussions by specialists in these fields. Webb had not utilized all the backlog of available material as of 1930 when he was writing his book, but in a pioneer work, in which he was feeling his way, it would have been most unusual had he, or anyone, done so. The more important point is that in relation to the rank and file of the historical profession, or to the recognized leaders in the profession, he was much in advance, and even by 1942 in the great plains conferences, there was little evidence that other historians had gone beyond him. The really disturbing aspect of the matter was that there was little advance by either Webb or the others beyond the position of 1930, a sort of great plains orthodoxy seemed to have crystallized, and in the meantime the backlog of basic scientific thinking had been accumulating at an unusual rate. In these conferences there is no evidence that there was any recognition of the essential incompatibility of the Turner frontier and sectional tradition with the Webb regional approach. Had these conferences examined adequately the bases of regionalism as a historical method, the contrast should have become clear, as well as the more important conclusion that the study of history by this method brings out relationships in a significant manner that cannot be brought out by any other method. But, emphatically, no one method is so complete in itself that it should be adopted to the exclusion of all others.

Webb, Shannon, and most of the members of the Sky Top conference appear to have been thinking of regionalism in terms of geographical determinism. Wirth (p. 181) challenged the principle of determinism, calling attention to Lucien Febvre's theory of possibilism, and showed how Webb's treatment was inadequate to permit its application. He should have been more inclusive by applying his remarks to Shannon and to the other members of the conference. The omission, by Webb and Shannon, of the soil question, was pointed out by Colby as one of the most serious defects in the work of both men. Webb was excused because the works of Marbut and Kellogg had appeared after his book was published. Webb might well be excused for the omission, but not on that ground. Hilgard had distinguished between eastern and western soils some forty years earlier. Marbut had presented his treatment of western soils, together with a map as early as 1923, and had published his views on soil classification in 1922 and 1928. Scarcely any historians were dealing with the new soil literature in 1931 or in 1939. Shannon made an effort to remedy the defect in his book of 1945, devoting a part of the first chapter to soils. In this, as in the case of others doing pioneering work, Shannon's handling of the new field was only partially satisfactory. He was not acquainted with soil literature as a whole, and depended for guidance too largely on the official version of the United States department of agriculture according to Marbut and Kellogg.

The idea of possibilism might be applied profitably to the issue of adaptations to regionalism, and contribute toward clarification of some of the matters at issue in the Sky Top conference — railroads, fencing, windmills, revolvers, and cattle industry — as well as others of a similar sort — housing and fuel, tillage and crops. In suggesting this approach to a better understanding there is no intent to make commitments to any philosophical system — only to state the problems in a form that is adaptable to objective treatment. The idea of possibilism might be considered in two different applications: The same technique applied to more than one kind of environment where it possessed a different cultural value in each; and more than one technique applicable in the same environment. In other words, it is often if not always probable that there is more than one way of doing things to achieve the same or similar ends. In the first instance it was possible for a new technique to be invented, developed, or utilized in several different environments, and in each serve a somewhat different purpose, sometimes highly significant and important, but not always, and yet combine into the cultural complex in a manner that would not serve an equal or equivalent function. It is those techniques that possess an unusual rôle that become both significant and important in imparting a regional peculiarity to the manner in which they are combined in the cultural complex. To the extent that more than one possibility exists the element of determinism is eliminated. The particular method in vogue may be the outcome of tradition in which the trend is explainable only on the basis of historical development, but in other instances, a choice from the list of possible methods may be exercised consciously by individuals, groups, or communities. In either case, the understanding of the fact of alternatives discredits not only geographical determinism, but also discredits attempts to rationalize the imposition by compulsion of one particular formula chosen by arbitrary decision of authoritarian procedures. In modern history the two points of view have often been placed in sharp contrast; the English method of "muddling through" as it has often been called, permitted a high degree of flexibility in meeting difficult situations, in contrast with authoritarian methods so often imposed upon the continent of Europe.

A windmill might be used in any country, but in a country in which wind of any substantial velocity was a rarity it would never be of much importance. In the central North American grassland, however, where absence of wind was the exception, the windmill was both significant and important. Within a few months after the first settlers arrived and set up newspapers in Kansas, they began to discuss the wind and the opportunity afforded to put it to work cheaply, in a country short of fuel, for power purposes. It is beside the point to ignore this fact of immediate recognition of both importance and significance, and to argue merely that windmills did not come in quantity until the 1880s. To place windmills within the financial reach of the average pioneer farmer awaited cheap steel, mass production methods, as well as mechanical refinements of machine design. It is essential to distinguish the existence of a certain device or invention as fact, and the difference in circumstances of utilization and diversity of uses as fact.

The importance and significance of communication were quite properly emphasized by Webb in connection with the grassland culture. He was concerned most conspicuously with horse culture in connection with the primitive Indian occupation, the pioneering period of the white invasion, and the organization of the cattle industry. It was a man-on-a-horse relationship. The railroad came in for a share of attention, but Shannon challenged its regional significance. The statement of the problem might be broadened with profit to the discussion.

In terms of so-called transcontinental communications, from the Mississippi valley or country eastward to the Pacific coast, 1849-1869, the most important medium and route, in terms of volume of business, was the water route around Cape Horn; secondly, the interrupted water route by way of the Isthmus of Panama; and of least importance, the animal-powered land routes. Of the animal-powered forms of communications, if the emigrant wagon and its household frieght are considered, the wagon was the most important, with the stage coach second. In terms of strictly commercial business and operations as a common carrier, the stage coach must be given first place as the carrier of passengers, mail, baggage, express, and other light freight; the freight wagon makes a poor second, and the pony express rates mostly for its spectacular value. By such tests, how were animal-powered communications a medium of major importance? They were vital, of course, but only under a particular set of circumstances, as a means of communications from water-served bases (later rail served bases) to the interior where there was no water (or later rail) competition. Animal-powered communications were too expensive to be used except as a last resort, and the extent to which the stage coach and the wagon train came to symbolize the West is a conclusive demonstration of the fact that, except the Missouri river, there were virtually no water communications available, in the interior, west of the Mississippi river.

The heart of the grassland communications problem, in a regional sense, was more largely centered on the transportation of bulk commodities in and out of the region than in the matter of personal movement, but the modes of transportation that combined both were still more important. If a single symbol of the grassland was to be chosen it should be the wheel; in the case of the freight wagon, the stage coach, the farm wagon, and buggy, drawn by animal power; and the railroad, automobile, truck, and tractor, driven by mechanical power. Thus the wheel acquired a peculiar regional significance because of the virtual absence of water communications, because of the importance of bulk commodities in its essentially commercial as distinguished from its self-sufficiency economy, because of the extent of regional interdependence, and because of the possibilities of application of mechanical power. All together these constitute an aspect of culture that is peculiarly regional. Water was the primary dependence of man for communications, on a world basis, until the mid-nineteenth century when the advent of mechanically-powered wheel communications superseded it for most inland services. In the second quarter of the twentieth century, both water and wheels are being superseded, in part, by air communications. The occupation of the grasslands of the world by modern civilization during the late nineteenth century was associated particularly with mechanically-powered wheel communications, and the arctic is being brought into the orbit of modern civilization by air communications. In North America the wheel pointed the advance westward, while air power points it northward.

In the case of fuel for the grassland, it was early realized that coal was essential. The preparatory period before the actual invasion began was one in which coal was displacing wood, in the forest country, in industry, in transportation, and in city heating, and developed there the coal burning stoves and locomotives. The attitude of the grassland man toward coal was quite different from that of the forest man, who had used both coal and wood and to a substantial degree, outside of cities, still might make a choice.

The housing problem presented much the same situation. The use of sawed lumber, shipped from some distance, was quite general in the forest region, and in the grassland the same tradition prevailed, the difference being the shipping from longer distances and the consciousness that there was limited choice. In most respects the exceptions only tend to emphasize the prevalence of the rule. The sod house frontier myth (Dick, 1937, and Hafen and Rister, 1941) must be rejected because there was no frontier line beyond which sod houses prevailed, and the extent to which sod houses were used at any time or place was limited. After World War I, there was an interest in reviving in America the pisé, or rammed earth, method of utilizing earth as building material on the spot, and during the depression stabilized earth houses were built, using either oil elumsion or cement as the stabilizer. All three of these methods of constructing earth houses were used only in a limited degree. With few exceptions, the prevailing house type would be more accurately described as the sawed house, rather than the sod house.

In restating the fencing problem much that has been said of the problem of fuel and housing applies. New research by Danhof (1944), and Hayter (1939), superseded or supplemented much of what Webb had to say. Danhof pointed to the fencing problem as insistent in some places, during the late colonial era, in the forest region. Experiments with substitutes occurred in several environments and with different materials. Wire was an early competitor, even barbed-wire was not at first a satisfactory adjustment, and Danhof concluded that Bessemer steel was the technological essential in providing a cheap, tough wire for both barbed and woven fencing, and that there were no regional implications in its origins. This was correct. The regional significance of barbed wire lay in its use, and the manner in which it fitted into the cultural pattern of the grassland economy.

The point made by Danhof, relative to Bessemer steel and fencing, is the same one the present author has made with a broader application. There could be no true machine age until steel had displaced iron. In other words it was not sufficient for iron to compete with wood, because wood was superior for so many uses. It required cheap steel to effect the real transition in substantial degree from wood to metal.

One word more, however, about the origin of barbed wire. Like so many innovations claimed by America, and by the West, the first barbed wire patents were probably not even American. According to A. M. Tanner (1892, and 1893, and earlier), the first barbed wire patent was that of Leonce Eugene Grassin-Baledans, in France, July 7, 1860, and the second was that of Louis Francoes Jannin, another Frenchman, in 1865, both of which antedated the first American patent, that of Hunt and Smith, in 1867. The third French patent was that of Gilbert Gavillard, August 27, 1867, which was similar to the second American patent, that of Michael Kelly, February 11, 1868.

The problem of weapons and the grassland would profit by a restatement. Webb's attention given to the Colt revolver was a significant attempt to study historically the impact of technology upon society. The particular aspect of that impact which he chose to examine was the interrelation of the revolver and the great plains, but in so delimiting his subject, he was not claiming that this was the only impact this weapon had made upon society, nor was he denying that the use made of it by the city under-world was not important, or was less important. It may appear pretentiously pedantic to elaborate upon such distinctions in defining the problem, but as the present author reads the record, most of the argument that has been aroused by Webb's revolver thesis has arisen out of disagreement upon the scope and limits of the problem.

In his book, The Sharps Rifle (1943), Smith chose three firearms as the significant ones in American history, because of major development in design and relation to history; the Kentucky rifle, the Colt revolver and the Sharps rifle. Webb chose to deal with the second of these in contrast with the first, and in relation to Indians, horse culture, and open country. Incidentally, the research of Haven and Belden (1940) superseded much of Webb's historical details, identifying the Walker-Colt episode as 1846-1847 and delivery of the model during 1847, and they designate the period 1846-1860 as that of the establishment of the supremacy of the Colt revolver.

Contemporary recognition of the problems as just defined is available, the present author having collected several valuable items as a by-product of other research. Boynton and Mason (1855) visited Kansas during the summer of 1854 in the interests of immigration promotion and reported on their experience at Fort Riley:

> We were surprised, incredulous, almost offended, when a young officer .... deliberately asserted, that our mounted men, though armed with revolvers, were in general not a match in close combat, for the mounted Indians, with their bows and arrows. But his explanations [of Indian methods of warfare] were satisfactory, and I shall henceforth regard these wild warriors as a formidable foe, even for those who are armed with the most effective weapons of modern times.

The army officer's explanation emphasized that few of the dragoons were trained horsemen, and that the horses were raw recruits as well, while the Indians and their horses were perfectly coordinated leaving both hands free, the arrows were effective at thirty yards, the Indians rode outside their horses, shooting from under the neck, they were almost invisible, and by circling tactics, yelling and frightening cavalry horses, rendered "any certain aim with the revolver impossible, while his arrows are discharged at horse and man more rapidly than even a revolver can be fired."

As bearing upon the Webb-Shannon differences, several points are not settled by this account. The Colt repeating rifle or the Sharps breech-loading rifle was not considered. Apparently, neither was in general use in 1854. Of course, the Spencer, Henry, and Ballard repeating rifles had not yet been invented. The aim of the revolver was bad, but there was no comparison as to how much worse other weapons of 1854 would have been in fighting the same Indians. The implication was that the rifle was not the equal of the revolver as of 1854. Furthermore, there was no commitment on how effective an experienced frontiersman would have been with the same equipment. Some positive conclusions seem warranted. The revolver was recognized as one of the most effective, or as the most effective, weapon of modern times as of 1854. In this account, the revolver had displaced the rifle. The dragoons, as fighting men, were not the equals of the Indians, even with weapon advantage. The Indians did fight in relatively close combat, within thirty yards. The rapidity of arrow shots was greater than the revolver, the arrows were effective, and in horsemanship the Indian was superior to the dragoon. This episode emphasizes the importance of exact statement of the questions at issue. The exact date is essential because of the rapidity of change in the weapon situation. Much of the significance of the Boynton-Mason account turns on the issue of men rather than on the issue of weapons.

The Sharps rifle was based on the Hall rifle, its inventor, Christian Sharps, received his basic patent in 1848 and a second patent in 1853. Trained in the Harpers Ferry arsenal in the technology of interchangeable parts, the design of his mechanism reflected that fact. Equipped with a sliding breech-block and designed for handling self-contained paper cartridges, it proved the most successful rifle of the decade of the 1850s. A Kansas newspaper item of February 1856, probably a disguised advertisement, claimed ten shots per minute, and said that the first lot had been sent by the federal government to the troops on the Texas frontier. A story in a second paper said that "the small carbine [is] now in use by the U. S. Mounted men." In testifying before a federal grand jury in Kansas in 1856, Dr. A. J. Francis said that arrangements had been made to place in the hands of "every reliable Free-State man a Sharps rifle and a brace of Colt's revolvers." In criticism of the policies followed in the Territory of Kansas a resident declared that "One regiment of mounted men, with Sharps rifles, stationed in Kansas, would have kept the peace without any difficulty ...."

The Spencer repeating rifle was based on the Sharps, the inventor had been employed in the Sharps plant, and he had used as many Sharps parts as possible in his new gun. Thus the influence of the Sharps rifle carried over into the repeater age. In 1865 the Ballard repeating rifle was attracting attention, one instance being given from central Kansas, that one man, armed with a Ballard rifle and plenty of ammunition had defended himself, successfully, for three hours against a party of twenty Cheyenne and Sioux. During the same year a newspaper article claimed the Ballard would fire fifteen shots per minute, was "particularly adapted to the Plains" and quoted General Conner as saying "'I consider them the best arm in the world for Indian fighting.'" Later references cited the virtues of the Henry repeating rifle, and still later the famous Winchester '73.

It is clear that these contemporaries were thinking of the weapons, among other purposes, in connection with fighting Indians, but the significance lies more largely on a broader basis. Whether used for Indians, for border warfare, or for hunting, these weapons provided adaptability to the requirement of rapid movement, horse culture, and open country, where firepower was determined by flexibility and rapidity of fire rather than deadrest accuracy.

The range cattle industry was one of the subjects of difference between Webb and Shannon and it turned on two issues; a matter of verbalism in the wording, and a failure of both to define altogether clearly the problem. The second is the really important one. A range cattle industry in any proper sense was impossible in the East because it was a forest region, or stated in terms of regional deficiency, a grassless region. With a very few exceptions of limited areas, the maintenance of a grass cover for grazing purposes in that region was done artificially. In a natural grassland, grass was the native vegetation and did not have to be maintained by artificial methods or the cultivation of tame grasses. That was one of the most difficult lessons that forest man had to learn on entering the grassland (Malin, "Bluestem Pastures", 1942, p. 23-26; Winter Wheat, 1944, 33-34, 80-82; Geog. Rev., 37 (1947) 241-250. The range livestock industry of the grassland was founded on relatively unlimited quantities of permanent, native grass. The grass was what made it a permanent industry, rather than a passing phase of a process of settlement as was true of the limited pastoral activities of the East. Native grass, and not chronological priority or the method of organization, was the determinant (within the meaning of the law of minimum) which made the range livestock industry (not just cattle) unique within the experience of Anglo-Americans. In 1857 William Gilpin had pointed out this fundamental distinction as related to native grass and livestock "spontaneously supported by nature as is the fish of the sea."

The experience of T. C. Henry, one time resident of New York state, Alabama, and Kansas, provided a clear cut example of contemporary appreciation of these basic facts as of the 1860s and 1870s. Henry was born and reared in New York, but went to Alamaba after the Civil War to make his fortune by raising cotton with freedman labor. He decided in, 1867, that he could not prosper on seven cent cotton that cost twenty cents, and transferred his operations to Abilene, Kansas. Speaking before the Dickinson county fair in 1870, he described the process of making a farm in the East; cutting the forest to let in light, digging ditches to provide drainage "in order that the earth might bring forth grass," and the constant renewal of grass for pasture and meadow. As to the South, he pointed out that "owing to the fact, startling in its importance, that no valuable variety of grass has ever been grown there," even more than to the political and social factors, was to be ascribed "the present prostration and comparative poverty of those states," and that so long as nothing adequate to remedy this defective feature was supplied, not even the corrections and reforms of the war could secure to them the degree of prosperity and welfare that was afforded the states of the north where grass could be cultivated successfully as a part of the agricultural system. Isolated exceptions were found by Henry in the South as : "Today the happiest and most flourishing section in the entire South is eastern Tennessee. The single fact that clover is grown there, and cotton cannot be, accounts for the great difference." In Kansas, Henry admonished his neighbors that day to quit talking about Kansas as deficient in terms of eastern forest standards, and turn attention to "our exclusive advantages," one of which was native grass (Cf. Malin, Winter Wheat, 1944, p. 32-35).

Another aspect of the desirable broader approach to the history of the range livestock industry relates to the general subject of stocking the ranges. The Texas aspect of the matter had been greatly over-emphasized in nearly all the literature, as well as the cattle factor, with the result that there are no satisfactory histories of the range sheep industry available that are at all comparable to those devoted to cattle. The story of the stocking of the range with both cattle and sheep may start chronologically with the Spanish approach to the area from the southward into the whole area from eastern Texas to the California coast. When Captain Cooke marched west through New Mexico and Arizona in 1846 and drew his map, he marked on the area south of the Gila river in southern Arizona and northern Mexico "numerous herds of wild cattle from San Bernadino to the point where the San Pedro river is left."

Irrespective of how the area is broken down into subdivisions for convenience of historical treatment, that broad and relatively unbroken Spanish front should be kept clearly in the picture — the separation from each other of Texas, New Mexico, Arizona, and southern California in the cattle and sheep histories is primarily a delimitation of convenience which is relatively artificial. From the hide, tallow, and horn trade, southern Texas developed limited beef outlets, including in the late 1840s some overland drives which became fairly substantial across eastern Kansas and western Missouri when interrupted by the Civil War. The resumption after the war, and the enlargement of the drives north is an oft told story, becoming a process of fattening and maturing as well as marketing. The cattle business began in extreme southeastern Texas and spread northwestward into the interior rather slowly; by 1870 to a diagonal line from about 98° at the Red river to a point somewhat west of 100° at the Rio Grande (Richardson, 1944); or about 101° 30' by 1880 (Gordon, 1884); and by 1883 the invasion of the high plains was under way (Texas Live-stock Journal). The quarantine against Texas fever was begun prior to the Civil War, at the Kansas line, but by 1883 the first line of defense was in northern Texas led by the cattlemen's associations which closed the trails or at least brought them under restriction. In 1883 they took a stand to try to establish the line at the Texas and Pacific railroad. The filling-in of western Kansas by farmers was only a secondary factor in that decade. The discovery of the nature of the disease (1889), and establishment of federal-state inspection and eradication programs (1890), gradually brought order into the industry. Railroad building into the cattle country changed the procedure, especially after 1887, from drives to rail shipments to northern grass such as the Kansas Bluestem Pastures or to market and corn belt feed lots. The shipments to the Kansas Bluestem Pastures began in April, lease contracts expired October 1. Movement out of the pastures to market, as grass fat beef or later to corn belt feed lots for full feeding, usually began late in July, or in some cases young cattle might be held a second year for maturing (Malin, "Bluestem Pastures", 1942).

The southern Pacific coast cattle trade began also in terms of hides, tallow, and horns (Dana, 1840), and after American annexation came the cattle boom, and the sheep boom. The Pacific Northwest developed its resources partly from emigrant livestock and partly from California, cattle and sheep being trailed on a large scale eastward to the northern plains after the Civil War, the Texas blood not being much in evidence in Montana herds in 1860. By the mid-eighties cattle were shipped by rail. The best study of this aspect of the northern grassland cattle movement is found in articles by Oliphant (1932, 1933, 1946). Wentworth (1942) gave a good presentation of sheep trails, and Towne and Wentworth (1945) present the sheep industry as a whole, but further similar work is needed to provide systematic treatment of the cattle trails. A newspaper item in the spring of 1879 estimated 100,000 head of cattle would be moved east, from Oregon and Idaho that season. The following year one firm that had been in the business seven years was trailing 23,000 head from Oregon to the Yellowstone country, divided into three herds moving one day apart, and starting about April 25. (Other data, Burlingame, 1942; Briggs, 1940; Osgood, 1929).

The stocking of the range from the East is possibly the most important, certainly not less so than the Texas source. Chronologically first was the livestock that was moved with emigrants to Oregon, and California. In the very nature of the situation, information as to the volume and natural increase from this source is quite vague. A second source is the border livestock business which supported the overland trade and powered the wagon trains over the Santa Fé Trail and other wagon trails into the interior, about 250-300 oxen or mules for a train of twenty-five wagons. In 1857 Kansas City claimed to be "the stock market for the territory west. There are more sold here than at any point west of the Mississippi river, and more work oxen than at any other single point in America." The plains trade made the Missouri mule famous. The cattle of the freighting firm of Russell, Majors, and Waddell wintered on grass in the season of 1855-1856 as far as one hundred miles southwest from Kansas City and Leavenworth. The breeding stock necessary to provide the volume of work oxen needs systematic attention. So long as the wagon train was the foundation of the transportation system, the advancing frontier of settlement carried this livestock industry deeper into the grassland.

The volume of cattle and sheep driven commercially to California or other points in the West is difficult to evaluate, but the frequency of local news items mentioning such herds and flocks makes certain of a substantial flow during the 1850s and later. A Fort Scott, Kansas, dealer in work and beef cattle advertised in 1857 mentioning that he was located on the California trail "where much emigration and stock pass every spring for California." With the establishment of ranches on the plains this source of stock was available, and was used. With the boom in the early eighties, stock dealers were reported in 1883 as canvassing the South below the Texas fever line, collecting cattle from states as far east as the Carolinas and Florida, moving them to the Arkansas frontier from which they were driven into the plains. The same year this movement by stock trains westward from the Mississippi valley states into the plains was the subject of special comment, in part, these cattle were being taken to the plains to improve the quality of the range herds. The volume of such movements, the methods of collecting, and the routes of movement from year to year have not received systematic attention.

In the long run, the most important sources of range cattle were the pure bred herds, some being shipped direct from Great Britain, but more from stock farms of the Mississippi valley. The first shipments were Shorthorn bulls to cross on Texas and other range herds. The Shorthorn was typically a tall grass or tame grass animal, and upon the short grass plains did not prove satisfactory. At the end of the seventies, Scotch black polled cattle, first Galloways, and later Aberdeen-Angus, were tried out and boomed in the early eighties, but did not hold their own. The Hereford, popularly called Whitefaces, gained ground rapidly after 1876 but were handicapped at first by deficiency in the hind quarters. In 1881, Gudgel and Simpson, breeders at Independence, Missouri, imported Anxiety IV, a Hereford bull of unusual conformation and potency. The Anxiety IV blood strain immediately gained recognition as one of the most significant of the breed. Herefords quickly became the dominant range breed because of their vitality, rustling abilities, early maturing, and fattening qualities, and over 75 per cent of the range cattle in the 1930s were Whitefaces. In this story is to be found one of the most important of all the adaptations to environment, yet the biologists have given no explanation wherein lies the secret. Also, the history of the pure bred herds that have provided bulls for the range since the 1860s has never been written, and it is this phase of the industry that the present writer ventures as the most important of all the sources which stocked the range; not in numbers, but in quality which produced prime beef superior to, and more uniform than, that produced upon the farms of the corn belt. Captain W. J. Tod, of Maplehill, Kansas, Captain Dan Casement, of Juniata, and R. H. Hazlett, of Hazford Place, are other Hereford men of national reputation. Captain Tod began his operations with George A. Fowler, of the meat-packing family, which suggests the importance of that industry in promoting livestock improvement. The railroad, commission men, and banking interests all played a vital rôle in the range industry as here presented.

The story of sheep is different, but possesses many features of similarity. The early phase of the industry turned on wool production so the Merino breeders of New England in particular prospered in growing breeding stock. With the shift to meat production as well as to wool, the longer wool breeds, Shropshire, Hampshire, and Southdown were preferred. Arizona and Idaho lambs and sheep were shipped in large numbers to mature or finish in the region east of the Rocky Mountains, illustrating a significant regional interdependence, each region serving one stage in the process.

The feature of regional interdependence of the livestock industry in all its phase is so important that it warrants some repetition and special emphasis. The legend has been built up by livestock propaganda, especially by the cattle interest, that the plains should have been left to the cattle men and had that been done, drouth periods would not have brought disaster. The utilization of the high plains of Texas makes the issue particularly clear as the cattleman's invasion occurred during the boom of the eighties and very nearly simultaneously with the small farmer invasion farther north. Both soon met disaster. Adverse winters, especially the disastrous one of 1885-1886 in the central area and 1886-1887 on the northern plains which crippled the industry, and in conjunction with the next decade of drouth, and world depression, the recovery was slow. Lying in southern latitudes the grass dried up early in any case, but dry seasons caused shortage. Shipping cattle north for grass reduced the numbers to what the Texas grass would probably support. Unexpected drouth conditions always forced additional shipments, sometimes only after much damage was done, the cattle arriving in Kansas or other northern pastures in weakened condition and sometimes with heavy losses in transit. It should be clear that there are two aspects of adjustment involved; larger numbers could be bred on the plains ranges than could be matured, the maturing and fattening being done in region more suitable; and the interregional economy provided flexibility in making adjustments to the unpredictability of climate.

The depression of the 1930s, and World War II, worked a number of changes in the livestock economy. One change was a sharp emphasis upon calves and young cattle in interstate movement. Another change was the tendency to ship Texas and southwestern cattle to California markets. A third was even more important. During the 1930s the South turned extensively to the growing of Kudzu, a legume which served for grazing and for hay and seemed to provide that section, for the first time, with a possible equivalent of native grass as the basis for a livestock industry. Richmond, Virginia, developed ambitions as an eastern cattle center. The net result was for the South and the Atlantic coast states to initiate a movement for reconstruction of railway rates from the southern range livestock states eastward. Hearings were held before the Interstate Commerce Commission early in 1941. The Kansas Bluestem Pasture country and the packing cities of the North protested the rate changes. The intervention of World War II obscured the possible development of this interregional realignment, but should the South draw any substantial portion of the southwestern cattle trade from the North it would disturb an important interregional relationship which began a hundred years earlier.

Chapter Sixteen

METHODOLOGY FOR HISTORY OF SOCIAL CHANGE:

POPULATION STUDIES

Some examples of methodology for the study of the history of social change are described in this and the following chapters. The first aspect is the study of the population problems. The initial study in the series was published in 1935, some other phases being treated from time to time thereafter (1936, 1940 [3 articles], 1944). The second aspect was agricultural studies focused upon the comparative analysis of internal differences from area to area and internal changes in time from the first recorded census enumeration after pioneer settlement to the last ones available when the study was closed (1936, 1940 [3 articles], 1942, 1944). The third aspect was the reconstruction in some kind of narrative form of the history of the process by which the population worked out possible adjustments to the exigencies of environment and circumstances (1935, 1936, 1940, 1944). In this phase, attention was focused especially, and necessarily, upon these procedures as concrete and realistic samples of the folk process of history in action.

Out of these investigations grew the consideration of the problem of the community as a historical concept, and one which must differ necessarily from the concepts held by sociologists. The latter discipline is not bothered by the time factor, which intervenes to prevent the historian from formulating similar definitions. Closely related, and growing out of the community problem, is the city or urbanization, an equally difficult concept to reduce to a workable definition. Historical samples of these problems appear in incomplete form in two published studies, Kinsley, and Abilene (1933,1944).

In the presentation of all these studies the present author is keenly aware of the handicap of verbalism and ideologies already firmly established. The language does not possess words that are satisfactory to designate the process of change in a frontier as it was transformed in the course of time into the society of the 1940s. The words "mature", "develop", "evolve", or any similar ones are freighted with a traditional ideology associated with the idea of progress, or the Darwinian evolution, from an alleged lower to a higher form, which also involves progress, or the organismic concept of the life cycle of birth, youth, maturity, old age, and death

The word "stabilized" does not express adequately the process, because factors intervene which disrupt the general tendency toward stabilization with age and the achievement of an appropriate harmony of the folk culture with the environment. Furthermore, there is a limit beyond which stabilization cannot go. Equilibrium is always unstable. Irrespective of the words used, it is imperative to deal with the social process as one of indeterminate and continuous change — an open system.

An aspect of methodology in Jenny's <u>Factors in soil formation</u> is suggestive for history. After challenging organismic ideology and the concept of maturity in soils, he discussed the question of when parent materials became soil, and gave the answer, whenever soil forming factors begin to operate. Also, soil already formed would be reduced to the status of parent material whenever a change occurred in the soil forming factors, initiating a new sequence in soil formation. In time, soil was parent material in relation to the future, and product of the operation of soil forming factors in relation to the past. Change in society may be viewed similarly in relation to society forming factors. As social history, the American frontier represents a relatively short span of historic time. The period occupied by the settling-in process was highly unstable, especially the first years, after which a relative equilibrium might be achieved. Major modifications in any existing factor, or variable, or the introduction of a new variable, must necessarily upset the unstable equilibrium, initiating a new sequence of succession toward reestablishment of a new unstable equilibrium in its areal setting. New settlers who arrived from time to time in any community represented major changes in the factors operating in a community, and changes of variable proportions. Power farming, during the second quarter of the twentieth century, represented the introduction of a new independent variable, highly disturbing to equilibrium. In either case, a new sequence in re-forming society was initiated. At any given instant, in relation to the future, the state of society is parent material, and in relation to the past, it is the product of the operation of society forming factors. And in any case, within the circle of known facts, the social process is one of indeterminate and continuous change — an open system.

Population studies

Population studies in the sense of historical demography have received little attention from American historians. Turner's frontier approach, as expounded in his essays, included the growth and development of a frontier community into a mature society. In practice, however, Turner did not develop a specific methodology for tracing the process of such development, or for reducing the process to a problem, or series of problems stated in an operational form, to which objective quantitative methods could be applied. He did not take particular communities or restricted geographical areas and follow their evolution through a period of time with a view to presenting systematically the growth of the principal features of the culture complex. Without such more or less standardized quantitative methodology it was impossible to measure in an objective manner what happened to the population on any particular frontier, and what changes occurred as it grew older. It was even more impossible to study the population of different frontiers comparatively, either as frontiers or as societies developing through periods of time, in order to determine specifically the continuities and the differences, either of time or of geographical environment, or of relations to the technological development of the background society. The most elaborate project that has been undertaken, but not completed, which was making some contribution in that direction, was Joseph Shafer's Wisconsin Domesday project. In the case of Webb's The Great Plains, he did not address himself to that problem at all as his attention was occupied with other aspects of the regional approach.

The present author has undertaken pioneer studies in historical demography, and the adaptation of the agricultural system to Kansas as a specific area of the central grassland. Some attention has been given to the town, to the facilities of entertainment, and to strictly social life. The most fruitful contributions thus far, both as respects methodology and historical conclusions, have been in the first two departments, population, and agricultural studies. The basic procedure is described most fully in "The turnover of farm population in Kansas," (1935), and "Local historical studies and population problems" in The Cultural Approach to History, edited by Caroline Ware (1940).

The sampling method is the basis of operations, using a township or community area of fixed geographical boundaries as the unit small enough that it can be manageable in the entirety of its social behavior. Several such samples can be thrown together for statistical purposes, but handled separately for narrative purposes. They can be combined as seems desirable on different kinds of classifications to emphasize community age, type-of-farming area, rainfall belts, soil types, immigrant or native populations, or other aspects that may be of interest. The sources of the data are the state and federal census records. Other records that might be used are local records of land titles, or tax lists, and probate court cases.

The procedure for studying population behavior was to list the farm operators from the agricultural schedule of the census, and then gather the data from the population schedules for each operator, his wife, and his family, by name. As Kansas took a decennial census from 1865-1925 inclusive, the federal and state census enumerations provided lists at five year intervals to 1885, after which period the federal census records were closed to investigation. For the later period the state census was depended upon exclusively. After 1925 statistical rolls of farm operators were available in Kansas for each year, but not general population data. The census list of farm operators for the first enumeration after settlement afforded the first base for comparison, and each list thereafter was compared with the base list to determine the individual operators remaining within the census area, but not necessarily upon the same farm. In turn, each census list was used as a base list for comparison with subsequent lists. Upon the death or retirement of an original operator, a son was counted as representing the family succession in the area. Tables were then prepared showing the total number of operators at each base enumeration interval, and the number remaining at each successive enumeration; another table was prepared in terms of percentages persisting; and then graphs were constructed, plotting comparative curves for each base year.

    The conclusions from these population studies are illuminating, and a few of them are summarized here; but the more complete treatment of the published portions are to be found in the original monographs. In the population turnover studies, there were variations from sample to sample, but there was a general uniformity of the curves illustrating farm-operator persistence, irrespective of whether the community was settled just prior to the census of 1860 or of 1895. The rate of turnover of population (or persistence as one may choose to call it) was associated primarily with community age; very high turnover during the pioneering period, a period of relative stabilization at low levels of persistence, and lastly a period of a high degree of stabilization, especially after 1915 or later. Of course, there were limits beyond which stabilization with age could not go. New independent variables entered from time to time, such as mechanical powered farming, following which new adjustments must be made. The turnover pattern of communities of a comparable age persisted to a remarkable degree, irrespective of depressions or of drouths, and there was very little difference as respects rainfall belts of 35 inches or over and 20 inches or less. Among the most stable found were some communities in the high plains where rainfall was less than 20 inches. New settlers were always more unstable than the old residents. Thus, with every replacement by new settlers, the movers were weeded out, leaving the more permanent element. It is not safe, however, to jump at conclusions as to why some moved and others stayed, or to assume that the best stayed and the worst moved. In most samples and periods there was a tendency toward greater stability during depressions, and mobility during periods of prosperity.

A second group of conclusions has to do with the total numbers of operators in each sample. The behavior was not uniform. On the pioneering frontier, numbers usually declined during depression and drouth. As already pointed out, the rate of loss, or turnover, was not greater necessarily when a drouth, or depression, year was used as the base from which to measure, and, on the contrary, was generally less. The real issue was the flow of replacement population. With a consistently high rate of loss, complete depopulation would have resulted if new settlers had not arrived promptly. An increase in the total number of farm operators meant that the flow of replacement population exceeded the losses. This usually occurred in the pioneer period, during booms. The decline in the total numbers of farm operators meant that the flow of replacement had diminished or ceased. This occurred in the pioneer period during drouth and depression periods. The rate of turnover in either case, based on the whole number of any particular census list, fluctuated little. In other words, the frontier was not a safety valve in the sense in which that theory has been used by the Turner school of history.

No satisfactory method has been devised for tracing what became of those who moved. The local newspapers record that many returned to the East, moved to towns, went to the mines in the mountains, or worked for the railroad, during depressions, but sufficiently exact data are not available from such sources to afford reliable quantitative treatment. Significant increases in numbers occurred frequently in older samples during depressions, in part, at least, reflecting an urban-rural movement. In most respects the depression of the 1930s was no exception to the general behavior pattern. The newest settled areas suffered the highest rate of turnover and net losses, but the conspicuous characteristic was the stability regularly associated with older communities, the propaganda to the contrary notwithstanding. There were fewer farmers on the move, proportionally, than in any previous depression in Kansas. Again it should be emphasized that replacement population had ceased to flow into the region and that, rather than instability, accounted for population losses.

A third group of conclusions relate to what proportion of the total population of a sample, at any given date, were descendants of the operators of any prior date Allowing for wide variations in individual samples, it is clear that the high rate of turnover of operators during the pioneering phase left a very small proportion of the original settler descendants among the farmers fifty to seventy-five years later. The assumption so often made that the original settlers determined the character of a community is unsafe as a generalization. After a substantial stabilization of the community had been achieved, even though the turnover might still be relatively high, the proportionate influence of the subsequent population was much greater

The principal foreign-born groups of farm operators in Kansas who were sufficiently concentrated to dominate the communities were Germans, Swedes, and Bohemians. On the whole, they were highly persistent, but the second and third American-born generations reflected much the American pattern of behavior (1935, 1940). With the cessation of a large replacement population from the mother countries, these communities rapidly lost much of their distinctive character. As a rule, negroes and Jews did did not appear conspicuously in farm population. Negro colonies were established during the period of the Civil War and reconstruction migrations, but the population mostly drifted to the cities. Partly, the explanation lay in inability to carry on independent farming enterprise; partly, it was failure to adapt to environment; but, more largely, the answer probably lay in the cessation of the flow of replacement population to compensate for the high turnover characteristic of all newly established enterprises.

Studies of the internal migration of the United States are handicapped by the fact that federal census enumerations recorded only the state of birth and the place of residence at the time of the enumeration. Also, all the later federal enumerations were closed to investigation, except to federal employees, so only the printed mass statistics were available. Kansas census records, 1875-1925, recorded the additional data of the state from which the individual moved to Kansas. As the Kansas census and the early federal enumerations are open for examination of individual names, it was possible to make more satisfactory studies than have ever been made elsewhere. Even much of the migration intervening between birth and the removal to Kansas can be pieced together, for persons with families, by noting the birth states of the children. In connection with local studies, data on these matters were printed in 1936, 1940, and 1944 as pertained to Edwards, Dickinson, and Saline counties. The present author reported at the annual meeting of the American Historical Association, in December, 1939, on the broader aspects of interstate migration of native Americans direct from the state of birth to Kansas. This was published in abstract in The Cultural Approach to History (1940), edited by Caroline Ware, without the detailed statistical table which is now printed. The sample areas were grouped from east to west across the length of the

Interstate migration of native farm operators to Kansas Men Direct from the state of birth

Percentages are figures on the total native farm operators as a base

	1875		1885		1895		1905		1915		1925	
	No.	%	No.	%	No.	%	No.	%	No.	%	No.	%
I. First tier counties												
Three townships:												
Total farm operators	442		535		489		541		605		569	
Total native	345	100.	429	100.	413	100.	468	100.	539	100.	539	100.
Native direct	135	39.13	184	42.89	192	46.48	195	41.66	202	37.47	160	29.68
Contiguous	42	12.17	66	15.39	65	15.73	88	18.8	111	20.59	107	19.85
Non-contiguous	93	26.96	118	27.50	127	30.75	107	22.86	91	16.88	53	9.83
II. Second tier:												
Total farm operators	467		611		616		629		596		566	
Total native	330	100.	458	100.	465	100.	520	100.	541	100.	534	100.
Native direct	132	40.	253	55.24	253	54.40	258	49.61	212	39.18	162	30.33
Contiguous	14	4.24	37	8.08	33	7.09	45	8.65	48	8.87	52	9.73
Non-contiguous	118	35.76	216	47.16	220	47.31	213	40.96	164	30.31	110	20.60
III. Extreme territorial												
Frontier 1860:												
Total farm operators	393		695		615		645		714		688	
Total native	305		539		511		534		634		629	
Native direct	160	52.45	277	51.39	307	60.07	312	58.42	275	43.37	184	29.25
Contiguous	4	1.31	17	3.15	41	8.02	45	8.42	65	10.25	56	8.90
Non-contiguous	256	51.14	250	48.24	266	52.05	267	50.00	190	33.13	128	20.35
IV. First post-War												
Frontier 1865-70:												
Total farm operators			251		245		252		248		234	
Total native			139		126		130		165		189	
Native direct			53	38.13	37	29.36	63	48.46	52	31.51	40	21.16
Contiguous			6	4.31	20	15.85	24	18.46	22	13.33	22	11.64
Non-contiguous			47	33.32	17	13.51*	39	30.00	30	18.18	18	9.52

POPULATION STUDIES

V. Second post-war											
Frontier mid-70s:											
Total farm operators	268		632		687		744		881		700	
Total native	218		438		421		508		700		624	
Native direct	65	30.	174	39.72	213	50.6	269	52.95	307	43.85	198	31.73
Contiguous	2	.91	11	2.51	27	6.41	63	12.40	106	15.14	74	11.85
Non-contiguous	63	29.09	163	37.21	186	44.19	206	40.55	201	28.71	124	19.88
VI. First recovery											
Frontier 1878-79:											
Total farm operators			194		134		207		250		281	
Total native			159		106		169		203		213	
Native direct			75	47.17	52	49.05	86	50.9	97	47.76	79	37.08
Contiguous			12	7.54	13	12.26	27	16.00	39	19.21	37	17.37
Non-contiguous			63	39.63	39	36.79	59	34.90	58	28.55	42	19.71
VII. West of 100°:											
Total farm operators					726		581		833		916	
Total native					515		439		711		817	
Native direct					186	36.11	185	42.14	345	48.55	316	38.66
Contiguous					65	12.62	72	16.40	175	24.61	199	24.35
Non-contiguous					121	23.49	113	25.74	170	23.94	117	14.31

Omitted: Kansas born, foreign born, and native indirect migration.
* This figure of 13.51% is out of line with probabilities, but no satisfactory explanation has been found.

Interstate migration of farm operators from state of last residence Men

Percentages are figures on basis of internal migrants, native and foreign born

		1875		1885		1895		1905		1915		1925	
		No.	%	No.	%	No.	%	No.	%	No.	%	No.	%
I.	Total farm operators	442		534		489		541		604		569	
	Kansas born	1		32		105		159		253		318	
	Foreign born direct to Kansas	18		40		32		34		52		18	
	Internal migration to Kansas	423	100.	462	100.	352	100.	333	100.	296	100.	228	100.
	From adjacent state	214	50.5	215	46.5	158	44.9	179	53.7	169	57.1	162	71.0
	From non-contiguous state	209	49.5	245	53.5	194	55.1	154	46.3	127	42.9	66	29.0
II.	Total farm operators	466		611		616		628		596		566	
	Kansas born	1		25		88		169		271		324	
	Foreign born direct to Kansas	27		53		57		59		40		24	
	Internal migration to Kansas	438	100.	523	100.	470	100.	400	100.	284	100.	217	100.
	From adjacent state	104	23.7	123	23.5	115	26.6	106	26.5	92	32.4	87	40.0
	From non-contiguous state	334	76.3	400	76.5	355	73.4	294	73.5	192	67.6	130	60.0
III.	Total farm operators	393		694		614		644		715		689	
	Kansas born	0		7		36		94		227		341	
	Foreign born direct to Kansas	24		46		42		60		48		38	
	Internal migration to Kansas	369	100.	637	100.	536	100.	478	100.	446	100.	257	100.
	From adjacent state	60	16.2	107	16.8	106	19.7	102	21.3	161	36.1	95	36.9
	From non-contiguous state	309	83.8	530	83.2	430	80.3	376	78.7	285	63.9	162	63.1
IV.	Total farm operators			250		245		257		248		234	
	Kansas born			2		10		28		96		138	
	Foreign born direct to Kansas			75		96		103		68		41	
	Internal migration to Kansas			167	100.	138	100.	128	100.	84	100.	54	100.
	From adjacent state			65	38.9	58	42.0	46	36.0	29	34.5	32	59.2
	From non-contiguous state			102	61.1	80	58.0	82	64.0	55	65.5	21	40.8

V. Total farm operators	270		632		687		744		885		701	
Kansas born	0		8		10		64		241		345	
Foreign born direct to Kansas	19		85		153		142		131		45	
Internal migration to Kansas	251	100.	535	100.	524	100.	538	100.	509	100.	296	100.
From adjacent state	60	24.3	132	24.6	134	25.7	177	32.9	207	40.6	119	40.2
From non-contiguous state	191	75.7	403	75.4	390	74.3	361	67.1	302	59.4	177	59.8
VI. Total farm operators			194		134		207		250		281	
Kansas born			0		1		13		57		96	
Foreign born direct to Kansas			2		5		15		34		55	
Internal migration to Kansas			192	100.	127	100.	177	100.	159	100.	126	100.
From adjacent state			56	29.1	55	43.3	71	40.1	69	43.4	59	46.8
From non-contiguous state			136	70.9	72	56.7	106	59.9	90	56.6	67	53.2
VII. Total farm operators					723		538		833		913	
Kansas born					9		33		157		274	
Foreign born direct to Kansas					71		37		57		48	
Internal migration to Kansas					638	100.	511	100.	620	100.	554	100.
From adjacent state					347	54.3	276	54.0	351	56.6	359	64.8
From non-contiguous state					291	45.7	235	46.0	269	43.4	195	35.2

Defects in the census data leave some unaccounted for.

state in seven successive frontiers. The most significant points of emphasis were that the native migration to Kansas, direct from the state of birth, increased in proportion after 1875, both 1895 and 1905 being high, and thereafter the proportion declined, and furthermore, the part of that migration that moved into Kansas from adjacent states was small in proportion to that coming from non-contiguous states. These general conclusions confirmed fully the more specific ones relative to Edwards (two samples), Dickinson, and Saline counties (1936, 1940, and 1944).

In the second statistical table is presented, for the first time, the data on interstate migration to Kansas from the state of last residence. The conclusions drawn from these data support those presented in 1939 at the American Historical Association meeting, and printed in 1940. The migration was not from one frontier to the next adjacent states but from non-contiguous states, except for the farthest west frontier, which meant mostly that the settlers made the long jump from some state east of the Mississippi river to Kansas. In the case of the west tier of counties of Kansas forming the seventh group, two factors are sufficient, probably, to explain that behavior; the building of the Burlington, and Rock Island railroads into the area from Nebraska, and Iowa; and secondly, after 1895 the change in all migration was setting in, especially by 1915. The decisive aspect of the table as a whole is that, for the census dates 1875-1905, the probabilities were that very nearly three of every four, or four of every five, of the migrant operators listed on those dates had migrated from a distance to Kansas, and that the adjustments necessary to so marked a change in environment were substantial. Conspicuous also was the number of Kansas born operators in the enumerations of 1915 and 1925, a fact which contributes to an explanation of stabilization of population to its environment. Three points relative to migration need to be emphasized because of the tradition in Turner circles that one frontier supplied the population for the next. The Goodrich (1936), and Thornthwaite (1934, 1936) studies in migration employed an inadequate methodology, but made the best use possible of the mass statistics from the printed federal census enumerations. Thornthwaite's (1934 p. 10) conclusion that the trans-Mississippi states received settlers chiefly from eastern states, which had been settled forty or fifty years earlier, is so vague as to be virtually meaningless. He did not define the word "settled." Shannon (1945) followed Thornthwaite in part, concluding that only the occasional family made the long jump from non-contiguous states. The Shannon conclusions are so far out of line with the facts derived from the Kansas census data that the error must be emphasized. Whether or not studies based upon an adequate methodology would reveal different conclusions for states north or south of Kansas cannot be forecast with certainty, but the probabilities are that they would differ little from the Kansas results.

The age of the population of the frontier has been the subject of much inconsistent or contradictory treatment. One contention is that the frontier was settled by men who had made several successive removals. If so, then the frontiersman must have been anything but young. Another extreme held that the frontier was composed conspicuously of young couples who were just starting out in life. Paxson, (1930, p. 29-31) defined the frontier in terms of a cycle, from the coming of the cabineer until his first born, in turn, married and set out on a new frontier. The present author, in 1936, 1940, and 1942, presented quantitative data on the subject for Kansas and for the first time removed this phase of the frontier problem from the realm of merely speculative generalization. In the samples analyzed, the young couples just beginning life were conspicuously in the minority. The accompanying table of median ages gives the most complete body of data on that aspect of the age question, both for comparative areas in Kansas, and for successive stages in community age, 1860-1925. It should be noted that children and other members of the communities were eliminated, the data applying only to farm operators and their wives. This segregation gives the results statistical meaning. The men were conspicuously middle-aged, and in only two entries in the table did the median fall below 35. The age distribution clustered rather closely around the median.

Median age of farm operators and wives by rainfall belts: Kansas

Date	35— inches		30-35 in.		25-30 in.		20-25 in.		20—inches	
	Men	Women	Men	Women	Men	Women	Men	Women	Men	Women
1860	37.0	34.0	33.0	31.0	32.8	31.2	-	-	-	-
1865	39.0	35.0	37.0	34.0	35.0	29.0	-	-	-	-
1870	40.0	35.5	38.0	35.0	35.5	32.5	-	-	-	-
1875	44.5	38.0	39.0	35.6	35.0	33.0	36.0	34.0	-	-
1885	45.0	39.0	41.0	37.0	40.0	36.0	40.0	36.0	-	-
1895	43.0	38.0	43.0	38.0	43.0	38.0	42.0	39.0	40.0	37.0
1905	44.8	41.0	43.0	39.0	44.0	39.0	42.0	36.0	44.0	39.0
1915	45.0	41.0	42.0	39.0	42.0	37.0	39.0	37.0	44.0	38.0
1925	45.6	40.5	43.0	39.5	41.5	38.5	40.6	37.5	42.5	37.0

Note: The same sample townships and counties were used in this as in the other phases of the population studies.

After the publication of "The turnover of farm population in Kansas" (1935), the USDA, through its division of farm population and rural life, undertook a study of farm population in Kansas. A. D. Edwards was in charge, and he applied largely the methodology just described. The present writer cooperated fully; explained the procedures and illustrated them by materials published and unpublished, pointed out errors and difficulties that had been encountered in the experimental work, and suggested improvements. Edwards's study covered Haskell county as a whole, with an intensive concentration on one township sample, and was published as Social Science Research Report No. 7, January, 1939. In evaluating the report, distinction should be made between the research results and the policy conclusions, the latter of course conformed in general with the departmental policy. The important conclusion is that the methodology in the hands of another investigator produced essentially the same research results as obtained by its originator. The present writer emphasized to Edwards the importance of studying the replacement population in the community, from census to census. The older population had made some adjustments, but the new population was composed of beginners. He made this differentiation and presented the data and graphs (pp. 17-23) showing the quantitative difference between these two segments of the operators under examination. The new, or replacement, population always followed approximately a new frontier type of curve, and the old population followed a stabilized society type of curve. He found also, what the present writer pointed out, that periods of prosperity seemed to show, for the most part, greater instability than periods of drouth and depression.

It is important that anyone organizing a research project of any size, according to this method, should master fully the possibilities of the punch card and statistical machine methods of compiling and computing data The study is less liable to result in errors, — but the all-important factor is labor-saving that may mean the difference between failure and completion of the project

The kind of population studies thus briefly described are in progress only for Kansas. The full value of these can be brought out only through comparative studies by substantially the same method for other areas states to the north, to the south, and to the west of Kansas, and especially samples from eastern states, and for social groups other than farm operators. So far as can be determined by preliminary surveys, no other state has quite so complete a record of data essential to the purpose as appears in the Kansas census, but such federal census materials as are available, and county records, seem to indicate that similar studies are feasible for other areas. For Kentucky, with its long axis running east and west in the general direction of population movement, it would seem that a county basis of study would have to be devised. Louis Warren's effective use of county record in his book, <u>Lincoln's parentage and childhood</u> (1926), indicates that in those records lies an opportunity to make valuable population studies. The most effective thing that could be done to make possible the tracing of migration within states, as well as between states, would be the making of indices of the names in the federal census records. The Kansas State Historical Society has made a beginning of such an index for its early census enumerations, but it would have to be done for all of the states. For the early years when numbers were relatively small such a project would not be prohibitive. If there were state enumerations between the federal recording dates, as in Kansas, that would make possible a check on the residence of each individual every five instead of every ten years. Without comparative studies, both as to time and area, it is impossible to evaluate in any quantitative and objective manner the process of the peopling of the United States and the changes in internal composition and structure through the years from the first pioneer settlement to the twentieth century.

The second aspect of the question is equally important. The center of interest of the studies described has been farm operators, wives, and families. But what was the behavior of other groups; town people, the professions, different types of business, the extractive industries, processors, marketing and distributive businesses, personal service businesses (barbers, blacksmiths, livery stables, garages, restaurants, real estate, and insurance), the general store, etc.? Were they more, or less, stable than farm operators? Preliminary investigations would indicate that the personal service businesses were far more unstable, probably the most unstable of all the professions. In the population turnover study of 1935, a comparison was made with a sample of college students. Their persistence over a period of six years, toward a college degree, provided a curve of the same conformation as the farm operator curve. Such graphs worked out comparatively are imperative to any real knowledge of historical development and change in population structure, and should be of outstanding importance to students of governmental policies, whether designed for over-all application to society, or to benefit particular groups. As suggested in 1935, the basic behavior pattern revealed in these farm population turnover studies is not peculiar to farm operators of Kansas, but seems to indicate "that the problem is primarily one of group behavior, apart from specifically assignable accidents of farm life" — a large percentage of losses during the early stages, with a gradual stabilization later.

Chapter Seventeen

AGRICULTURAL STUDIES

     In studies of agriculture in the grassland the sampling method as described for population studies was extended and adapted for the purpose in hand, using the same communities. Because of this concentration of the present author's own research in the central portion of the grassland region, consideration of other portions is omitted except as to some statements of general matters. For the Prairie Peninsula of Illinois, or for the barrens of Kentucky, where Anglo-Americans first met the grassland problems, there have been no comprehensive studies of either population behavior or agricultural adaptation that deal explicitly with the grassland problem as such. The Illinois Centennial Commission history of that state (1920) did not meet the problem of the prairie and the Bidwell and Falconer History of Agriculture in the Northern United States, 1620-1860, (1925) recognized it only in a general way. Poggi, The Prairie Province (1934) was a geographical study with a section on settlement which was inadequate as history. Studies by ecologists of the area were intensive and valuable. Even without new research, a competent synthesis of existing studies from all these related fields is much to be desired. Illinois was a leader in ecological, geographical, and historical studies in separate fields, but the several disciplines did not get together for cooperative synthesis that might have brought the data from all the sources to bear in an explicit and comprehensive presentation of the grassland problem and its impact upon Anglo-American behavior.

     The problem of irrigation in the lowest rainfall areas of the grassland and desert are likewise omitted from special consideration. Three aspects only of the problem are selected for emphasis; the importance of the synthesis of geographical, ecological, soil, and other scientific material with the historical; and the comparative study of the whole irrigation problem within the different areas of North America and among the other continents with similar problems. Sauer's type of approach from the standpoint of historical geography is particularly important. Comparative studies should reveal the differences, and limitations, in different parts of the world, and break down the provincialism that pervades so much of irrigation history. The final point is to keep the record straight relative to federal reclamation work under the Newlands Act of 1902, and its amendments, and the social myth built up by propaganda under that system. The work of irrigation was inaugurated and expanded in such haste as to bring a large number of the projects to a point of imminent collapse when they could not pay out; partly because of blunders, miscalculations, and incompetence, and partly because of a misunderstanding of the whole problem of American rural population behavior in its historical setting. The fact of governmental sponsorship and treasury support was no guarantee of success or of avoidance of large scale hardship to the individual settlers. The relative failure of so many of these projects, as revealed during the 1920s, should have served as a warning of the high degree of failure in other government-sponsored settlement projects of the 1930s. The farm population studies reviewed in the preceding chapter point to a general principle, that in any group of people a large proportion (although variable in each sample group) will always fail to complete the plan, the largest losses appear near the outset, the rate of loss declining sharply after the initial test period and then tapering off to a relative stability among the surviving few. Each new increment added must in turn experience this selective process. This principle of group behavior was suggested in 1935, and further studies seem to confirm it.

As the advancing population reached Kansas in 1854, the problem of agriculture under the prairie environment was discussed. Missourians, and some Indians among the Immigrant Tribes from the East, had a limited experience prior to the opening of the territory. The recognition was quite general that the settlers were facing a new environment, one sufficiently different that modifications in crops and methods were accepted as inevitable (Malin, Winter Wheat, 1944, Chapters 1-3). The question was not whether changes were necessary, but what nature and extent of changes would be required. The long-term answer to that question was determined by the several factors of climate, topography, and soil. From east to west the moisture problem was most conspicuous. As the rôle of irrigation was negligible for the country as a whole, almost to the Rocky Mountains, the decreasing available moisture westward emphasized what came to be called dry-land farming as the settled area advanced toward the mountains.

Various traditions have become relatively stereotyped in dealing with the size of farms. The group of land reformers are disposed to adverse criticism of land policy on the ground that it did not fit conditions (Report President's Commission on tenantry, 1937, p. 5). Into this pattern is fitted the formula that the original land unit was too small, and that immediately upon patents being issued the process of consolidation into economical, larger units began. Statistical studies of farm size from the printed census figures are at best of limited value and often are positively misleading. First there was no uniform or adequate definition of what constituted a farm and in pioneer days no definition could have been applied consistently. Students of the land problem should digest carefully the introductory explanations of the federal census, especially for 1870, and 1880. Furthermore, the very nature of mass statistics serves to conceal, rather than to reveal, actual changes and their significance. One kind of tendency may be cancelled by another. So long as there was unoccupied land held by the government or non-resident owners, it was used largely by the community as commons (1942,1944). Even if the pioneer had received larger acreages, under the terms of the governmental land policy, he seldom had the capital to finance adequately even the traditional quarter-section farm— the buildings, fences, machinery, horsepower, and manpower. Whether he could have marketed, profitably, large production is open to question. The early years were usually marked by the severest struggle to cultivate even small acreages.

Such limited discussion by resident farmers of the late nineteenth century as has been made a matter of record in the area of the present author's research, leads mostly in one direction, emphasis on more efficient management of a smaller farm rather than enlargement of the unit (Winter Wheat, 1944). This general trend, to which there were only occasional exceptions, is represented by a letter to the editor of the Junction City (Kansas) Union, February 20, 1862, in which the argument warned farmers against taking too much land, because of costs of taxes, improvements, breaking sod, fencing, building, and hiring help. The letter writer insisted that eighty acres was enough, and that excellence and profit in farming depended upon the operator doing his own work, with small expense; the income would be small, but sufficient, and the farmer would remain independent, and no outsiders would intrude upon his domestic circle.

This limitation to eighty acres was in part a sound argument from the standpoint of the small farmer's inability to finance adequately larger farms, even with the advantage of cheap or free land, and in part it illustrated the easterner's misconception, even when on the spot, of the size of farm necessary to support a family in the vicinity of the 97 meridian. In respect to the first point no adequate answer to financing has been forthcoming, and on the second point only experience with different types of farming could answer for any given time or place, and no answer on the matter of size could be final because of the changing conditions, especially those accompanying mechanization. Another factor in the farm-size problem was the element of speculation in land. The concept of "actual settler" or the permanent farm home was largely a myth as most owners of land, irrespective of whether they were resident or non-resident, bought land on the assumption that it would be resold soon at a profit (Malin, Mobility and history, 1943). To a larger degree than any historian has yet been willing to admit, the much discussed periods of agricultural prosperity and depression of history have been more accurately reflections of land prices rather than reflections of profits and losses on actual farm operations. It is probable that, over a period of years of fluctuating weather and prices, only the better managers made profits out of farm operations. These matters become clear only when farms are studied individually at different times and in terms of acreages and yields for samples in the several farm-size groups.

The size of the farm possesses significance only in terms of utilization, and changes in use are reflected necessarily in size. New transportation facilities often change farm sizes, inducing such shifts as livestock raising to grain farming. The rise of cities provides new kinds of markets and induce dairying and truck farming; cannery contracts may determine the crops. A farm of ten to twenty-five acres may be a substantial size of intensive truck, fruit, poultry, or irrigation farm, but insufficient in size for grain or livestock. All such situations are lost in mass statistics printed in the federal census. Before statistical studies of farm-size can have any particular meaning, the data must be classified into groups of comparable kind, because only similar or substantially similar things can be compared statistically if results are to possess validity. In the selected samples used in the present author's studies, the history of each sample and its changing internal structure were studied to determine the kind of farm program prevailing and the nature of changes. Only under such known conditions do studies in farm-size possess much significance.

First, a sample community in eastern Kansas will illustrate several of the aspects of the problem under discussion. Kanwaka township, Douglas county, Kansas, is an upland mixed farming community occupying the ridge between the historic towns of Lawrence, and Lecompton. The accompanying tables and discussion present farm sizes, and crop and livestock programs over a period of eighty-five years (1855-1940).

When the state was first settled the prevailing method of acquiring land from the public domain was the preemption system by which a settler might buy a quarter section, prior to the offering of the land at public auction, at the minimum price of $1.25 per acre. Under this land system the predominant size of farms was 160 acres with a few 80s. In 1860, 90 per cent of all farms in Kanwaka township were quarter sections.

Size of farms: Kanwaka township, Douglas county, Kansas

(Figures in percentage, except where otherwise indicated)

Date	1860	1865	1870	1875	1880	1885	1895	1905	1915	1920	1925	1930	1935	1940
No. of farms	79	100	109	143	152	125	159	168	138	149	169	154	176	162
Under 3 acres			} 6.4	} 1.3	} 1.3	0.8	} 2.5	} 2.9	} 3.6	4.0	1.18	1.3	4.54	2.44
3-9 "											2.36		2.84	1.85
10-19 "											0.59		0.57	0.00
20-49	6.3	13.0	24.8	2.8	3.9	0.8	3.1	5.4	4.3	4.0	1.78	2.6	2.27	1.85
50-99			25.7	21.7	28.4	13.6	22.0	20.8	21.7	16.8	19.52	17.5	16.47	14.19
100-174	89.89	75.0	25.7	48.9	36.2	47.2	45.9	42.9	38.4	33.6	41.42	40.0	38.00	38.27
175-259	3.9	} 5	} 12 17.4	25.9	28.3	36.8	24.5	26.8	29.0	37.6	18.93	} 36.4	15.90	16.66
260-499		} 7					1.9	1.2	3.0	4.0	11.83	1.3	15.34	17.90
500-999					} 1.9	0.8					1.18		1.14	3.08
1000-4999				0.7							0.0		0.0	0.0
5000+											0.0		0.0	0.0
No data											1.18		2.84	3.70
160 or less	96.2	88.0	82.6	73.4	69.8	62.4	73.5	72.0	68.0	58.4	66.85	61.4	64.69	58.60

Note: The difference in the total number of farms in this table and in the table giving numbers of farms is explained by the fact that for some years the data were missing on one or more farms. The percentages were figured, therefore, on the basis of the number of farms with specific data on the assumption that error would be less serious because the no data farms were probably scattered random throughout the range of the census list. The problem was most serious for the years 1920 and 1930 and 1940 when no formal Kansas census was taken and agricultural statistical rolls had to be used for the present purpose.

Douglas county, Kanwaka township, (Mixed farming upland)

Number of farms in each size-group

Date	1860	1865	1870	1875	1880	1885	1895	1905	1915	1920	1925	1930	1935	1940
No. of farms	80	100	110	143	152	128	160	168	138	163	169	160	176	162
1-19 acres	0	0	7	0	2	1	4	5	5	6	7	2	14	7
20-49 "	0	0	27	4	6	1	5	9	6	6	3	4	4	3
50-99 "	5	13	28	31	43	17	35	35	30	25	33	27	29	23
100-174	71	75	28	70	55	59	73	72	53	50	70	64	67	62
175-259	1	5	13	20	25	23	20	23	24	37	32	34	28	27
260-499	2	7	6	17	18	23	19	22	16	19	20	22	27	29
500-999					2		3	2	4	6	2	2	2	5
1000-4999				1	1	1								
No data	1		1			3	1			14	2	6	5	6

Livestock: Kanwaka township, Douglas county, Kansas

(The numbers represent the average of the farms having each type of livestock, not the average of all farms)

Date	1860	1865	1870	1875	1880	1885	1895	1905	1915	1920	1925	1930
Milk cows												
40 acres						1.00[1]	1.75	4.82	4.6	3.33	1.4	0
80 "		3.46	2.16	5.66	3.66	3.06	2.86	4.06	4.95	5.52	3.04	4.25
160 "	2.25	4.9	4.36	3.84	4.92	6.02	4.64	6.20	4.79	3.82	4.80	5.5
320 —	2.95		5.26	6.45	6.43	7.82	6.05	8.5	6.38	4.0	5.4	8.62
500 —	5.66	8.75	9.9	7.22	8.8	120.00	4.66	14.00	5.25	5.75	6.0	14.00
				15.00	8.00							
Other cattle												
40 acres						0	1.00	15.4	5.75	3.33	8.33	5.5
80 "	5.66	3.53	5.66	6.0	8.15	3.37	2.5	5.11	14.04	6.95	6.5	6.8
160 "	5.44	8.6	5.69	7.61	8.45	8.32	11.36	7.94	11.90	5.4	12.4	13.91
320 —	7.33	21.91	10.48	13.64	15.3	17.13	24.25	21.2	10.51	16.2	22.6	22.84
500 —			16.93	17.00	22.41	50.00	72.66	38.00	55.25	63.66	32.0	20.5
				72.00	54.6							
Swine												
40 acres						18.00	6.33	5.4	6.66	1.5	6.5	4.0
80 "	17.25	3.18	8.12	8.0	9.33	6.53	11.11	10.44	6.5	14.76	8.4	12.7
160 "	13.7	4.76	7.68	5.46	10.00	9.7	15.41	15.93	12.75	11.43	14.7	15.7
320 —	11.66	6.73	9.08	6.46	16.3	19.13	24.18	23.27	14.4	9.5	24.1	22.87
500 —			17.11	8.74	21.95	100.00	41.66	17.00	9.25	9.25	32.0	29.00
				2.00	36.00							

Note:
1. Only one form is represented

Once the land was in private hands the readjustments in the size of farms were continuous. During the periods of rural depression the size decreased, or in other words, the number of farms of 160 acres or less increased. In prosperous times the smaller farms were consolidated in part into units of 160 acres or larger.

The period showing the greatest sub-division into small farms of less than a quarter section came in the years of depression immediately following the Civil War. In Kanwaka township, in 1870, 25 per cent of the farms were 40 acres, another 25 per cent were 80 acres, and a third 25 per cent were quarter sections. There was an increasing number over 1865 of larger farms (half sections or three quarters) but the proportion was small,— only 17 per cent of all farms.

In the boom period of the 1880s, the pendulum swung in the opposite direction, when in 1885 nearly two out of every five farms (37.6 per cent) were approximately a half section or larger. Less than half were 160-acre farms. Forty-acre farms practically disappeared and only a few 80s survived.

The depression of the 90s broke up the big farms, restoring a moderate number of 80-acre and 40-acre operators, and it was not until 1920, as a result of the World War boom, that the large farms again came back in approximately the same proportion as in 1885. After 1920 there was no long-time trend in either direction, but short sharp swings of the pendulum about every five years. In 1925 and 1935 the small farms multiplied; while in 1930 and 1940 the large farms appeared in increasing numbers. One exception must be made to this generalization, however, inasmuch as after 1925 the 80-acre farms declined continuously in numbers. From the standpoint of the political objective of preserving the traditional family-size farm, the statistics did not indicate any decisive success. As in earlier periods of history, economic and technological, rather than political factors, seem to control the size of farms. The larger the farm, and the more elaborate the equipment necessary to operate it, the more difficult it is for a single family to own.

During the early pioneer days it made little difference what size the farm might be as one man could care for only a limited number of acres of crops. The early census enumerations did not record the acreages in each crop; only the production. The census of 1860 reported on the crop of 1859 which was the most favorable crop raised in early Kansas history. The average production of corn per farm in Kanwaka township that year was 600-700 bushels. With yields of 40 to 50 bushels per acre this meant that an average farm may have had 10 to 15 acres of corn, but a substantial number could have had only 2 to 5 acres. Nearly every farm produced white potatoes, some as few as 10 bushels, the highest was 300, but the average was 68 bushels. These were the two universal crops, corn and potatoes, but in addition to these about one-third of the farms produced a few bushels of oats, and a third of them some wheat, possibly from one to five acres in each case. Few farms raised both oats and wheat. Probably the total cultivated area of each farm did not exceed 15 to 20 acres. The year 1864, reported in the census of 1865, was not a good crop year and army service had drained the country of labor, so farm production suffered. The average farm produced only 200-300 bushels of corn. By 1870, there was evidence of expansion of cultivated acreages somewhat in proportion to the size of the farm. Corn, potatoes, and oats being the standard crops; only a few farmers raised wheat, which was mostly of the spring variety. Among the more occasional crops, during the decade of the 1860s, were barley, rye, buckwheat, and sweet potatoes, but they were raised by only a few farmers and the acreages were small. Buckwheat was rather generally raised in 1859, but practically disappeared in later years.

In the census of 1875, for the crop year 1874, crop acreages were listed for the first time. Corn and potatoes were still the only crops raised by all farmers, but about two-thirds of the farms raised oats, and one of every six raised winter wheat.

The census of 1930, for the crop year 1929, showed only corn and potatoes in the four 40-acre farms; corn and potatoes on practically every one of the twenty-seven 80-acre farms, with winter wheat on seven, oats on fifteen, and sorghum or kaffir corn on fourteen. There were sixty-four 160-acre farms with corn and potatoes as the universal crop combined with oats on forty-nine farms, sorghum or kaffir corn on thirty-five, and winter wheat on twenty-nine. The group of larger farms of 320-480 acres (2-3 quarters) numbered fifty-five, again with the corn-potato combination, and forty-two with winter wheat, forty-one with oats, and thirty-eight with sorghum or kaffir corn. Only two farms contained over 500 acres.

In respect to corn it is notable that the acreage devoted to that crop had reached the optimum suitable to each size of farm, the census of 1930 showing almost exactly the same average number of acres of corn, for each size-group of farms, as that of 1875. The average corn crop in both years for an 80-acre farm was 24 acres; for a 160-acre farm 37 acres, and for a half-section farm 48 acres in 1875, but only 43 acres in 1930. The oats crop had not been so definitely stabilized. In 1875 the average acreage in oats for the three sizes of farms was 5, 9, and 11 respectively, while the 1930 figures were 8.6, 10.4, and 13.5 acres. The wheat crop was much less stabilized, the acreages being 3, 13.6, and 8, in 1875, and 11, 22, and 31 in 1930. The potato acreage in 1875 range from one-half to two acres, while in 1930 the almost uniform report was one-quarter acre. The average total acreage devoted to the four principal crops combined in each size group had doubled approximately in the interval between 1875 and 1930.

In addition to these basic crops there had been a number of supplementary or experimental crops introduced from time to time, some of which soon dropped out while others gained a permanent place in the crop program. The decade of the seventies is usually known as the period of the granger movement which grew out of a number of rural grievances; high railroad rates, monopoly, and low prices being most discussed. Other and possibly more serious difficulties grew out of a decade of erratic weather, drouth, heat, and winds, with attending dust storms and blowing soil, chinch bugs, and grasshoppers. The chinch bug menace became so serious during the decade of the seventies that every possible means of combating it was resorted to. As chinch bugs thrived especially in spring wheat, that crop was abandoned and some farmers advocated discontinuing winter wheat. The problem of one or more substitute crops in eastern Kansas was made more insistent by overproduction on corn and its consequent low price. Such a substitute crop had to meet a number of tests, besides adaptability to climate and soil, especially a sure cash market, and small bulk and weight in proportion to value, in order to stand high transportation costs to market. Hemp, castor beans, and flax, each received attention, especially flax. A St. Louis paint and linseed oil company offered to loan seed to eastern Kansas farmers, and the Kansas Pacific railroad carried on a publicity campaign to encourage experimentation. Some promoters went a step further, urging the building of mills in Kansas to crush flax seed, making the oil at home, thus promoting home industry, employment, home markets, and the saving of transportation costs. In consequence of such agitation, some flax was raised for a number of years, but it never attained an important or permanent place in the crop system. Some sweet sorghum had been raised for molasses as early as the territorial period, but the making of sugar from sweet sorghum attracted much attention during the eighties and nineties. As a sugar crop, sorghum was a failure, but it became a permanent part of the agricultural system as a forage or seed crop along with the non-saccharine varieties, kaffir in the nineties, and feterita and sedan grass after 1910, and more recently, atlas sargo, and other improved varieties.

Until the 1880s, the Kanwaka township farmer depended primarily upon native grass for pasture and hay, but a few planted timothy, clover, and millet. In another ten years timothy was the dominant tame grass, supplemented by a little clover. By 1905 the field was divided among four rivals, timothy, clover, blue grass, and alfalfa, but alfalfa was not generally grown until after 1905, when timothy dropped out for the most part. The latest of the newcomers were sweet clover and lespedesa, during the last fifteen years, 1925-1940. In spite of eighty-five years of agriculture in Douglas county, a substantial acreage of native grasses remained, and where given anything like a fair chance they survived in vigorous condition. The early settler began worrying about the grass problem very soon after settlement, predicting the early disappearance of native grasses for both pasture and hay. They were thinking, naturally, in terms of eastern forest clearings, not in terms of the western prairie-plains grass region, and they did not appreciate the vigor and recuperative power of Kansas bluestem and its associates.

Farm power in early Kansas was supplied by horses and oxen, and in 1860 the numbers were about equal in Kanwaka township. By the close of the Civil War oxen were going out of use, rapidly, there being only one ox to six horses. By 1870 only a few oxen remained. Prior to 1905 there were only a few mules, and there was no general use of mules until about 1915. Prior to World War I this part of Kansas represented, quite literally, a man and horse power era. Any degree of mechanization came only in the very last years.

It is clear from the review of the crop program of the early years that Douglas county was a part of the corn belt, raising that crop almost to the exclusion of others. The only profitable way to market corn was in the form of livestock and their products. Sheep raising did not gain an important foothold during any period, so the livestock business centered around cattle and hogs. Although hog production fluctuated widely, there is little indication of an increase in the number of hogs per farm after the 1870s. The census statistics are not an altogether accurate indication, however, partly because of deficiencies in reporting, but partly also, because more efficient feeding, and the marketing of younger animals, yielded a larger amount of pork from the same corn acreage. Not all farms, especially the small farms, had all three types of livestock, milk cows, beef-cattle, and hogs. The following averages apply only to such farms as had the particular class specified. If the averages were based on the whole number of farms in the group the figures would be much reduced.

Milk cows always occupied a substantial place in farm planning, such 80-acre farms as had cows having 3-5; the 160-acre farm 4-6, and the larger farms averaging about 6-10. After 1870, however, there was no permanent increase in average numbers of cows per farm. The increases in production were derived from better cows and more efficient management. In the beef-cattle division the numbers fluctuated inversely with the prosperity of field crops. During years of prolonged depression, such as the late 1870s, cattle increased. During the boom years of the mid-eighties, cattle declined. In 1880 the 160-acre farms reported an average of fifteen head of cattle per farm in addition to milk cows, but five years later the average had dropped to eight. During the depression of the nineties the average increased to eleven per 160-acre farm, declining again by 1905 to eight. Taking the last sixty years as a whole, however, there was no clear long-time trend in numbers in either direction. On the other hand, as in the case of hogs, there were substantial changes during the last years in the methods of feeding and in marketing younger animals which made comparisons for recent years uncertain. There is no record available, however, of feeders, either cattle or sheep, transient livestock, finished for market on Kanwaka corn.

     Probably the most significant changes in the livestock picture during the last sixty years, were the improvements in quality, and emphasis on pure bred or high grade animals. During the 1880s William A. Harris built up his herd of Scotch Shorthorns, at Linwood, Kansas. Although his famous farm was just outside Douglas county, Harris was closely identified with Lawrence. He was one of the pioneers in the United States in developing the Scotch strain of Shorthorns, and was rated one of the nation's foremost breeders. During the same period the firm of Shockey and Gibb was among the foremost American importers of Hereford cattle, and gave Lawrence national publicity in that field. The depression of the nineties was disastrous to most of the breeders of fine stock. Although a number of fine herds of thoroughbred cattle were identified with the county later, probably none occupied so distinguished a position nationally as the Harris and Shockey cattle.

     In days when the farmers were suffering from over-production of agricultural politicians, sensational statements often appeared on the subject of long-time depletion of soil fertility. As proof of the destruction of farm lands it is customary to cite instances of fabulous crop production in the early days compared with later yields. Much, if not most, of such propaganda should be discounted. There is good reason to doubt whether many of these reports of extraordinary yields are either reliable or representative. Early Kansans were realestate boomers, and because of the reputation of the state for drouth it was customary to exaggerate. Measurements of acres and of production were faulty, — usually they were only rough guesses. Only small tracts of the best land were first opened up on each farm, and under favorable conditions they should have given high yields, higher than in later years when expansion of acreage under the plow could be accomplished only by cropping less fertile land. Of course, it should be emphasized that the definition of soil fertility is itself a debated question, and one of those that cannot be proven positively one way or the other. Neither can the question of productivity be settled, in part for the simple reason that there are no long-time comparative records that are adequate, and in part because production capacity depended so largely upon machinery and methods of culture. Without question, some land was abused, but on the other hand much land was more efficiently farmed as the years passed. It is not possible to strike a satisfying average for a whole county, and much less for a whole state. Modern machinery, especially power machinery, provided more efficient handling of land than was possible under a horse-power agriculture. Even if maximum yields per acre were not increased above those under more primitive methods, there were not so many failures. The greater stability insured higher average production over a period of years. The problem of soil fertility was not so much one of restoring past depletion as of guarding against losses from erosion and from the more intensive utilization which was being exacted of the land and which was becoming more intensive with the passing years.

Other samples have been presented elsewhere, for the early years of Dickinson and Saline counties, between $97°$ and $98°$, in <u>Winter Wheat</u> (1944), and for parts of Edwards county, near $99° 30'$, fragments appear in separate articles (1935, 1936, and 1940). The reader must consult these publications for the details and the local setting. In those samples, the introduction and proving of hard winter wheat and the grain sorghums through the 1880s and the 1890s was reflected conspicuously in the nature of the farm program. Corn and livestock gave way to wheat and to the sorghums in their proportionate place even where actual corn acreages were not yet reduced. In Geary and Riley counties, just to the eastward of Dickinson county, the corn-hog-cattle combination persisted generally, with the addition of the sorghums and alfalfa, the transition line dividing corn and wheat being fairly sharply defined between the Bluestem Pastures region and the central hard winter wheat region. The hill and limestone rock topography entered into the sharpening of this transition, and to a lesser extent the soils, and farther west the outcrop of the Dakota sandstone strata.

Besides the Kanwaka township, Douglas county, sample, tables are printed for five other samples, each representing a different type of farming area, or a different sub-area. The commercial corn-belt sample, Walnut township, Brown county, shows the clear predominance of the 80, 160, and 240-acre (3 eighties) farms until 1920 when, for the first time, there were more half-section farms than eighties, and thereafter the proportions remained fairly stable, with the quarter-section farm of historical tradition as the most numerous size, the 240-acre unit second in number, and the half-section third.

The behavior pattern was substantially different from the Kanwaka sample in the mixed farming area. To the westward and southwestward of Douglas, and Brown counties, lies the Bluestem Pasture region. The nature and variety of farming, corn, sorghums, alfalfa, and some wheat, combined with the commercial summer pasture operations, produced so complicated a situation that statistical tables are an inadequate method of presentation. For large areas of upland, the pastures prevailed, while the broader river bottoms permitted farming operations of substantial proportions. Some reference has been made already to the manner in which the pasture practices interlocked with the southwestern range cattle industry, the corn belt, and grass-fat market. There are also stock farms devoted to the production of pure-bred cattle, hogs, horses, and sheep. Pending the completion of a full-scale historical treatment of the Bluestem Pastures, already in progress, the present author's introductory essay (1942) gives the best presentation of that region (See also, Doll, Kansas Experiment Station Bulletin 294, 1941).

The table for Buckeye township, Dickinson county, repeats part of the table in Winter Wheat (1944), but gives the full chronological range 1860-1940. The quarter-section farm remained the largest size-group, with the next two large size-groups about equally represented in numbers. The list for 1940 indicated a slight break in the pattern, but only time can determine whether or not it was a change of trend. One degree westward on the 98 meridian, in the southcentral Kansas hard winter wheat belt, is Vinita township, Kingman county. The first census after settlement was 1880. By 1895 the half-section farms and the 80-acre farms were second and third in number to the 160-acre farms, which represented half of the whole number. The clear change of trend to the larger farm was not registered until the enumeration of 1905 (federal 1900 might indicate it, if available), but not until 1920 or forty years after the first settlement record did the half-section farm take the lead. 1930 indicating the extreme trend toward large units. The 1940 census again raised questions, the half and the quarter-section groups being equal.

On the 100 meridian, Highpoint township, Ness county, offers much of interest, but the most of it cannot be read from the statistical table. Occupying an area particularly subject to uncertainties, livestock occupied an important part in its economy, and utilized non-resident owned land during its early history. The reported size of farms often possessed little meaning even after 1905 when the census revealed a sharp shift to larger units. By 1920 more intensive settlement revealed even more clearly the fact that three-quarters to a full section-and-a-half farms were approximately the units that had actually been utilized irrespective of the way in which records were kept. From 1920 through 1940 there was little readjustment in size.

Wallace county, lying against the Colorado line, the 102 meridian, near the northwest corner of Kansas, was studied as a whole. It is typical of what is usually thought of as level grama-buffalo grass plains, with a soil, fine-grained, heavy, dark, and windblown. It is that beautiful kind of plains country that incites forest man to denounce it as the place where one can look farther and see less than any other place in the world. First settled near the end of the nineteenth century, the normal holdings were quarters and half-sections. The decade, 1895-1905, worked a readjustment into larger holdings, and then the resettlement period 1905-1915 reverted partly to nominal quarter-section farms. By 1920 the emphasis was again on sizes of three quarters and up, continuing either the livestock tradition or yielding to wheat under the influence of mechanized farming. The trend to very large holdings represented in 1940 was probably abnormal, the figures in the last three enumerations being influenced by corporation farming and the intervention of the legislature to outlaw that form of business organization in agriculture.

Some conclusions should be clear from all these samples, especially the point that was made in the introduction of this discussion. In the early and middle years for all the samples, the possible income to the operator was small, very small, irrespective of prevailing prices of farm products, except for the few large farms. Few farms, even under favorable crop conditions, could have produced much to sell. Even survival requirements placed a heavy emphasis upon management, bare subsistence, and self-denial. Only the large farms, well managed in specialized agriculture, involved a cash income large enough that wealth could be accumulated out of farming operations, over a long period of years, as distinguished from value derived from enhancement in land prices resulting from the growth of settled communities. Men possessed with that sixth sense of business management often did make their fortunes out of both farming operations and buying and selling of land, but such was not the experience of the average small farmer. Periodic depressions, and especially the prolonged agricultural depression after 1920, bore down with particular weight upon American agriculture because they were periods of deflation of land values when the burden of financing agriculture fell exclusively upon the income from farm production. Nothing else could do more toward clarifying the whole farm problem, historical and contemporary, than to isolate successfully those two factors in their bearing upon the agricultural situation. As the present author has pointed out elsewhere, (Winter Wheat, 1944), irrespective of how low an income was received during the early stages of agricultural development, on the basis of capital investment there were few, if any, productive industries that were expected to yield, and which did produce, so high a rate of return as agriculture. In fact, it is a mistake to treat early American agriculture, and especially pioneer agriculture, as a capitalistic enterprise in that sense, because the farmer's principal investment was labor, not capital. That fact, kept clearly in view, tends to clarify the farm problem as twentieth century agriculture finds itself more and more in a position where it must operate for the first time on a cost accounting, strictly capitalistic, basis as any other productive industry, and must reduce its cost of production to a point where it can compete on the world's markets on a cost basis. Furthermore, the shift of farming operations to larger farm units coincided, not only with mechanization, which came only with an effective low-cost mechanical power after World War I, but it coincided closely with the high level of farmer stabilization emphasized in the farm population turnover discussion. It is not intended that this statement is to be interpreted as a cause and effect relationship, merely the two are closely associated chronologically, all these independent variables interact together upon the whole rural scene.

    The importance of the evils and abuses agitated by the historic farmer movements, and by agricultural politics (in the whole middleman chain of services) in relation to a possible rural prosperity, have been much exaggerated. They too must be reduced to perspective among the several independent variables, and even if all such alleged evils could have been fully remedied, they could not in themselves have made the farmer prosperous. As an example, an 80-acre Kanwaka township farmer grew for his principal crop, on the average, 24 acres of corn, yielding 30 to 40 bushels per acre, or 720 to 960 bushels. If he sold half of it for cash and was "robbed" by the middlemen of 25 per cent of a supposed just price of 40 cents per bushel, the total loss would be $36.00 to $48.00 per year. Although those sums were appreciable in the economy of the last third of the nineteenth century, they should be a warning that such income differences on his leading crop would not explain adequately the difference between farm prosperity and depression. And such an approach to the problem of the farmer in his historical setting may contribute a better comprehension of how small the operations of the small farmer really were, and why a few dollars in money meant so much to him. In overall effect, the hazards of weather on crops and prosperity were greater than the hazards of price. To ponder agriculture in this perspective brings the historian face to face with the stark realism of rural life.

Brown county, Walnut township, (Commercial corn-belt county)
Number of farms in each size-group

Date	1860	1865	1870	1875	1880	1885	1895	1905	1915	1920	1925	1930	1935	1940
No. of farms	35	145	239	130*	210	240	222	244	226	197	217	177*	226	197
1-19 acres	1	1	0	0	9	11	5	5	3	1	4	5	7	4
20-49 "	5	4	13	8	20	10	6	3	7	3	1	5	5	1
50-99 "	4	26	71	38	62	61	58	49	34	16	14	12	18	18
100-174 "	20	37	75	50	83	92	94	111	108	97	102	77	89	86
175-259 "	3	18	23	15	16	26	34	32	38	45	58	39	54	50
260-499 "	2	12	21	12	16	23	21	28	26	25	22	35	27	28
500-999 "		2	4	5	3	7	4	3	1	5		3	3	6
1000-4999 "			2							1				
No data		45	30	2	1	20		13	9	4	16	1	23	4

Note: *Evidently the census roll is incomplete for 1875, and 1930.

Dickinson county, Buckeye township, (East Central winter wheat region)
Number of farms in each size-group

Date	1860 whole county	1865 whole county	1870 Grant twp.	1875 Buckeye twp.	1880	1885	1895	1905	1915	1920	1925	1930	1935	1940
No. of farms	12*	80*	47*	75	78	112	100	100	131	122	116	105	111	98
1–49 acres		2		1					1	1	2	2	2	2
50–99 "		6	8	33	30	37	19	18	21	17	12	14	12	11
100–174 "	7	45	13	31	31	43	47	53	59	70	48	52	40	35
175–259 "	2	4	7	5	11	11	12	10	15	18	24	18	24	19
260–499 "	3	7		3	7	11	16	15	15	12	27	18	21	24
500–999 "		1		1	1	8	5	4	1	1	1	1	0	4
1000–4999 "	1					1							1	1
No data			19	1		1	1		19	3	2			

Note:
* The data for 1860, and 1865 are for the whole county; and those for 1870 are for Grant township, which includes Buckeye. The data for 1875 and later are for Buckeye alone.

Kingman county, Vinita township (South Central winter wheat region)
Number of farms in each size-group

Date	1880	1885	1895	1905	1915	1920	1925	1930	1935	1940
No. of farms	80	84	106	86	80	76	81	68		71
1-49 acres			1	4	0	2	0	0		1
50-99 "	4	4	19	8	6	2	3	3		6
100-174 "	67	58	54	23	32	17	19	11		24
175-259 "	1	5	10	19	8	13	12	10		8
260-499 "	8	9	20	25	25	30	31	29		24
500-999 "		1	1	6	6	10	9	11		7
1000-4999 "		1	1	1	1	1	1	1		1
No data		6			2	1	6	3		

Ness county, Highpoint township (100° winter wheat-livestock)
Number of farms in each size-group

Date	1880	1885	1895	1905	1915	1920	1925	1930	1935	1940
No. of farms	111	103	56	96	120	158	160	134	149	147
1-49 acres				1	1	0	1	1	1	1	2
50-99 "				0	0	1	3	1	2	1	4
100-174 "	96	81	27	29	6	13	13	6	17	16
175-259 "	1	4	0	2	5	7	7	5	4	4
260-499 "	12	7	26	42	69	85	78	80	78	77
500-999 "	2	5	2	17	22	39	49	34	35	35
1000-4999 "		3		4	11	6	3	5	3	9
No data		3		1	6	4	8	1	10	

Wallace county, Size of farms, (West line wheat-livestock region)
Number of farms in each size-group

Date	1895 No.	1895 %	1905 No.	1905 %	1915 No.	1915 %	1920 No.	1920 %	1925 No.	1925 %	1930 No.	1930 %	1935 No.	1935 %	1940 No.	1940 %
No. of farms	360		246		314		312		270		387		475		319	
0-19 acres	0	.0	0	.0	0	.0	2	1.2	0	.0	3	.75	2	.0	1	.0
20-49 "	3	.83	1	.40	2	.63	2	1.2	2	.5	1	.25	4		0	
50-99 "	6	1.67	2	.81	6	2.0	3	1.8	4	1.0	2	.6	4		2	
100-174 "	273	75.8	80	32.5	133	42.3	58	18.6	40	10.8	34	8.8	48	10.1	20	6.2
175-259 "	2	.55	3	1.22	3	1.0	10	3.1	12	3.2	11	2.8	18	3.8	11	3.4
260-499 "	62	17.2	52	21.1	64	20.4	86	27.5	112	30.2	100	25.8	124	26.0	67	21.0
500-999 "	5	1.4	34	13.8	30	9.5	57	18.2	113	30.6	121	31.3	132	27.8	88	27.6
1000-4999 "	4	1.1	41	16.6	30	9.5	56	18.0	76	20.6	86	22.2	91	19.1	111	34.8
5000- "			17	6.8	6	2.0	10	3.1	5	1.3	10	2.5	6		11	3.4
No data	5	1.4	16	6.5	40	12.7	28	9.0	6	1.5	19	5.0	47	10.0	8	2.5

Note: The large number of farms deficient in acreage data diminish the meaning of the percentage figures. An important element of disturbance was the acquisition of a large acreage by farming corporations, which underwent forced liquidation when corporation farming was outlawed in 1933.

Chapter Eighteen

LAND TENURE, OPERATOR TURNOVER, AND FARM ORGANIZATION

In an essay on agricultural policy under the title "Mobility and history", (<u>Agricultural History</u>, 1943) the present author discussed several aspects of the land problem and of the farm from the standpoint of business organization. A rethinking of the whole land tenure problem was advocated. The accompanying table provides data on some aspects of that matter, using land tenure statistics drawn from twenty-three sample communities used in the other population and agricultural studies. The Kansas agricultural census of 1920 listed land tenure under three heads; owner, part owner and part renter (owner-renter), and renter. The procedure used in the original study of population turnover was applied to these three classes of operators, listed according to tenure, with a view to determining the behavior of each, 1920-1935. It is not possible to determine whether or not the operators occupied the same land throughout that period, only that they remained within the same census unit-area and were therefore still a part of the same community.

The results are tabulated in the accompanying table. In twenty of the twenty-three samples the owner-renters were more stable in the community (not necessarily on the same piece of land) than other forms of tenure. In seven of the twenty samples, the stability of the owner-renter was approximately twenty or more points higher in the percentage scale than the closest other type of tenure. In another seven of the twenty samples, the stability of the owner-renter was about ten points higher than the closest other type. In another six samples, although higher, the differences between the owner-renter and the other form of tenure were slight. In respect to the owner group in seventeen of the twenty-three samples the owners were more stable than the renters. In twelve of this seventeen the owner persistence was substantially higher than the renter, but in the other five it was slight. Put in a different form, in only twelve cases, or about 50 per cent of the whole twenty-three was there a substantial difference. On the other hand, in three samples of the twenty-three, the renters were substantially more stable than the owners, in one, more stable than either owner or owner-renter. In only two samples was the owner more stable than the owner-renter. The second major conclusion, to supplement that at the head of this paragraph, is not that the owner was more stable than the renter, but the fact revealed by the statistics that the supposedly greater stability occurred in only about 50 per cent of the cases and in even those it was slight. The real point is that there was little difference in respect to community stability as between owners and renters. Again, the distinction may be emphasized that it is persistence in the community, not on the same farm, that is being measured. No records exist that would determine exactly the land occupied for the period under consideration or for any earlier period. These conclusions do not change any facts, but they do modify the approach to the problem represented by those facts.

Percentage of farm operators of 1920 present in 1935 in the same community

Township	County	Owner-renter	Owner	Renter
Center	Doniphan	76.1	54.7	53.
Eudora	Douglas	54.5	51.1	32.7
Kanwaka	Douglas	61.2	54.9	27.5
Valley	Linn	43.4	40.4	15.3
Walnut	Brown	70.	56.6	51.2
Agnes City	Lyon	68.4	59.5	29.7
Pike	Lyon	51.4	48.4	36.4
Reading	Lyon	73.5	51.6	29.
Macon	Harvey	66.6	57.1	53.7
Wayne	Edwards	81.2	52.6	41.4
Big Creek	Russell	63.3	52.5	34.
Highpoint	Ness	51.8	51.	36.5
	Barber	46.6	37.8	22.2
Sinclair	Jewell	58.8	53.4	37.2
	Hamilton (whole)	37.5	32.2	16.3
	Wallace (whole)	56.2	44.3	12.6
	Cheyenne (4 twps.)	56.9	46.7	29.4

		Owner-renter	Renter	Owner
Vinita	Kingman	70.8	51.4	23.5
Center	Decatur	70.5	47.0	37.5
Grainfield	Gove	71.4	32.2	23.8

		Owner	Owner-renter	Renter
Alexandria	Leavenworth	60.2	57.	40.8
Long Island	Phillips	45.	34.6	23.2

		Renter	Owner-renter	Owner
Walnut	Saline	46.8	45.8	32.5

Statistics may show what happened, but they do not explain why. In the course of examination into the problem of an explanation, the situation was submitted to farmers, bankers, and grain men. In no case was anyone aware of the owner-renter form of tenure being adopted deliberately as a matter of considered policy, but in every instance the same explanation was given, along with illustrations drawn from the community known to the commentator. It offered a flexible method of expanding or contracting operations to suit conditions. Expansion was possible by renting such additional land as would make machinery units operate to capacity and reduce unit costs, without incuring debts for purchase of additional land, with its tax and interest burdens. When conditions did not seem to warrant large operations the lease would not be renewed, the investment risks on the additional land being carried by the owner rather than the operating farmer. It was an object lesson in business organization for bringing together into one operating unit larger amounts of capital than one man could contribute, for dividing the risks among two or more persons, and for affording opportunity for the participants to enjoy a certain diversification of investment. It calls attention to one of the conspicuous aspects in which agriculture was deficient in comparison with city business organization, but illustrates how the practices of farm operators were evolving out of experience a sounder body of business practices. In conclusion there is one defect in the data that obscures the fullest analysis of the situation. The census data did not record the amount of land owned and rented and therefore no conclusions are possible, whether or not there was a possible optimum ratio determinable as between the owned and the rented portions.

The "actual settler" concept and the "family size" farm are in large part social myths which were more closely associated with propaganda than with history. The idea of tenantry was branded with a social stigma. There is an increasing, though not yet a very substantial, tendency to recognize that tenantry is not necessarily a mark of pathology, but to some extent at least, it is a necessary adjustment to commercial agriculture in a machine age with high capital requirements. Tenantry became a form of rural business organization in which more than one person contributed the capital and shared the risks, and as such might serve as one of several sound and desirable methods of conducting a business (Malin, Mobility and history, 1943). Ladd Haystead, agricultural editor of Fortune (December, 1945), presented excellent case studies of tenantry under certain circumstances as a positive advantage both individually and socially.

Salter (1943) presented an important monograph on a Wisconsin township settled in the 1850s and 1860s by a highly stable population. Much of the land was passed along within the membership of the resident families. When no heirs existed, or when one heir was unable to finance the purchase of the rights of other heirs, the land had to be sold out of the family. Tenantry and mortgages were essential factors in the succession of title, either within the family or when sold out of the family, and the better the land and improvements, the more necessary were these factors in the tenure problem.

As a matter of historical development, a part of the range livestock industry came to operate on essentially an owner-renter basis, the government becoming the landlord through the medium first of the forest reserve ranges (1905-   ) and later of the public domain ranges The stockman owned his home ranch and grazed his livestock during a regulated season under lease upon the public range. The Bluestem Pastures of Kansas, in private ownership, occupied a somewhat similar relation to the southwestern cattlemen. So much has verbalism and the stigma of a name confused thinking that men who denounced tenantry and pledged themselves to the preservation of the family-size farm, at the same time demanded the Taylor Grazing Act and the extension of the leased grazing system of the forest reserves to the whole of the public lands — an owner-tenant system.

Along with the re-thinking on land tenure in terms of business organization is the associated problem of farm management. The present author emphasized management in his Winter Wheat (1944), and Renne, in reviewing the book (J.F.U.& L. Econ., February, 1945) took issue, emphasizing group action and area diversification rather than management of the individual farm. The point that Renne overlooked was that in either case the issue was management, and that he was only arguing for a particular kind of management. Again, Haystead has presented something of the management side in his Meet the Farmers (1944), and in Fortune (May, and December, 1945) — not community planning, but planning for the individual farms under the direction of professional farm-managerial service. The family-size farm myth assumed tacitly that all farmers were endowed equally with managerial ability, an assumption long abandoned in connection with the industrial city factory worker. There were several possible methods of achieving division of labor in agriculture as in industry, and a number of adjustments were in evidence, but only a beginning had been made in exploring the extent and efficiencies of available alternatives.

Chapter Nineteen

THE COMMUNITY AND THE CITY

The community

Out of the studies reviewed in the preceding chapters emerges a dynamic concept of the community, one that differs necessarily from that of the sociologists who think of the community as a unit of contemporary social organization. Historically speaking, the community does not fit so neatly into any pattern. In eastern Kansas before the time of railroads, and farther west prior to the entrance of a railroad on the scene, there was often something of an orderly sequence of developments. Settlers were scattered, acquaintances were wide, often as extensive as a conventional county, and there had not yet emerged any fixed centers of organization. Individual settlers felt free to meet at different places and to participate in activities that might bring them together. Later this larger area became more differentiated, centering around a local trading center for some activities, or a schoolhouse or other convenient place for union religious, or social gatherings. With the settling-in of numbers, the major religious denominations set up separate services, school district lines were drawn, and families became identified specifically with particular social interests and groups. With the establishment of a village, the differentiation between town and county began. Where newspapers were established that covered such an area, even in the correspondence columns it is often possible to trace fairly definitely such processes of evolution. A more complicated sample community is traced for another purpose, in the final chapter on Dutch Henry's Crossing in the present author's John Brown and the Legend of Fifty-six (1942), but incidentally it illustrates much of the community evolution as a process

Many Kansas communities had their origin, however in the building of a railroad, with townsite promotion planned at intervals along the line, together with county seat ambitions. Such communities had a different history and one less likely to follow a normal pattern of development. Often a local promoter or group dominated these communities or attempted to do so. As the local newspapers are the best source for community development in Kansas (the Kansas State Historical Society has a complete file of practically every newspaper printed in the state after 1875), newspaper coverage is emphasized as a measure of community unity and organization. Of course, the quality of the newspaper depended upon the editor, but systematic analysis of the newspapers reveal more than most casual users of such sources realize. An important limitation must be recognized, however, because a stage was reached in every community when the newspaper no longer afforded adequate coverage. Apparently, the dividing line is to be found in social differentiations and complexity of interests, especially those centering around the distinctions between town and country. In the present author's book Winter Wheat (1944), he pointed out that the break came about the time of the boom of the late 1880s; the collapse of the boom brought a partial restoration of the old relationship, but after the opening of the new century, the newspaper no longer served the historian in the manner of the earlier day. In Kinsley, Edwards county (1935), the change occurred about the same time, in eastern Kansas the changes occurred earlier, but in such ambitious cities as Kansas City, Missouri, Leavenworth, Atchison, Lawrence, and Topeka, Kansas, this integrated community aspect of the newspapers vanished very soon after establishment, because political and city rivalries occupied the center of editorial interest.

Still another aspect of the community problem is its relation to the railroad. This matter is dealt with incidentally in an Edwards county study in beginnings (1940), and in Winter Wheat. The railroad was not only the principal taxpayer, it was sometimes virtually the only one, and naturally it exercised an interest in local officials who spent the money, and to a degree it became the guardian of the welfare of the communities served by the road. It is notorious that most pioneer communities were in need of a guardian against the corruption of the local county court house rings. During the pioneer period most of the traffic was incoming freight, which must pay the operating charges both ways until there was something to ship out. Abuses charged against the railroads have been exploited at length and far beyond the facts, but the present author knows of no community study done in such a manner as to illustrate the constructive side of the railroad as the indispensable institution which made possible the general settlement of the grassland.

The city

The ubiquitous problem of the city in the twentieth century demanded attention from all directions and resulted in attracting more than a little speculation as to the rôle of the city in the history of civilization. The English geographer, Vaughn Cornish (1923), formulated his theory that great capitals occupied a forward position in respect to the vital expanding or challenging interest of a state, and that the factors determining its geographical location were a granary, a crossroads, or a fortress, or combinations of them. In a federal state, such as the United States, there were two capitals: New York, the economic capital; and Washington, the political capital. Whittlesey, in The Earth and the State (1939), gave his own individual turn to the idea of capitals. N.S.B.Gras (1922, 1925), an economic historian, developed the historical evolution and rôle of metropolitan areas and directed his students in studies of particular samples. "The industrial city; center of cultural change" was the theme of R. E. Turner (1940). He was too much preoccupied with the economic interpretation of history, but the discussion was both an important and a significant aspect of the larger problem. The most complete sociological presentation of the twentieth century city was that of McKenzie (1933), The Metropolitan community, in which the approach was ecological. Park (1929) formulated the idea of "Urbanization as measured by newspaper circulation." He gave it a purely twentieth century application as sociologist with an economic emphasis. There is another aspect of the newspaper, however, that is highly significant if the emphasis is centered in that institution as a collector and disseminator of intelligence, not just news. For historical purposes the newspaper is too recent a medium to serve as an independent variable in general historical interpretation. The suggestion is here proposed that while the city serves all the functions suggested in the foregoing citations, its most significant and dynamic rôle in history is that of intelligence center for the accumulation and dissemination of intelligence, using that word in the most inclusive sense embracing the fine arts, education, recreation and amusement, and all forms of intellectual activity including printing, and reproduction of ideas. It should be noted that the city is not necessarily a prime creative force except that in its accumulation of numbers of people it may include its proportionate share of creative individuals, and under some circumstances it may draw within its orbit more than its proportionate share.

In early colonial America, cities occupied coastal positions, exchange centers primarily between their immediate hinterland and Europe. Quebec, for the

St. Lawrence valley; Philadelphia, Boston, and New York, leaders for the middle and upper Atlantic coast; Charleston, Norfolk, and Savannah for the southern Atlantic coast; New Orleans, Mobile, and Pensacola for the Gulf coast. With the penetration of the interior by white settlement, these coastal cities became rivals as gateways. Quebec, and Montreal remained outlets for the Great Lakes, and the St. Lawrence basin; New York, Philadelphia, and Baltimore were rivals for water and turnpike connections across the mountains. In an era of water communications, New Orleans became the forward city of the Gulf slope, the interior of North America, south of the Great Lakes, facing that direction rather than eastward. Behind New Orleans in the interior were Memphis, St. Louis, Cincinnati, and Pittsburgh on the river, and Cleveland, Toledo, and Chicago on the lakes with artificial water way ambitions for river connections. About the middle of the nineteenth century a redistribution of urban power came about under two diverse influences; the fact that the United States became a two-front nation, and the rise of rail communications as a factor in internal tensions. The turn of the tide in the rivalry of rails over water after the middle 1850s made New York the forward city serving the interior from the eastward. Rails laid the basis for the rise of Chicago as the leading interior city, rivaling St. Louis with its water and rails, the southern river cities failing to keep pace with the expanding volume of urban services.

 After the acquisition of the Pacific coast, centering on the year 1846, the fact of being a two-front nation changed further the outlook of the interior cities which faced west as well as east, but there was the constant threat of an isthmian canal which might perpetuate the influences of New Orleans on a rail and water basis. The first Pacific railroad project was that of Asa Whitney, in 1843, by the northern route from the Great Lakes. At the southern extreme the lower Mississippi river cities looked westward by way of the El Paso route secured by the Mexican war and the Gadsden Purchase. New Orleans, Jackson, Memphis, or even Cairo might have connected with such a project. Memphis had a more desirable route directly westward to Santa Fé. Douglas had thrown in his lot with Chicago and from 1845 was interested in a railroad by way of the South Pass. Benton's interest was in St. Louis and the central route. Cincinnati was quite effectively cut off from direct access to the western trade and cultural influence, but was unwilling to admit defeat as is evidenced by the Boynton and Mason expedition, and book of 1854, and subsequent participation in the Kansas controversy. When the panic of 1857 crashed its hopes, Toledo thought it was on the point of realizing its dream of cutting through between St. Louis and Chicago by means of a direct rail connection to Hannibal, Missouri, and to St. Joseph on the Missouri river.

In historiography, this theme of the sectional and city rivalries for the eastern terminus of the Pacific railroad and its relation to the Kansas-Nebraska Act was the principal contribution of F. H. Hodder in a number of papers beginning in 1912. The problem of the key cities of the plains is a theme that had not yet been developed in historiography, a line of rival cities lying west of the Chicago-Mississippi river line. As already pointed out, Gilpin (1857) thought there would be one key city and that it would develop in the Independence-Kansas City area, and Van Horn, of the Kansas City Enterprise, was equally convinced of the inevitable dominance of his city. A conspicuous group of geopolitical prophets at the beginning of Kansas history were convinced that the key city of the plains would lie at the most western possible point of available coal for fuel and power, and of agriculture, and that while water navigation was desirable, the basis of communications in the new region was rails. The most westerly points considered seriously were Council Grove on the Santa Fé trail, and sites in the vicinity of Fort Riley, near the junction of the Smoky Hill, and Republican rivers, the head of possible river navigation. These considerations determined the location of the first Kansas capital at Pawnee, at what was thought to be the east edge of the Fort Riley military reservation, with Governor Andrew H. Reeder and other territorial officers as stockholders. Political controversy growing out of the slavery issue destroyed the project, but nearly a dozen other cities were planned on sites close by, Junction City making the strongest bid for some realization on its ambitions. The reasons alleged for going as far west as this 97 meridian line, the western edge of the oak-hickory timber-tall-grass region was that it was the point beyond which agriculture was thought to be impossible and therefore there could be no threat of some city farther west cutting off its forward position facing the plains and the Pacific coast as Independence had cut behind Franklin, Missouri, on the Santa Fé trail, and Westport in turn had cut off Independence, and Kansas City expected to cut off St. Louis.

The dreams of most of these planners of the future did not materialize, and much of the planning is important to the understanding of the era and the concept of regionalism involved. Over-optimism about the extent and quality of coal deposits was one of the fatal defects of much of this planning, as well as the failure to realize the long road necessary to bring agriculture into harmony with the new low-rainfall grassland. The Missouri river cities, Kansas City, St. Joseph, and Atchison, maintained largely their dominance, for example in the wholesale hardware trade in western Kansas, until the second decade of the twentieth century, and St. Louis did not bow to the inevitable until about 1910 when the leading wholesale hardware firm of that city established a branch at Wichita. By the next decade Wichita, Hutchinson, and Salina were well established as the wholesale distributors for the country farther west, as well as performing a part of the marketing and processing services for their territory. But so far as large cities were concerned, those of near metropolitan proportions, Denver was the first one to qualify west of Kansas City, thus fulfilling Gilpin's prophecy for both locations. Farther west, Salt Lake City was the next metropolitan center on the road to San Francisco.

As Lake Superior and Lake Michigan lie athwart the northern route from the Pacific eastward, Chicago occupied a double terminal position for that route and the north central route by the South Pass. On the latter route, Omaha grew on the Nebraska side of the state line, its concentration of wealth contributing to the state it served, which was unlike Kansas City whose wealth derived from Kansas was diverted to Missouri. West of Omaha, Cheyenne was the only sizable city on the route to Salt Lake City. On the northern route, Chicago, Milwaukee, and Duluth on the lakes and the Twin Cities of Minnesota divided the eastern terminal influence, but between Minneapolis and St. Paul, and the Pacific coast cities, there was no important city. In this situation is to be found much of the basis for discontent of that region.

Along the southern approaches to the plains grew a line of moderate-size cities, Tulsa, Oklahoma City, Fort Worth, Dallas, Houston, and San Antonio. Between that line and the Pacific coast there were four cities of moderate local importance, each limited by geographical factors; El Paso, Albuquerque, Tucson, and Phoenix. Except for these, however, the long jump was made from the plains-edge to the Pacific coast. In considering the problem of the grassland region it is imperative to think of these cities in larger terms than economics or politics. The more concise test of their adequacy as regional cities serving regional needs lay in the study of them as intelligence centers, and it is there that regional deficiency was most conspicuous. The cities took over, almost bodily, the forest man's concepts. Only in isolated instances did urbanization make adjustments which harmonized with grassland-desert regionalism.

The rail unifications that bound the Mississippi line of cities with the Atlantic seaboard seemed on the brink of realization in 1856 and 1857. The panic of 1857 halted the process which was not resumed until the New York Central consolidation in 1869, twelve years later, and others waited nearly another decade. In 1869 the first Pacific railroad was completed, with three others by 1883. It was this internal communications revolution that laid the foundation for the communications pattern of three-quarters of a century duration, even motor coach and truck traffic following much the same model. The dominance of New York was extended over the whole country as far west as the continental divide, and even to a large degree westward of that line, not only in economic matters, but even more completely in those things that fall under the head of accumulation and dissemination of intelligence.

In respect to the Trans-Mississippi West, Ladd Haystead, in his book *If the prospect pleases* (1945), argued that the continental divide was becoming the new regional dividing line, everything west looking to the Pacific coast cities, while everything east looked in that direction. So far as land communications dominated, that conclusion seemed sound. The present writer has pointed out (Space and history, 1944) however, that air communications were working a new redistribution of power among cities, one in which east-west lines and cities based on such relations must give way largely to north-south lines in circumpolar relations and the cities serving that new orientation. On the basis of either of these lines of orientation, however, it is clear that no city, or cities, of sufficient importance, have developed in the Trans-Mississippi West to serve as independent intelligence centers, and thereby, to qualify in any peculiar regional sense as a center, or centers, of an indigenous grassland culture.

Chapter Twenty

HARMONIZATION OF CULTURE

Folk process and science

The original study in the series which has just been discussed, was a history of Wayne township, Edwards county, Kansas, a part of the present author's home community, and was presented as a high school commencement address in May 1933 (Printed in the Lewis Press, June 1-July 6, 1933). The experience of this community did not conform with the traditions of the Turner school, and for comparative purposes a second study was made using Kanwaka township, Douglas county. These samples represented differences in area and time, eastern and western Kansas, settlement beginnings as of 1854, and 1877, the one prior to rail transportation and the other identified with the railroad settlement. The results of the two samples were in substantial agreement, and led to the addition of new samples and finally to the generalized study of 1935 on population turnover. The expansion of the population studies into the agricultural studies began also with Wayne township and its farmers' club, 1886-1893, which discussed first hand the local problems of adaptation of crops, tillage methods, machinery, and farm management (Agricultural History, 1936). It was here that the idea of adaptation as a folk process emerged as representing a major factor in regional history. The discussion at the club meetings served the function of systematizing to a certain degree the folk experience of the community. The methodology by which these studies was pursued was developed piecemeal, by experiment, and standardized as the scope of the project expanded. It was only after the work was fully developed that the fact became evident that there was a close parallel between this methodology as applied to history, and that of the ecologists and other sciences, and that the problem of regionalism in history required a new kind of synthesis of the sciences of biology, climatology, geology, geography, and pedology. The present author received his introduction to biology, (laboratory, museum, field, and theory) from Charles Sylvester Parmenter, of Baker University. The extent of that debt has been brought into sharper focus as this book has developed. So far as the general point of view on the West was derived from anyone, the present writer's first obligation was to Frank Heywood Hodder. It is within this background that Webb's The Great Plains made its contribution.

Even though there are resemblances between the methodology applied here to history and the methodology of the ecologists and other scientists, there is objection to labeling it an ecological interpretation of history. Brand names have a way of making commitments beyond the facts, and tend to place the stamp of statics and orthodoxy upon any body of methods so named. Methods should be kept in their proper place, not as ends in themselves, but as tools to be used, modified, or discarded as seems desirable. A reluctance to accept the term cultural approach to history has already been expressed (1940). The similarities between the historian's problems of methodology and those of the several sciences are evident. The soil scientist, Jenny, used the method of independent variables; the geographers, the idea of the "circle of facts"; the ecologist, Gleason, the individualistic concept of the plant association; Tansley, the concept of the ecosystem; the microbiologists stress the direct method of studying organisms in the soil rather than in isolation; and the Bridgman group in physics, the importance of stating questions in a form that could be subjected to operational proof. In all these instances there were common objectives which are equally of interest to the historian; to eliminate the subjective element of personal opinion or philosophical interpretation, to avoid the distortions of single factor interpretations, to escape the errors of verbalism, and to arrive at objective and positive conclusions. No particular brand of methodology is advocated, only that the methodology be adequate.

Highly significant to this study is the historical relation of the folk process to science. Amidst the highly mobile population of the United States, and in consequence of the short period of occupation of the continent, especially the western portion, the folk foundation of culture never matured to the degree that it did in the older cultures of Europe or Asia, where the sense of folk culture was ever present. Furthermore, the machine age intervened as an independent variable to initiate a new cycle in the indeterminate process of readjustment. Science may be treated as still another independent variable introduced into the complex process of culture formation. In America its influence was more direct than elsewhere. The initial settlements, and preliminary experimentalism, occurred prior to the time when science had much of anything to offer, and when much or even most of what was proposed in the name of science was wrong as applied to the problem in hand. As time passed a larger backlog of basic scientific knowledge became available, and more experience was acquired in the technology of its application. An increasing degree of coordination of folk knowledge with scientific knowledge became possible. This did not occur without conflict, however, as the farmer exhibited hostility toward the impractical aspect of science, and the scientist often, not only disregarded, but ridiculed, folk knowledge. Of course, in the use of the term folk knowledge, there must be a clear differentiation from mere superstition, or what is often contemptuously called the old almanac-type of tradition. One of the most revealing conclusions from these studies has been the soundness of the leadership evidenced in the progressive experimentalism of this folk-process and the importance to the most modern science of reconciling its conclusions with such earthly practical experience in the hands of the common man. It is proper to propose the question of just how folk knowledge and science differ, or whether, after all, they are only different stages or aspects of the same thing; the unorganized folk process, the more formal trial and error experimentalism, and finally the planned research based on theoretical principles and objectives. In this age of science and technology, government became peculiarly the agency for implementing these independent variables, especially in agriculture where internal organization of the industry was still primarily upon an individual footing. Organized action is always attractive because of the power that may be wielded in getting things done. This applies to private organization as well as government. Such organized action can initiate innovation but it can also stifle changes. It tends to run in a cycle of initiation, stabilization, conservation of gains, and finally stagnation. Therein lies the particular danger of the tendency of government to dominate society and through it power to smother folk initiation. Some of the most important leaders in government aid to the farmer in helping himself have been the first to admit that the net effect of their best endeavors was the opposite.

Although largely overshadowed by the rise of formal organized science, and often held in contempt by the scientist, in times of stress the folk process reasserted itself. During the drouth of the 1930s, the plains farmer came forward with much of the originality that was implemented by the farm equipment companies and governmental agencies. The machinery exhibits at the Dodge City fair in 1936 were a conspicuous demonstration of that fact. Again, in October, 1945, the extension service of the Kansas State College of Agriculture gave recognition to the war-stimulated resourcefulness of the people by holding an exhibit at the State Fair grounds. The machinery was arranged in a circle, one half of the circle consisting of implements made or improved by the farmers themselves, and in the other half of the circle was arranged the new implements being offered by manufacturers duplicating much of the farmer-made machines. A similar parallel exhibit showed devices for the home.

With increasing frequency in biological literature the plea appeared in the decades of the 1930s and 1940s for the recognition by specialists of the point of view of the old-fashioned naturalist, the broad synthesis and perspective which is the only means of bringing out the full value of the contributions of the specialist. The point of view of reconciling the folk process and science looks in the same direction, and is even more fundamental to the larger problem of the evolution of the basic ideas of civilization.

The twentieth century was so conspicuously an age of science that a new obligation was imposed upon history to study the impact of science upon society and of society upon science, to study technology and its interrelations with society, a sort of an ecological study of these as social factors. Other men had done somewhat similar things, but Webb's attention to the revolver, to fencing, and to the windmill, were valuable examples of that method, and there are many other possible examples of technology that need similar treatment. The present author has given an example in the development and utilization of the lister idea in its several ramifications as a significant and important aspect of regional technology, and has pointed out the possibilities of similar studies dealing with agricultural machinery for other regions (1944). One of the particular objects of this book is to direct attention to the significance of synthesis of science, technology, and history, (in this instance it is limited to regionalism) and it has been presented because of the conviction that in this direction lies the most important immediate field of activity for the historian. The backlog of basic science and technology has accumulated faster than the historian has evaluated it in relation to the history of society.

As pointed out in chapter eleven, few scientists are trained in history and social science, and likewise, few historians and social scientists have training in science. Either they must each undertake to acquire a competence in the other's field or they must cooperate in studying the history of science, and technology, and society (Shryock, 1943, 1944). Few graduate school training programs recognize this problem. At the same time, however, that this field is pointed out as the most insistent among several in need of cultivation, the limitations of a science interpretation of history are indicated. Science embodies only a segment of the cultural heritage of ideas, and a strictly science interpretation of history could result only in distortion. It is more important in the long run to have an interpretation of history in terms of ideas, not merely scientific ideas. The greatest obstacle to this is the difficulty of stating the problem of ideas as the basis of history in a meaningful operational form, but that task should constitute a challenge rather than a deterrant.

The discussions in the chapters on the rôle of science are not to be interpreted as denying or minimizing the validity of science. Far from it. They are only an insistence that science did not possess, and never will possess, the final answers to any of these questions. The same applies to folk knowledge. The primary purpose is to call attention explicitly and clearly to the fact that this problem of harmonization exists, and to insist upon recognition of it. Science itself is dynamic and subject to continuous, indeterminate change and enlargement of knowledge. By whatever kind of agency they may be administered, the shifting impersonal findings of science, and their applications, must be harmonized with the deep-rooted cultural experience of the race, or in other words, with the folk process as the dynamic expression of human beings living together in society. Either operating alone tends to defeat itself; the folk process to become rigid, and science to lose contact with the reality of the human equation. The historian must never permit the "illusion of finality" to intrude into his thinking in respect to science or to the folk process. In this respect as in others history is an open system.

Culture and environment

The Anglo-American had brought with him from Europe the forest complex as a characteristic of his cultural heritage, and upon establishing himself in North America, learned from the Indian the corn complex. This combination of forest and corn became a unique feature of North American culture within the temperate climate range of corn. Corn was adapted to a wider variety of uses than any other crop for both man and animals. For man it was used in the green stage for roasting ears, or cut, or dried; in the ripened stage, for bread (various forms), mush, and hominy (various forms). For animals, the plant itself made excellent forage, and the grain was the richest of all the grains grown under similar climatic conditions. For the forest man, it was the easiest crop to grow, the surest of yield, and produced the greatestest food value per acre and per man hour of labor of any available crop. More than any other cultivated crop, corn was the unique foundation of American agriculture, and the whole round of practices associated with preparation of the soil, cultivation, harvesting processing, and methods of preparation of food for men and animals were interwoven into the culture of the people. Even the urbanization process did not eliminate it, and the forest-corn complex exercised a powerful and unconscious influence over those who were no longer connected with agriculture and were totally unaware of its origin and meaning.

Upon coming out of the forest into the tall grass section of the grassland country, forest man found that corn was even more prolific than on the less fertile forest soil, and pushed the corn belt as far north as low temperatures permitted and as far west as dry heat permitted. The tall grass prairie of the middle latitude put the finishing touches to the corn complex in the Corn Belt centering in Illinois and Iowa.

The northward movement of corn culture is a story that belongs to the history of the northern grassland; the shortened growing season, the menace of early frost that might ruin the crop, cut it short, or produce a crop of "soft" (immature) corn. The focus of the problem to which the research of the present author is directed relates to the westward extension of corn culture into the low-moisture region and the limits of corn growing under such conditions, the transition to other and more adaptable crops, the sorghums (forage and grain) as substitutes for corn and used for much the same purposes, and the adaptation of completely different crops, especially hard wheat, the making of a grassland wheat culture or wheat-livestock culture in contrast with the eastern forest-corn culture. The most difficult change was in this last named aspect, because it struck at the heart of the forest-corn culture complex in all its ramifications. Many years were required for the settlers to learn under the impact of the new environment, and it proved impossible to teach the eastern people of farm and city, whose only contacts were indirect and whose interests centered in other things — in things they considered really important. They had the numbers that were important politically to determine policy and public opinion, along with an intolerant attitude of contempt for the minority interests of the grassland which they did not understand and about which they did not care to learn. The unkindest cut of all was when the livestock interests of the grassland joined forces with the forest corn culture of the industrial city in opposing the development of an effective diversified utilization of the grassland resources.

The first settlers in Kansas raised corn as their principal crop and continued the practice as the frontier of settlement moved westward. The Populist movement was as much as anything else a revolt of forest-corn culture against the grassland-wheat environment, which was unconsciously expressed in 1890 in Mrs. Mary "Yellin" Lease's famous admonition to Kansas farmers to raise less corn and more Hell. In eastern Kansas, corn was then and remained largely the dominant crop, but not so farther west As Miller (1925) pointed out, the heart of Kansas

Populism was in the central third of the state where the farmers had already made a partial success in farming and had accumulated moderate wealth which was threatened. It is an excellent illustration of the historical principle that the revolts of that kind did not come from downtrodden people, but from people on-the-make. It is necessary to go farther than Miller had gone and point out that the occupants of this grassland area had not yet achieved a harmonization of their culture with environmental factors. Agricultural succession in the ecologist's usage of that term had not yet achieved an unstable equilibrium. The corn and wheat cultures were still in conflict in the decade of the nineties and success in the growing, the milling, and the baking of hard winter wheat was required to establish firmly the victory in the central part of the state (Malin, Winter Wheat, 1944), or as Shaler (1891, 1894) put it, to reconcile culture with environment. Farther west, as in Edwards county on 99° 20', corn was still being raised in about the same acreage to the end of the century, but hard winter wheat accounted for the increased acreage (1935, 1936). Not until World War I could it be said that wheat acreage definitely exceeded corn acreage in Kansas as a whole. During the first quarter of the twentieth century the sorghums won out over corn in the western part of the state, and in the second quarter century were fully established under the influence of improved varieties produced by scientific plant breeding. Nutrition studies show that some of the improved varieties of sorghums were equivalent to corn in the rations of several kinds of livestock.

The map of types-of-farming areas in Kansas (Hodges, et al 1930) is particularly significant in interpreting agricultural adaptation. With the exception noted later, the eastern third of Kansas consisted of the mixed-farming areas, the commercial corn-belt and the Bluestem Pastures region and was bounded on the west by the 97 meridian. Such wheats as were grown in this section were soft winter wheats. The central third of the state, between 97° and 100° was the hard winter wheat belt, and west of 100° was hard winter wheat, sorghums, and livestock. The breaks between the three groups of regions at 97° and 100° coincide closely with the original vegetational divisions of tall grass prairie, mixed-grass prairie-plains, and further confirms the fundamental character of those boundaries as natural division zones for native, and agricultural vegetation. This clear three-fold division seems to be fully valid, however, only for Kansas as a whole, because the north tier of Kansas counties was designated on the type-of-farming map as secondary corn belt to a point west of the 100 meridian, and in the three westernmost counties along the northern boundary a substantial corn acreage persisted. These northern counties belonged to the Nebraska corn belt because they were far enough north for the effective rainfall to permit corn growing farther west than elsewhere in the state of Kansas. North of the Platte river in Nebraska lay the sand hills which again interrupted the otherwise regular east-west distribution of both the native and the agricultural vegetation. Still farther north into the Canadian plains, the native vegetation of the northern nuclear grass area was of northern species of grasses, especially the needle grasses (genus Stipa) and modern agriculture was correspondingly different, — hard spring wheat and associated crops.

The southwestern part of Kansas is also transitional, and the basis of this was brought out most clearly by superimposing rainfall, length of growing season, and wind maps, making a single composite. The rainfall belts run north and south; the temperature belts northeast to southwest, except in southwestern Kansas where they shift to an east-west direction; and the high wind zone centers in a circular area in the Panhandle of Texas with a somewhat lesser velocity sweeping southwestern Kansas. The convergence of these three factors of rainfall, temperature, and wind on that area and southward into central Texas results in a somewhat different native vegetational distribution and crop distribution from central Kansas. Still farther south the southern grass nuclear area of mesquites (genus Hilaria), the southern gramas, and associates, dominated the southern extremes.

Reconciliation of culture with environment was a theme, both significant and important, in the thinking of Shaler (1891, 1894). He cited the centuries of accumulated folk experience in Europe and in Asia, and the want of it in America, especially in the Mississippi valley. The agricultural difficulties of the 1890s were in part, according to his interpretation, a consequence of this lack of reconciliation. The present author would use the term harmonization. The type-of-farming areas, as they were mapped fifty years later on the basis of actual farm practices, represented a greater degree of harmonization and in consequence a more stabilized cultural system, using the term culture in the anthropological sense. Of course, in using the terms harmonization, equilibrium, and stabilization, there is no intent of implying any "illusion of finality" or of determinism. There should be no mistake that there was possible one and only one crop or combination of crops, or one and only one kind of machinery in use, or one and only one method of utilizing it. Always, or at least the exceptions would be rare, there was more than one combination of crops, machines, and usage, not only possible, but actual. From time to time new factors or independent variables were added, such as that of mechanical power, which initiated a new phase of the readjustments of unstable equilibrium. The maintenance of harmonization of all factors was an indeterminate process.

Chapter Twenty-one

AN OPEN SYSTEM

     There were many ways of illustrating how society was changing its foundations and outlook during the century closed by World War II. None explained everything, none were complete explanations of anything, yet each well chosen contrast contributed something to a better understanding of civilization in transition. In spite of the interpretations of anthropologists who divide the history of culture into the stone age, the bronze age, and the iron age, the fact remained that civilization was founded primarily upon wood, a forest civilization, until the nineteenth century. Following that line of thought the nineteenth century was conspicuously a minerals age. The emphasis was upon coal and iron and other minerals deposited in the earth. At the prevailing rate of population growth and utilization, the forest age could think of its basic material as replaceable, but the minerals age was utilizing a body of basic materials that were irreplaceable, and at the accelerated rate of population increase and of utilization in the nineteenth century, there came to be a growing conviction of calamity in the exhaustion of forests and of minerals, and of soils that were capable of producing forests and food for the support of man. It was such consideration that started Malthus, Marsh, and Reclus, and many other nineteenth century worthies, to speculating on the future of man and the world's resources. By the end of the nineteenth century the closed world idea had crystallized, expressed most vividly in America by Turner and in the global sense by Mackinder — the passing of the frontier, closed space, the geographical pivot with the monopoly of power, the capacity to produce, the struggle for the most valuable space and remaining resources, the end of opportunity, and other pessimistic ideas reduced to popular catch words and phrases.

     In terms of science, these views were given apparent support by the theories of cosmogony associated with astronomy and geology of the nineteenth century, which assumed that the universe began as a molten mass and was cooling rapidly into a dying world. A further assumption held that the source of energy was heat released by the chemical process of combustion. On the earth, among living things, the single source of food supply was the energy of the sun acting through photosynthesis in plants. Whatever the validity of the theories of the sun's energy, there could be no immediate effect upon the food supply of the world, but nevertheless the dying world idea fitted into the growing pessimism about population, food supply, raw materials, and the idea of a closed world. Furthermore, the idea of fixity of species was only partially modified by the influence of Darwin, and Wallace, as change could occur only at a glacial pace and largely within the limits of natural determinism. Plants and animals were replaceable, but the emphasis was upon doubt of capacity to keep pace with population.

The industrial and communications revolutions, which have been given a particular definition and emphasis elsewhere in this book, were based upon mechanically powered machines, and they in turn were dependent primarily upon coal and its associates as fuel for the generation of the essential steam and electrical current, and as explosive force. Not until the mid-nineteenth century, however, could it be said that the shift from wood to coal for fuel had become general for steam engines in Europe and somewhat later in America; likewise for heating and cooking, and for the operations of the metallurgical industries. The full force of both revolutions was felt only in the last quarter of the nineteenth century when cheap Bessemer and open-hearth steels became standards — the steel age. The situation made its contribution to the imperialistic revival of the last quarter of the nineteenth century and to the ruthless drive for control of these irreplaceable raw materials. Uncertainty about the future was the inevitable outcome of a civilization based upon exhaustible and irreplaceable resources.

In respect to both categories, the replaceable and the irreplaceable, the conviction of inadequacy culminated in the closed space philosophy of the twentieth century geo-politics; that the whole surface of the globe was discovered, explored, and occupied, and in the future, the nations could only order their living within this closed space and its limited supply of raw materials. In a direct sense World War I and World War II were the inevitable consequences of the operation of this body of ideologies.

The beginnings of the occupation of the North American grasslands became an important event of the second quarter of the nineteenth century but was only a phase of a general world-wide movement of population into the grasslands; the steppes of southeastern Europe, Asia, India, southern Africa, Australia, and South America. By the twentieth century, at least a partial utilization of these areas, the grassland 40 per cent of the land surface of the earth, was an accomplished fact. The occupation of the grasslands coincided in point of time with the minerals century and its emphasis on substitutes for wood. To what extent cause and effect operated between the two phenomena may remain a moot question, but there is no doubt that minerals contributed largely to the rapidity of the development: the steel plow, the railroad, iron and steel farm machinery, iron and steel wire (smooth, barbed, and woven), the drive well, and the steel windmill. Without these the occupation of the grasslands would have been slower and it might have been only partially possible. The competition for the world's grasslands as a source of bread and meat supply fitted into the same turn of the century outlook on the future of exhaustible and irreplaceable materials.

Somewhat later chronologically, but in part paralleling the preceding ideological sequences, another kind of outlook was in the making — an open system. In biology the rediscovery of Mendelian principles of hybridization opened the door to new concepts of genetics; chromosomes, genes, and mutations. Ecology emerged in terms of the interrelations of plants, animal life, and environment, a developmental concept of plant, animal, and soil. The chemical balance-sheet theory of soil fertility was displaced by the new theory of soil formation as an open physical system with a capacity for the replacement, indefinitely, of the raw materials and food of civilization. Theoretically, at least, a civilization should feel a greater sense of security under a system of replaceable raw materials.

The twentieth century saw also the broadening of the basis of civilization by the utilization of the light metals, aluminum, and magnesium, as well as the numerous alloys, adding indefinitely to the total supply of available metals, and to the wider distribution geographically of these irreplaceable minerals. Further broadening of the range of materials was found in plastics derived from both plant and mineral sources, and the chemical treatment of wood which multiplied the available supply, widened the range of uses, and enlarged the possibilities of replaceability of forest products in terms of the soft woods.

Early in the World War II period scientists were asked to list the outstanding problems for post-war research (Robbins, 1944). The result put at the head of the list the analysis and study of human behavior; second, the general field of medical problems; third, the future sources of energy. On this last problem the Smithsonian Institution Report 1941, brought together a symposium summarizing the status of energy research; Abbott on solar radiation, Hottel on artificial converters of solar energy, and Lawrence on the new frontiers of the atom. Already the theories of the source of heat in the universe had undergone fundamental change under the influence of radium, radioactivity and the atomic theory. In 1944, Jeans brought out the fourth edition of his book,

The Universe around us, stating in the preface that the most important development since the third edition of 1931 was that the discoveries in nuclear physics provided an adequate explanation of the radiation of the sun. The successful beginnings in the disintegration of the atom opened a new epoch in civilization, energy released by physical processes in contrast with combustion through chemical processes. It is this fact that gives the term atomic age a significance beyond popular imagination. It changed, not only the theories of cosmogony which disposed of the dying universe ideas of the nineteenth century, but assured the world of virtually unlimited energy for heat and power before the time the supplies of coal would be running low. This new development invites what might be called an energy interpretation of history.

These scientific developments were revolutionary and they challenged the idea of closed space, but they did not destroy it and its associated ideologies. Such ideas, once implanted in men's minds, are the most difficult to eradicate. The idea of monopoly of the world's essential resources was countered by Mather (1942), who pointed out that geology of the world's continents emphasized, with few exceptions, the general distribution of essential minerals among all continents. The principal exception among minerals was tin, and in lesser degree radium, and nickel. On the assumption of a minerals monopoly, the struggle among nations for possession of certain known supplies was groundless. And Bowman, in a radio address July 15, 1945, emphasized that the earth was rediscovered again and again by new approaches, new crops became new instruments of power, diversity of the earth was man's opportunity. Regional interdependence was also the means of broadening the base of civilization and relatively untried opportunities opened in those directions. Orlando Park (1945) called upon ecologists to grasp the opportunity offered by the tropics as a region for study. Armstrong (1943) called the sea "the greatest potential source of raw materials," mineral, animal, and plant. Stefansson repeatedly called attention to the possibilities of the arctic.

The curse of the Mackinder-Turner idea of closed space and its heritage should be met head-on. In history, man found himself repeatedly faced with a similar conviction of closed space. It was technology, with its limitations, that was a conspicuous factor in such closed space dogmas. Alexander the Great was supposed to have wept because there were no more worlds to conquer. Whether or not the story was true is immaterial, but the attitude of mind typified is an important historic fact. Alexander had not occupied all global space, but he had reached the limits of the technology and the organizing capacity of his time. Rome reached a similar limit of expansion. The Mediterranean world spent centuries within the geographical limits beyond which it seemed unable to expand. The last two centuries of the medieval era were conspicuously a period of preparation — intellectual, technological, and experimental — for breaking out of the bounds of that closed world. The Portuguese became the leaders in the demonstration phase — Columbus and the discovery of America incidents — not of breaking open a closed world, but, more important, of a closed mind. The twentieth century conviction of closed space, as epitomized in Mackinder's teaching is again a matter of closed mind more than closed space, or a closed world. The world is no more closed in 1946 than in 1446, except as the minds of men were indoctrinated with the Social Myth and believed that they had no choice but between totalitarianism and annihilation, a sacrifice of freedom for an alleged security. The modern world was the victim of the determinism of a space geography interpretation of history although space was only one of the many independent variables operating as a former of history. Thoreau (1849) was doing more basic thinking than Turner and Mackinder when he wrote that "the frontiers are not east or west, north or south, but wherever a man <u>fronts</u> a fact...."

The factors of science entered the scene conspicuously during the depression of the 1930s and the World War II era, but there is question whether or not it was used as an independent variable in history. As a counter argument to the doctrine of Turner and Mackinder, science was sometimes presented as "the endless frontier," individualist opportunity in the Turner sense, and at times as the tool for implementing the totalitarian state. The report to the President of July 15, 1945, by Vannevar Bush, represented the first, and the writing of Russians, such as Marshak, and Vernadsky, the latter. The reply to both was that the outcome depended upon how science was used or misused. As Whitaker (1940) pointed out there was reason to question the assumption that man could remake nature "socially as well as technically" without risk of doing just the opposite of what was intended. The omnipotence of science as a Social Myth in either form went beyond the facts.

In an over-all sense, man was able to maintain a rate of discovery sufficient to reharness natural forces to new uses as rapidly as already known natural resources were exhausted. So far as science was able to predict on the basis of past experience, there would seem to be no reason to doubt man's ability to continue. In that sense the historian is justified in the hypothesis concerning the future, that the world is an open system of indeterminate change. This makes no commitments to the philosophies of either progress or degradation. In respect to them, the verdict is "not proven," but each individual is free to make his own choice in personal philosophies as distinguished from history.

BIBLIOGRAPHY OF LITERATURE

The bibliography is based on a plan of grouping materials from two or more chapters according to subject; at least, so far as such an arrangement is workable. Thus most plant ecology is listed in one place for Chapters 1, 2, 7, 10, and 11. The second list is for animal ecology and additional ecological material dealing with plants and animals together (Chapters 3, 10, and 11). Included in these lists are only a selection of titles from some twenty-five hundred, and the reader who desires additional literature on any subject will find it in the bibliographies of several of the more important works cited in these lists. Some important omissions were made necessary by deficiencies of library facilities.

Chapters 1, 2, 7, 10, and 11. Plant ecology

Adams. C. C. and G. D. Fuller. Henry Chandler Cowles, physiographic plant ecologist. Annals Assn. Amer. Geogrs., 30 (1940) 39-43.
Adamson, R. S. The classification of life forms of plants. Bot. Rev., 5 (1939) 546-561.
Aikman, J. M. Distribution and structure of the forests of eastern Nebraska. Univ. Nebr. Studies, 26 (1929) 1-75
—, Native vegetation of the [Shelterbelt] region. In, Possibilities of shelterbelt planting in the plains region. Pp. 155-174. Washington, 1935.
Albertson, F. W. Ecology of mixed prairie in west central Kansas. Ecol. Mono., 7 (1937) 481-547.
—, Prairie studies in west central Kansas. Trans. Kas. Acad. Sci., 41 (1938) 77-83; 42 (1939) 97-107; 44 (1941) 48-57.
Albertson, F. W. and J. E. Weaver. History of native vegetation of western Kansas during seven years of continuous drought. Ecol. Mono., 12 (1942) 23-52.
—, Effects of drought, dust, and intensity of grazing on cover and yield of short-grass pastures. Ecol. Mono., 14 (1944) 1-29.
—, Injury and death or recovery of trees in prairie climate. Ecol. Mono., 15 (1945) 393-433.
Aldous, A. E. and H. L. Shantz. Types of vegetation in the semi-arid portion of the United States, and their economic significance. Jour Agric. Res., 28 (1924) 99-128.
Allard, H. A. and W. W. Garner. Further observations on the response of various species of plants to length of day. USDA Technical Bulletin 727 (August 1940) 1-64.
Allard, H. A. and M. W. Evans. Growth and flowering of some tame and wild grasses in response to different photoperiods. Jour. Agric. Res., 62 (1941) 193-228.

Allen, J. A. The flora of the prairies. Amer. Nat., 4 (1870) 577-.
—, The fauna of the prairies. Ibid., 5 (1871) 4-9.
Arrhenius, O. Species and area. Jour. Ecol., 9 (1921) 95-99.
Arthur, J. C. Development of vegetable physiology. Bot. Gaz., 20 (1895) 381- .
Aughey, S. Improvement in western pasture-land. Science, 1 (1883) 335.
Baker, O. E. The agriculture of the great plains. Annals Assn. Amer. Geogrs., 13 (1923) 109-168.
—, Agricultural regions of North America, Part XI — The Columbia Plateau wheat region. Econ. Geog., 9 (1933) 167-197.
Bensley, R. R. Chemical structure of cytoplasm. Science, 96 n.s. (1942) 389-393.
Bentley, H. L. A report upon the grasses and forage of central Texas. USDA Div. Agrost. Bull., 10 (1898).
Bessey, C. E. A meeting place for two floras. Bull. Torrey Bot. Club, 14 (1887) 189-191.
—, The grasses and forage plants of Nebraska. Ann. Rpt. Nebr. Brd. Agric., 1887.
—, Plant migration studies. Univ. Nebr. Studies, 5 (1905) 11-27.
Bews, J. W. The world's grasses: Their distribution, economics and ecology. London, 1929.
Bidwell, P. W. and J. I. Falconer. History of agriculture in the northern United States, 1620-1860. Pub. No. 358, Carnegie Inst., Washington, 1925.
Blake, A. K. Viability and germination of seed, and early life history of prairie plants. Ecol. Mono., 5 (1935) 405-460.
Bollinger, C. J. The eastern boundary of the great plains in north central Oklahoma. Proc. Okla. Acad. Sci., 5 (1926) 123-124.
Booth, W. E. Revegetation of abandoned fields in Kansas and Oklahoma. Amer. Jour. Bot., 28 (1941) 415-422.
Braun, E. L. Glacial and post-glacial plant migrations indicated by relic colonies of southern Ohio. Ecology, 9 (1928) 284-302.
—, A history of Ohio's vegetation. Ohio Jour. Sci., 34 (1934) 246-.
—, The undifferentiated deciduous forest climax and the association — segregate. Ecology, 16 (1935) 514-519.
Bray, W. L. The ecological relations of the vegetation of western Texas. Bot. Gaz., 32 (1901) 99-123; 195-217; 262-291.
Briggs, L. J. and H. L. Shantz. The wax seal method for determining the lower limit of available soil moisture. Bot. Gaz., 51 (1911) 210-219.
—, The wilting coefficient and its indirect determination. Bot. Gaz., 53 (1912) 20-37.
—, The relative wilting coefficients for different plants. Bot. Gaz., 53 (1912) 229-235.
—, Relative water requirements of plants. Jour. Agric. Res., 3 (1914) 1-63.
—, Hourly transpiration rate in clear days as determined by cyclic environmental factors. Jour. Agric. Res., 5 (1916) 583-645.
—, Comparison of the hourly evaporation rate of atmometers and free water surface with transpiration rate of Medicago satina. Jour. Agric. Res., 9 (1917) 277-292.

Briggs and Shantz—(continued)
—, The water requirements of plants as influenced by environment. Proc. Second Pan-American Science Congress, 1917.
—, The water requirements of plants: I. Investigations in the great plains in 1910 and 1911. USDA Bur. Plt. Ind. Bull., 284 (1913) 1-48; II. A review of the literature, Ibid., 285 (1913) 1-96.
Bromley, S. W. The original forest types of southern New England. Ecol. Mono., 5 (1935) 61-90.
Bruner, W. E. The vegetation of Oklahoma. Ecol. Mono., 1 (1931) 99-188.
Burns, W., L. B. Kulkarni, and S. R. Godbole. Succession of xerophytic Indian grasslands. Jour. Ecol., 19 (1931) 389-391.
Carpenter, J. R. The grassland biome. Ecol. Mono., 10 (1940) 617-.
Cattell, J. Biological symposia. (A series of volumes devoted to current symposia in the field of biology). Lancaster, Penn. 1940.
Cain, S. A. Concerning certain phytosociological concepts. Ecol. Mono., 2 (1932) 475-508.
—, Foundations of plant geography. Ecology, 26 (1945) 101-102.
Chapline, W. R. and C. K. Cooperrider. Climate and grazings. USDA Yearbook, (1941) Climate and man. Pp. 459-476.
Clapham, A. R. Over-dispersion in grassland communities and the use of statistical methods in plant sociology. Jour. Ecol., 24 (1936) 232-251.
—, Studies in depth adjustment of subterranean plant organs. New Phytologist, 44 (1945) 105-109.
Clark, O. R. Interception of rainfall by prairie grasses, weeds, and certain crop plants. Ecol. Mono., 10 (1940) 243-278.
Claude, A. The constitution of protoplasm. Science, 97 n.s. (1943) 451-456.
Clements, F. E. Reports on ecological investigations 1914-1941. Carnegie Inst., Washington, Yearbook, 42 (1943).
—, Research methods in ecology. Lincoln, Nebraska, 1905.
—, Plant physiology and ecology. New York, 1907.
—, Plant succession; an analysis of the development of vegetation. Carnegie Inst., Washington, Pub. No. 242, 1916.
—, Development and structure of the biome. Ecol. Soc. Abstracts, 1916.
—, Plant indicators: the relation of plant communities to process and practice. Carnegie Inst., Washington, Pub. No. 290, 1920.
—, The phytometer method in ecology; the plant and community as instruments. Ibid., Pub. No. 356, 1924.
—, Plant succession and indicators. New York, 1928.
—, The relict method in dynamic ecology. Jour. Ecol., 22 (1934) 39-68.
—, The nature and structure of the climax. Ibid., 24 (1936) 252-.
—, Climatic cycles and human population in the great plains. Sci. Mo., 47 (1938) 193-210.
Frederic Edward Clements, by H. L. Shantz. Ecology, 26 (1945) 317-319.
Clements, F. E. and E. S. Clements. Climate, climax, and conservation. Carnegie Inst. Washington, Yearbook, 40 (1941) 176-182.

Clements, F. E. and R. W. Chaney. Environment and life in the great plains. Carnegie Inst. Wash., Supp. Pub. No. 24, 1936.
—, Climatic cycles and human populations in the great plains. Ibid., No. 43, 1938.
Clements, F. E. and V. E. Shelford. Concepts and objectives in bioecology. Carnegie Inst. Washington, Yearbook, No. 26 (1927).
—, Bio-ecology, New York, 1939.
  Review: Clements and Shelford, Bioecology, (1939).
    Hutchinson, Ecology, 21 (1940) 267-268. (See also Reviews of Weaver and Clements, Plant Ecology).
Clements, F. E. and J. E. Weaver. Experimental vegetation; the relation of climaxes to climates. Carnegie Inst. Wash., Pub. No. 355, 1924.
Clements, F. E., J. E. Weaver, and H. C. Hanson. Plant competition; an analysis of community functions. Carnegie Inst. Wash., Pub. No. 398, 1929.
Cook, C. W. A study of the roots of Bromus inermis in relation to drought resistance. Ecology, 24 (1943) 169-182.
Cooper, J. G. On the distribution of the forests and trees of North America, with notes on its geography. Ann. Rpt., Smithsonian Inst., 1858, 246-280.
Cooper, W. S. The fundamentals of vegetational change. Ecology, 7 (1926) 391-413.
—, Henry Chandler Cowles, Ecology, 16 (1935) 281-283.
Costello, D. F. Important species of the major forage types in Colorado and Wyoming. Ecol. Mono., 14 (1944) 107-134.
—, Natural vegetation of abandoned plowed land in the mixed prairie association of northeastern Colorado. Ecology, 25 (1944) 312-326.
Cottle, H. J. Studies in the vegetation of southwestern Texas. Ecology, 12 (1931) 105-155.
Cowles, H. C. The ecological relations of the vegetation on the sand dunes of Lake Michigan. Bot. Gaz., 27 (1899) 95-391.
—, The physiographic ecology of Chicago and vicinity,... Bot.Gaz., 31 (1901) 73-108; 145-182.
—, The plant societies of Chicago and vicinity. Chicago, 1901.
—, Chicago textbook of botany. 1911.
—, The causes of vegetative cycles. Bot. Gaz., 51 (1911) 161-183. Also, Annals Assn. Amer. Geogrs., 1 (1912) 1-20.
—, Persistence of prairies. Ecology, 9 (1928) 380-382.
—, (The issue of Ecology, July 1935, was dedicated to him, and all papers published in it were in tribute to him. Vol. 16, 281-534).
Conrad, H. S. Plant associations on land. In, Theodor Just (Editor) Plant and Communities. Notre Dame, 1939, 1-27.
Daubenmire, R. F. Plant succession due to overgrazing in the Agropyron bunchgrass prairie of southern Washington. Ecology, 21 (1940) 55-64.
—, An ecological study of the vegetation of southeastern Washington and adjacent Idaho. Ecol. Mono., 12 (1942) 53-80.
Deam, C. C. Grasses of Indiana. Int. Dept. Conserv. Div. For. Pub., 82, 1-256.
Egler, F. E. Vegetation as an object of study. Phil. Sci., 9 (1942) 245-260.

Eiseley, L. C. Pollen analysis and its bearing upon American prehistory; a critique. Amer. Antiq. 5 (1939) 115-139.

Elias, M. K. Tertiary prairie grasses and other herbs from the high plains. Geol. Soc. Amer. Spec. Papers, No. 41 (1942).

Erdtman, G. An introduction to pollen analysis. Chron. Bot. Co., 1943.

Evans, E. A. Carbon dioxide utilization in animal tissue. Science, 96 n.s. (1942) 25-29.

Eyster, H. C. Osmosis and osmotic pressure. Bot. Rev., 9 (1943) 311-324.

Fenton, E. W. A botanical survey of grasslands in the south and east of Scotland. Jour. Ecol., 19 (1931) 392-409.

Flory, E. L. Comparison of the environment and some physiological responses of prairie vegetation and cultivated maize. Ecology, 17 (1936) 67-103.

Franck, J. Some fundamental aspects of phytosynthesis. Sigma Xi Quarterly, 29 (1941) 81-105.

Freeman, O. W., J. D. Forrester, and R. L. Lupher. Physiographic divisions of the Columbia intermontaine province. Annals Assn. Amer. Geogrs., 35 (1945) 53-75.

Fuller, G. D. Evaporation and plant succession. Bot. Gaz., 52 (1911) 193-208.

—, Evaporation and stratification of vegetation. Ibid., 54 (1912) 424-426.

—, Postglacial vegetation of the Lake Michigan region. Ecology, 16 (1935) 473-487.

Fults, J. L. Somatic chromosome complements in Bouteloua. Amer. Jour. Bot., 29 (1942) 45-55.

Garner, W. W. Recent work in photoperiodism. Bot. Rev., 3 (1937) 259-275.

Garner, W. W., and H. A. Allard. Effects of relative length of day and night and other factors of the environment on growth and reproduction in plants. Jour. Agric. Res., 18 (1920) 553-606.

—, Photoperiodism, the response of the plant to relative length of day and night. Science, 55 n.s. (1922) 582-583.

—, Further studies in photoperiodism. Jour. Agric. Res., 23 (1923) 871-920.

Gates, F. C. Pines in the prairie. Ecology, 7 (1926) 96-98.

—, Grasses in Kansas. Rpt. Kas. St. Brd. Agric. Quarter ending December 1936.

—, Woody plants, native and naturalized in Kansas. Trans. Kas. Acad. Sci., 41 (1938) 99-118.

Gershenfeld, L. Ultraviolet light as a sanitary aid. Ann. Rpt. Smithsonian Inst., (1942) 209-225.

Gilmore, M. R. Uses of plants by the Indians of the Missouri river region. Bur. Amer. Ethnol. 33 (1919) 43-154.

Gleason, H. A. Some unsolved problems of the prairies. Bull. Torrey Bot. Club, 36 (1909) 265-271.

—, The vegetation of the inland deposits of Illinois. Bull. Ill. State Lab. Nat. Hist., 9 (1910) 23-174.

—, The structure and development of the plant association. Bull. Torrey Bot. Club, 44 (1917) 463-481.

Gleason,—(continued)
—, The vegetational history of the middle west. Ann. Assn. Amer. Geogrs., 12 (1922) 39-85.
—, On the relation between species and area. Ecology, 3 (1922) 158-.
—, Species and area. Ecology, 6 (1925) 66-74.
—, The individualistic concept of the plant association. Bull. Torrey Bot. Club, 53 (1926) 7-26.
—, Further views on the succession concept. Ecology, 8 (1927) 299-.
—, The individualistic concept of the plant association. In, Theodore Just, (Editor) Plant and Animal Communities, Notre Dame, Indiana, 1939, 92-108. (Proceedings of the Conference on Plant and Animal Communities, held at the Biological Laboratory, Cold Spring Harbor, Long Island, New York, from August 29 to September 2, 1938).
Gleason, H. A. and F. C. Gates. A comparison of the rates of evaporation in certain associations in central Illinois. Bot. Gaz., 53 (1912) 478-491.
Good, R. D'O. A theory of plant geography. New Phytologist, 30 (1931) 149-171.
Gordon, C. Report on cattle, sheep, and swine, supplementary to enumeration of live stock on farms— 1880. United States Tenth census (1880) 3: 951-1116 at 960. (Common grass names).
Gould, F. W. Plant indicators of original Wisconsin prairies. Ecology, 22 (1941) 427-428.
Gray, Asa. Review: Two new genera of Dioecious grasses of the United States. Trans. Acad. Nat. Sci. St. Louis, 1859, Vol. I, 431-442. In his Scientific Papers, 1: 112-115.
Griffith, D. The grama grasses: Bouteloua and related genera. Contributions from the United States Nat. Herbarium, 14 (1912) 343-428.
Grinnell, J. A revised life-zone map of California. Univ. Cal. Pub. in Zool., 40 (1935) 327-329.
Hall, E. Notes on some features of the flora of eastern Kansas. Amer. Jour. Sci.,100 (1870) 29-35.
Hall, H. M. and J. Grinnell. Life zone indicators in California. Proc. Cal. Acad. Sci., 4 ser., 9 (1919) 37-67.
Hall, W. L. The timber resources of Nebraska. Yearbook, USDA 1901, 207-216.
Hanson, H. C. Prairie inclusions in the deciduous forest climax. Amer. Jour. Bot., 9 (1922) 330-337.
—, Ecology of the grassland. Bot. Rev., 4 (1938) 51-82.
Hanson, H. C. and W. Whitman. Characteristics of major grassland types in western North Dakota. Ecol. Mono., 8 (1938) 57-114.
Hargitt, G. T. What is germ plasm? Science 100 (1944) 343-348.
Harper, R. M. Some undescribed prairie in northeastern Arkansas. Plant World, 20 (1917) 58-61.
Harvey, L. H. Floral succession in the prairie-grass formation of southeastern South Dakota. Bot. Gaz., 46 (1908) 81-108; 277-298.
Hisaw, F. L. Prairie vertibrates. Bull. Ecol. Soc. Amer., 6 (4: 1925) 20.
Hitchcock, A. S. Ecological plant geography of Kansas. Trans. Acad. Sci., St. Louis, 8 (1898) 55-69.

Hitchcock, —(continued)
—, Manual of grasses of the United States. USDA Misc. Pub., 200 (1935) 1-1040.
Kearney, T. H. Jr. Notes on grasses and forage plants of the southeastern states. USDA Div. Agrost. Bull. 1 (1895).
Kearney, T. H. and H. L. Shantz. The water economy of dry land crops. USDA Yearbook, 1911, 351-361.
Kramer, J. and J. E. Weaver. Relative efficiency of roots and tops of plants in protecting the soil from erosion. Univ. Nebr. Conserv. Surv. Div. Nebr. Conserv. Bull., 12 (1936).
Kramer, P. J. Absorption of water by plants. Bot. Rev., 11 (1945) 310-355.
Landauer, W. Shall we lose or keep our plant and animal stocks? Science, 101 (1945) 497-499.
Larson, F. and W. Whitman. A comparison of used and unused grassland mesas in the Badlands of South Dakota. Ecology, 23 (1942) 438-445.
Livingston, B. E. The relation of desert plants to soil moisture and to evaporation. Pub., 50, Carnegie Inst. Washington, 1906.
—, Evaporation and plant development. Plant World, 10 (1907) 269-.
—, Evaporation and plant habitats. Ibid., 11 (1908) 1-10.
—, A study of the relation between summer evaporation intensity and the centers of plant distribution in the United States. Ibid., 14 (1911) 205-222.
—, A quarter century of growth in plant physiology. Ibid., 20 (1917) 1-15.
—, Atmometers of porous porcelain and paper, their use in physiological ecology. Ecology, 16 (1935) 438-472.
Livingston, B. E. and F. Shreve. The distribution of vegetation in the United States, as related to climatic conditions. Pub., No. 284, Carnegie Inst. Washington, 1921.
MacDougal, D. T. A half-century of plant physiology. Ann. Mo. Bot. Garden, 19 (1932) 31-43.
McDougall, W. B. Plant ecology. Philadelphia, Second Edition, 1931.
McGuire W. W. On the prairies of Alabama. Amer. Jour. Sci. and Arts, 26 (1834) 93-98.
Martin, E. V. Studies of evaporation and transpiration, under controlled conditions. Carnegie Inst. Wash., Pub. No. 550 (1943).
Martin, E. V. and F. E. Clements. Adaptation and origin in the plant world. Pub. No. 521, Carnegie Inst. Wash., 1939.
Mason, S. C. A preliminary report upon the variety and distribution of Kansas trees. Eight Bien. Rpt. Ks. St. Brd. Agric., 1891-92, 259-274.
Maximov, N. A. The plant in relation to water: A study of the physiological basis of drought resistance. London, 1929. Review: The plant in relation to water. Jour. Ecol., 17 (1929) 414-416.
—, The physiological significance of the xeromorphic structure of plants. Jour. Ecol., 19 (1931) 273-283.
—, Plant physiology. New York, 1938.
Menzel, D. H. The nature of solar energy. Sigma Xi Quarterly, 28 (1940) 157-164, 180.

Moyer, L. R. The prairie flora of southwestern Minnesota. Bull. Minn. Acad. Sci., 4 (1910) 357-372.
Mueller, I.M. An experimental study of rhizomes of certain prairie plants. Ecol. Mono. 11 (1941) 165-188.
Mueller, I. M. and J. E. Weaver. Role of seedlings in recovery of midwestern ranges from drought. Ecology, 23 (1942) 275-294.
—,Relative drought resistance of seedlings of dominant prairie grasses. Ecology, 23 (1942) 387-398.
Nelson, A. The red desert of Wyoming. USDA Div. Agrost. Bull. 13 (1898).
Nichols, G. E. The interpretation and application of certain terms and concepts in the ecological classification of plant communities. Plant World, 20 (1917) 305-319; 341-353.
Norton, E. A. and R. S. Smith. The relationship between soil and native vegetation in Illinois. Ill. St. Acad. Sci., 24 (1931) 90-93.
Olmstead, C. E. Growth and development in range grasses. I. Early development of Bouteloua curtipendula in relation to water supply (with one figure), Bot. Gaz., 102 (1941) 499-519.
—,Growth and development in range grasses. II. Early development of Bouteloua curtipendula as affected by drouth period. Ibid., 103 (1942) 531-542.
—,Growth and development in range grasses: III. Photoperiodic responses in the genus Bouteloua. Ibid., 105 (1943) 165-181.
—,Growth and development in range grasses: IV. Photoperiodic responses in twelve geographic strains of sideoats grama. Ibid., 106 (1944) 46-74.
—,Growth and development in range grasses: V. Photoperiodic responses of clonal divisions of three latitudinal strains of sideoats grama. Ibid., 106 (1945) 382-401.
Pammel, L. H. Notes on the grasses and forage plants of Iowa, Nebraska, and Colorado. USDA Div. Agrost. Bull., 9 (1897).
Park, O. The measurement of daylight in the Chicago area and its ecological significance. Ecol. Mono., 1 (1931) 189-230.
—,Observations concerning the future of ecology. Ecology, 26 (1945) 1-9.
Parker, M. W. and H. A. Borthwick. Day length and crop yields. USDA Misc. Pub., 507 (1942).
Pool, R. J. A study of the vegetation of the sand hills of Nebraska. Minn. Bot. Studies, 4 (1914) 185-212.
Pool, R. J., J. E. Weaver, and F. C. Jean. Further studies in the ecotone between prairie and woodland. Univ. Nebr. Studies, 18 (1918) 7-53.
Pound, R. and F. E. Clements. The phytogeography of Nebraska. First edition, 1897. Second edition, 1900.
Review: C. MacMillan, Observations on the distribution of plants along shore of Lake of the Woods. Minn. Bot. Studies, 1 (1897) 949. In, American Naturalist, 31 (1897) 980-984.
—,The Vegetation regions of the prairie province. Bot. Gaz., 25 (1898) 381-394.
Ramaley, F. Sand hill vegetation of northeastern Colorado. Ecol. Mono., 9 (1939) 1-52.

Raup, H. M. Botanical problems in Boreal America. Bot. Rev., 7 (1941) 147-209; 209-247.
Rayner, M. C. The mycorrhizal habit in relation to forestry. Chron. Botanica, 6 (1940) 12-13.
Reed, H. S. A brief history of ecological work in botany. Plant World, 8 (1905) 163-170; 198-208.
Reitz, L. P. and H. E. Morris. Important grasses and other common plants on Montana ranges. Mon. Agric. Exp. Sta. Bull., 375(1939).
Riegel. A. Life history and habits of blue grama. Trans. Kas. Acad. Sci., 44 (1941) 76-85.
Robertson, J. H. A quantitative study of true-prairie vegetation after three years of extreme drought. Ecol. Mono., 9 (1939) 431-.
Rodgers III, A. D. John Torrey; a study of North American botany. Princeton, 1942.
—, John Merle Coulter: Missionary in Science. Princeton, 1944.
—, American botany, 1873-1892: Decades of transition. Princeton, 1944.
Roterus, V. Spring and winter wheat on the Columbia Plateau. Econ. Geog., 10 (1934) 368-373.
Runyon, H. E. Distribution of seeds by dust storms. Trans. Kas. Acad. Sci., 39 (1936) 105-113.
Ruthven, A. G. The faunal affinities of the prairie region of central North America. Amer. Naturalist, 42 (1908) 388-394.
Rydberg, P. A. and C. L. Shear. A report upon the grasses and forage plants of the Rocky Mountain region. USDA Div. Agrost. Bull. 5 (1897).
Sampson, A. W. Natural vegetation of range lands based upon growth requirements and life history of the vegetation. Jour. Agric. Res., 3 (1914) 93-147.
—, Effects of grazing upon Aspen reproduction. USDA Bull., 741 (1919).
—, Plant indicators— concept and status. Bot. Rev., 5 (1939) 155-.
Sampson, H. C. An ecological survey of the prairie vegetation of Illinois. Ill. Nat. Hist. Surv., 13 (1921) 523-577.
Sargent, C. S. Forests of North America. Tenth Census United States, Vol. 9 (1884).
—, The silva of North America. Boston, 1891, 1902.
—, Manual of trees of North America. Boston, 1905
Sarvis, J. T. Composition and density of the native vegetation in the vicinity of the northern great plains field station. Jour. Agric. Res., 19 (1920) 63-72.
—, Effects of different systems and intensities of grazing upon the native vegetation at the northern great plains station. USDA Agric. Dept. Bull., 1170 (1923).
Savage, D. A. Drought survival of native grass species in central and southern great plains. USDA Tech. Bull., 549 (1937).
Scarth, G. W. Stomatal movement: Its regulation and regulatory rôle. A review. Protoplasma, 2 (1927) 498-511.
—, Mechanism of the action of light and other factors on stomatal movement. Plant Physiology, 7 (1932) 481-.
—, Cell physiological studies of frost resistance: A review. New Phytologist, 43 (1944) 1-12.

Schaffner, J. H. Notes on the salt marsh plants of northern Kansas. Bot. Gaz., 25 (1898) 255-260.
—, Origin of timber belts. Ibid., 27 (1899) 392-393.
—, The spreading of the buffalo grass. Ibid., 27 (1899) 393-394.
—, Development of forest belts in the northwestern part of Clay county, Kansas. Trans. Kas. Acad. Sci., 20 (1906) 74-79.
—, The grasses of Ohio. Ohio St. Univ. Bull. Vol. 21, No.28 (1917).
—, The characteristic plants of a typical prairie. Ohio Nat., 13 (1913) 65-69.
—, Observations on the grasslands of the central United States. Ohio St. Univ. Studies, Contributions in botany, 178 (1926) 1-56.
Schmidt, K. P. Herpetological evidence for the postglacial eastward extension of the steppe in North America. Ecology, 19 (1938) 396-407.
Sears, P. B. The natural vegetation of Ohio:
    I. The virgin forest. Ohio Jour. Sci., 25 (1925) 139-149;
    II. The prairies. Ibid., 26 (1926) 128-146;
    III. Plant succession. Ibid., 26 (1926) 213-231.
—, The archeology of environment in eastern North America. Amer. Anthropologist, 34 (1932) 610-622.
—, Post glacial climate in eastern North America. Ecology, 13 (1932) 1-6.
—, Xerothermic theory. Bot. Rev., 8 (1942) 708-736.
Seifriz, W. The physiology of plants. New York, and London, 1938.
—, A symposium on the structure of protoplasm. Chronica Botanica, 6 (1941) 241-245.
—, The structure of protoplasm, II. Bot. Rev., 11 (1945) 231-259.
Shantz, H. L. Natural vegetation as an indicator of the capabilities of land for crop production in the great plains region. USDA Bur. Plt. Ind. Bull., 201 (1911) 1-100.
—, The natural vegetation of the great plains region. Ann. Assn. Amer. Geogrs., 13 (1923) 81-108.
—, Drought resistance and soil moisture. Ecology, 8 (1927) 145-157.
—, Plants as soil indicators. Yearbook, USDA 1938, 835-860.
—, The relation of plant ecology to human welfare. Ecol. Mono., 10 (1940) 311-342.
Shantz, H. L. and L. N. Piemeisel. Indicator significance of the natural vegetation of the southwestern desert region. Jour. Agric. Res., 28 (1924) 721-801.
—, The water requirements of plants at Akron, Colorado. Ibid., 34 (1927) 1093-1190.
Shantz, H. L. and R. Zon. Grassland and desert shrub. (1928) in USDA Atlas of Amer. Agric., 1926. Double-page colored map of natural vegetation of the United States (1923).
Shear, C. L. Field work of the Division of Agrostology. USDA Div. Agrost. Bull., 25 (1901).
Shimek, B. The prairies. Bull., Lab. Nat. Hist. St. Iowa, Vol. 6, No. 2 (1911) 169-240 and plates.
—, An artificial prairie. Ibid., No. 4 (1913) 35-42.
—, Papers on the prairie. Ibid., Vol. 11, No. 5 (1925).
—, The persistence of the prairie. Ibid., 11, 3-24.

Shimek, —(continued)
—, The prairie flora of Manitoba. Ibid., 11, 25-36.
—, Miscellaneous papers. Ibid., Vol. 14, No. 2 (1931), especially. The relation between migrant and native flora of the prairie region, 10-16.
    In Shimek's study "The Prairies" (1911) he was using the word soils in the sense then prevalent, which related to parent material in the geological sense, and the word xerophyte in the traditional sense prior to Maximov, The Plant in relation to water, 1929.
Shively, S. B. and J. E. Weaver. Amount of underground plant materials in different grassland climates. Univ. Nebr. Conser. Surv. Div. Nebr. Conser. Bull., 21 (1939).
Shreve, E. B. The daily march of transpiration in a desert perennial. Pub. No. 194, Carnegie Inst., Washington, 1914.
Silveus, W. A. Texas grasses. San Antonio, Texas, 1933.
Smith, A. D. A study of the reliability of range vegetation estimates. Ecology, 25 (1944) 441-448.
Smith, C. C. The effect of overgrazing and erosion upon the biota of the mixed-grass prairie of Oklahoma. Ecology, 21 (1940) 381-.
Smith, E. F. Plant pathology: A retrospect and a prospect. Science, 15 n.s. (1902) 601-.
Smith, J. G. Fodder and forage plants, exclusive of the grasses. USDA Div. Agrost., Bull., No. 2, revised 1900.
—, Grazing problems in the southwest and how to meet them. Ibid., No. 16 (1899).
Smithsonian Institution, Division of Radiation and Organisms. Photosynthesis studies. Ann. Rpts., 1929-1930.
Spaulding, V. M. The rise and progress of ecology. Science, 17 n.s. (1903) 201-210.
Standardized plant names. Prepared by F. L. Olmstead, F. V. Coville, and H. P. Kelsey. American Joint Committee on Horticultural Nomenclature. First edition, Salem, Massachusetts, 1933. Second edition, Harrisburg, Pennsylvania, 1942.
Steiger, T. L. Structure of prairie vegetation. Ecology, 11 (1930) 170-217.
Stevens, W. C. and F. E. Dill. Quercus macrocarpa in Kansas. Trans. Kas. Acad. Sci., 43 (1940) 185-197.
Stockman, W. B. Temperature and relative humidity data. USDA Weather Bu. Bull., October, 1905, 25-29.
Stoddart, L. A. The Palouse grassland associations in northern Utah. Ecology, 22 (1941) 158-163.
Tansley, A. G. Review: F. E. Clements, Plant succession; an analysis of the development of vegetation. Carnegie Inst. Washington, 1916. Jour. Ecol., 4 (1916) 198-204.
—, The classification of vegetation and the concept of development. Jour. Ecol., 8 (1920) 118-149.
—, Practical plant ecology; a guide for beginners in the field of plant communities. London, and New York, 1923.
—, Aims and methods in the study of vegetation. London, 1926.
—, Succession: The concept and its values. Proc. Intern. Cong. Pl. Sci., 1 (1929) 677-686.

Tansley, —(continued)
—, The use and abuse of vegetational concepts and terms. Ecology, 16 (1935) 284-307.
Taylor, G. Comparison of American and Australian deserts. Econ. Geog., 13 (1937) 260-268.
Tharp, B. C. Structure of Texas vegetation east of the 98th meridian. Univ. Texas Bull., No. 2606 (1926) 1-100.
Thimann, K. V. Action of light on organisms. Sigma Xi Quarterly, 29 (1941) 23-35.
Tolstead, W. L. Plant communities and secondary succession in south-central South Dakota. Ecology, 22 (1941) 322-327.
—, Vegetation of the northern part of Cherry county, Nebraska. Ecol. Mono., 12 (1942) 255-292.
Tracy, S. M. A report upon the forage plants and forage resources. USDA Div. Agrost. Bull., No. 15 (1898).
Transeau, E. N. On the geographical distribution and ecological relations of the bog-plant societies of northern North America. Bot. Gaz., 36 (1903) 401-420.
—, Forest centers of eastern America. Amer. Nat., 39 (1905) 875-.
—, The relation of plant societies to evaporation. Bot. Gaz., 45 (1908) 217-231.
—, The point of view in vegetation problems involving climate. Plant World, 12 (1909) 102-104.
—, Progress in the survey of the vegetational types of the north central states. Annals Assc. Amer. Geog., 20 (1929) 45.
—, Precipitation types of the prairies and forested regions of the central states. Ibid., 20 (1930) 44-45.
—, The Prairie Peninsula. Ecology, 16 (1935) 423-437.
—, The golden age of botany. Science, 95 n.s. (1942) 53-58.
Transeau, E. N., H. C. Sampson, and L. H. Tiffany. Textbook of botany. New York, 1940.
Turnage, W. V. and A. L. Hinckley. Freezing weather in relation to plant distribution in the Sonoran desert. Ecol. Mono., 8 (1938) 8 (1938) 529-550.
Van Dersel, W. R. Native woody plants of the United States, their erosion-control and wildlife values. USDA Misc. Publ., 303 (1938).
Vasey, G. Illustrations of North American grasses. 2 Volumes, 1890-1893. USDA Div. Bot. Bull., 12-13.
Vass, A. F. and R. Lang. Vegetative composition, density, grazing capacity, and grazing land values in the red desert area. Wyoming Agric. Expt. Sta. Bull., 229.
Vestal, A. G. A black-soil prairie station in northeastern Illinois. Torrey Bot. Club Bull., 41 (1914) 351-363.
—, A preliminary vegetation map of Illinois. Trans.Ill. Acad. Sci., 23 (1930) 204-217.
—, Strategic habitats and communities in Illinois. Ibid., 24 (1931) 80-85.
—, A bibliography of the ecology of Illinois. Ibid., 27 (1935) 163-.
Visher, S. S. The biogeography of the northern great plains. Geog. Rev., 2 (1916) 89-115.

Visher, —(continued)

—, The geography of South Dakota; a detailed discussion of the surface, resources, climate, plants, animals, and human geography, including the history of the area. So. Dak. Geol. Nat. Hist. Surv. Bull., 8 (1918) 1-117.

Wackerman, A. E. Why prairies in Arkansas and Louisiana. Jour. Forestry, 27 (1929) 726-734.

Waller, A. E. Professor John Henry Schaffner. Ohio Jour. Sci., 41 (1941) 253-286. Bibliography.

Ware, E. R. and L. C. Smith. Woodlands of Kansas. Kas.Agric.Expt. Sta. Bull., 285 (1939).

Warner, S. R. Distribution of native plants and weeds on certain soil types in eastern Texas. Bot. Gaz., 82 (1926) 345-372.

Weaver, J. E. A study of the vegetation of southeastern Washington and adjacent Idaho. Univ. Nebr. Studies, 17 (1917) 1-114.

—, Ecological relations of roots. Pub. 286 Carnegie Inst.Wash., (1919).

—, Root development in the grassland formation; a correlation of root systems of native vegetation and crop plants. Pub. No.292 Ibid., (1920).

—, Plant production as a measure of environment. Jour. Ecol., 12 (1924) 205-237.

—, Some ecological aspects of agriculture in the prairie. Ecology, 8 (1927) 1-17.

—, Root development of field crops. New York, 1926.

—, Who's who among prairie grasses? Ecology, 12 (1931) 623-632.

—, Quantity of living plant materials in prairie in relation to run-off and soil erosion. Univ. Nebr. Div. Conv. Dept. Cons. Surv. Bull., 8 (1935).

—, Competition of western wheat grass with relict vegetation of prairie. Amer. Jour. Bot., 29 (1942) 366-372.

—, Replacement of true prairie by mixed prairie in eastern Nebraska and Kansas. Ecology, 24 (1943) 421-434.

Weaver, J. E. and F. W. Albertson. Effects of the great drought on the prairies of Iowa, Nebraska, and Kansas. Ecology, 17 (1936) 567-639.

—, Major changes in grassland as a result of continued drought. Bot. Gaz., 100 (1939) 576-591.

—, Deterioration of midwestern ranges. Ecology, 21 (1940) 216-236.

—, Deterioration of grassland from stability to denudation with decrease in soil moisture. Bot Gaz., 101 (1940) 598-624.

—, Resurvey of grasses, forbs, and underground plant parts at the end of the great drought. Ecol. Mono., 13 (1943) 63-117.

—, Nature and degree of recovery of grassland from the great drought of 1933 to 1940. Ecol. Mono., 14 (1944) 393-479.

Weaver, J. E. and W. E. Bruner. Root development of vegetable crops. New York. 1927.

—, A seven-year quantitative study of succession in the grassland. Ecol. Mono., 15 (1945) 297-319.

Weaver, J. E. and F. E. Clements. Plant ecology, New York, 1929.
  Reviews: Weaver and Clements, Plant ecology, First edition; 1929
    Bates, Jour. Forestry, 27 (1929) 856-859.
    Brierly, Ann. App. Biol., 17: 398-399.
    Fuller, Bot. Gaz., 88 (1929) 452-454.
    Ramaley, Science, 70 (1929) 218-219.
    Vestal, Ecology, 12 (1931) 232-239.
  Reviews: Weaver and Clements, Plant ecology, Second edition;1938
    Braun, Ecology, 19 (1938) 486-490.
    Olmstead, Bot. Gaz., 99 (1937-1938) 957.
    Price, Jour. Forestry, 37 (1939) 270.
    Sears, Science, 89 (1939) 295.
Weaver, J. E. and J. W. Crest. Direct measurement of water-loss from vegetation without disturbing the normal structure of the soil. Ecology, 5 (1924) 153-170.
Weaver, J. E. and R. W. Darland. Grassland patterns in 1940. Ecology, 25 (1944) 202-215.
Weaver, J. E. and T. J. Fitzpatrick. Ecology and relative importance of the dominants of tall-grass prairie. Bot. Gaz., 93 (1932) 113-150.
—, The prairie. Ecol. Mono., 4 (1934) 109-295.
Weaver, J. E. and E. L. Flory. Stability of climax prairie and some environmental changes resulting from breaking. Ecology, 15 (1934) 333-347.
Weaver, J. E. and W. W. Hansen. Native midwestern pastures; their origin, composition, and degradation. Nebr. Conser. Bull., 22 (1941).
—, Regeneration of native midwestern pastures under protection. Ibid., No. 23 (1941).
Weaver, J. E. and W. J. Himmel. The environment of the prairie. Univ. Nebr. Conserv. Surv. Dept. Bull., 5 (1931).
Weaver, J. E., V. H. Haugen, and M. D. Waldon. Relation of root distribution to organic matter in prairie soil. Bot. Gaz., 96 (1935) 389-420.
Weaver, J. E. and J. Kramer. Root systems of quercus macrocarpa in relation to the invasion of prairie. Bot. Gaz., 94 (1932) 51-85.
Weaver, J. E. and W. Noll. Measurement of run-off and soil erosion by a single investigator. Ecology, 1935.
Weaver, J. E. and J. H. Robertson, and R. L. Fowler. Changes in true-prairie vegetation during drought as determined by list quadrats. Ecology, 21 (1940) 357-362.
Weaver, J. E., L. A. Stoddart, and W. Noll. Response of the prairie to the great drought of 1934. Ecology, 16 (1935) 612-629.
Weaver, J. E. and A. E. Thiel. Ecological studies in the tension zone between prairie and woodland. Univ. Nebr. Bot. Surv. Nebr. (n.s.) No. 1 (1917) 1-60.
Weaver, R. J. Water usage of certain native grasses in prairie and pasture. Ecology, 22 (1941) 175-192.
Webb, J. J. Jr. The life history and habits of buffalo grass. Trans. Kas. Acad. Sci., 44 (1941) 58-75.
Wenger, L. E. Buffalo grass. Agric. Expt. Sta. Kas. St. Coll. Agric. & App. Sci. Bull., 321 (1943).

White, C. A. Report on the geological survey of the state of Iowa ... made in the years 1866, 1867, 1868, and 1869. Two volumes. Des Moins, Iowa, 1870.

Whyte, R. O. Crop production and environment. London, Faber and Faber, 1946.

Williams, T. A. Grasses and forage plants of the Dakotas. USDA Div. Agrost. Bull., 6 (1897).

—, A report upon the grasses and forage plants and forage conditions of the eastern Rocky Mountain region. USDA Div. Agrost. Bull., 12 (1898).

Winter, J. M. An analysis of the flowering plants of Nebraska. Univ. Nebr. Conserv. Bull., 13 (1936).

Wood, J. G. The physiology of xerophytism in Australian plants. The stomatal frequencies, transpiration, and osmotic pressure of sclerophyll and tomentose-succulent leaved plants. Jour. Ecol., 22 (1934) 69-87.

Woodward, J. Origin of prairies in Illinois. Bot. Gaz., 77 (1924) 241-261.

Wright, A. H. and A. A. Wright. The habitats and composition of the vegetation of Okefinokee swamp, Georgia. Ecol. Mono., 2 (1932) 109-232.

Wulff, E. V. An introduction to historical plant geography. Authorized translation by Brissendon. Chronica Botanica, Waltham, Massachusetts, 1943.

Yarnell, S. H. Influence of the environment on the expression of heridity factors in relation to plant breeding. Science, 96 n.s. (1942) 505-508.

Zon, R. Forests and water in the light of scientific investigation, Appendix V of final Report of the National Waterways Commission, 1912, Sen. Doc. No. 469, 62 C. 2 s. Reprinted with revised Bibliography, USDA Div. Forest Service, 1927.

Desert

Brown, R. H. A southwestern oasis: The Roswell region, New Mexico. Geog. Rev., 26 (1936) 610-619.

Pickwell, G. B. Deserts. New York, 1939.

Kearney, T. H. et al. Indicator significance of vegetation in Tooele valley, Utah. Jour. Agric. Res., 1 (1914) 365-417.

Mallery, T. D. Changes in the osmotic value of the expressed sap of leaves and small twigs of Larrea tridentata as influenced by environmental conditions. Ecol. Mono., 5 (1935) 1-35.

Nelson, E. W. The influence of precipitation and grazing upon black grama grass range. USDA Tech. Bull., 409 (1934).

Shantz, H. L. and R. L. Piemeisel. Indicator significance of the natural vegetation of the southwestern desert region. Jour. Agric. Res., 28 (1924) 721-801.

Shreve, F. The rate of establishment of giant cactus. Plant World, 13 (1910) 235-240.

—, Establishment behavior of the palo verde. Ibid., 14 (1911) 289-296.

—, Changes in desert vegetation. Ecology, 10 (1929) 364-373.

Shreve, —(continued)
—, Rainfall, runoff and soil moisture under desert conditions. Assoc. Amer. Geogrs., 24 (1934) 131-156.
—, Problems of the desert. Sci. Mo., 38 (1934) 199-209.
—, Vegetation of the northwestern coast of Mexico. Torrey Bot. Club Bull., 61 (1934) 373-380.
—, Plant life of the Sonoran desert. Sci. Mo., 42 (1936) 195-213.
—, Lowland vegetation of Sinaloa. Torrey Bot. Club Bull., 64 (1937) 605-613.
—, The desert vegetation of North America. Bot. Rev., 8 (1942) 195-.
Shreve, F. and A. Hinckley. Thirty years of change in desert vegetation. Ecology., 18 (1937) 463-478.
Sumner, F. B. Biological problems of our southwestern deserts. Ecology, 6 (1925) 352-371.
Van Dyke, J. C. The desert. New York, First edition, 1901; Second edition, 1904.
—, The desert; further studies in natural appearances. New York, 1930.
Whitfield, C. J. Osmotic concentrations of chaparral, coastal sagebrush, and dune species of southern California. Ecology, 13 (1932) 279-285.
Whitfield, C. J. and H. L. Anderson. Secondary succession of the desert plains grassland. Ecology, 19 (1938) 171-180.
Whitfield, C. J. and E. L. Bentner. Natural vegetation in the desert plains grassland. Ecology, 19 (1938) 26-37.

Chapters 3, 10, and 11. Animal and general ecology

Adams, C. C. Postglacial origin and migrations of the life of the northeastern United States. Jour. Geog., 1 (1902) 303-310; 352-.
—, The southeastern United States as a center of geographical distribution of flora and fauna. Biol. Bull., 3 (1902) 115-131.
— On the analogy between departure from optimum vital conditions and departure from geographic life centers. Science, 19 n.s. (1904) 210-211.
—, Postglacial dispersal of the North American biota. Biol. Bull., 9 (1905) 53-71.
—, An ecological survey of northern Michigan. Prepared under the direction of Charles C. Adams.... Pub. by St. Brd. Geol. Surv., as part of Report for 1905, Lansing, Michigan, 1906.
—, An ecological survey of Isle Royale, Lake Superior. In, Rpt. Brd. Geol. Surv., for 1908. Lansing, Michigan, 1909.
—, Guide to the study of animal ecology. New York, 1913.
—, An outline of the relations of animals to their inland environments. Bull., Ill. St. Lab. Nat. Hist., 11 (1915) 3-32.
—, An ecological study of prairie and forest invertibrates. Ibid. Bull., 11 (1915) 33-280.
Allee, W. C. Note on animal distribution following a hard winter. Biol. Bull., 36 (1919) 96-106.
—, Animal aggregations. Quar. Rev. Biol., 2 (1927) 367-398.

Anderson, R. M. Summary of the large wolves of Canada, with description of three new arctic races. Jour. Mammalogy, 24 (1943) 386-393.
Bailey, V. The prairie ground squirrel or spermophiles of the Mississippi valley. USDA Div. Orn. Mam. Bull., 4 (1893).
—, The pocket gophers of the United States. Ibid., 5 (1895).
—, Revision of American voles of the genus microtus. USDA No. Amer. Fauna, Bull., 17 (1900).
—, A biological survey of Texas. Ibid., 25 (1905).
—, Wolves in relation to stock, game and the national forest reserves. USDA For. Serv. Bull., 72 (1907).
—, Harmful and beneficial mammals of the arid interior, with special reference to the Carson and Humboldt valleys of Nevada. USDA Farmers' Bull., 335 (1908).
—, Life zones and crop zones of New Mexico. USDA No. Amer. Fauna Bull., 35 (1913).
—, Revision of the pocket gophers of the genus Thomomys. Ibid., 39 (1915).
—, Biological survey of North Dakota. Ibid., 49 (1926).
—, Life history, distribution,... of grasshopper mice. USDA Tech. Bull., 145 (1929).
—, Mammals of New Mexico. USDA No. Amer. Fauna Bull., 53 (1931).
Baker, A. B. Mammals of western Kansas. Trans. Kas. Acad. Sci., 11 (1889) 56-58.
Bancroft, W. D. The universal law. Science, 33 n.s. (1911) 159-179.
Beed, W. E. A preliminary study of the animal ecology of the Nebraska game preserve. Univ. Nebr. Conserv. Surv. Div. Nebr. Conserv. Bull., 10 (1936).
Bird, R. D. A preliminary ecological survey of the district surrounding the entomological station at Treesbank, Manitoba. Ecology, 8 (1927) 207-220.
Bissonnette, T. H. Sexual photoperiodicity. Quart. Rev. Biol., 11 (1936) 371-386. (This paper includes a full bibliography of his publications, 1930-1936, 21 titles.)
Black, J. D. Mammals of northwestern Arkansas. Jour. Mammalogy, 17 (1936) 29-35.
—, Mammals of Kansas. 30 Bien. Rpt. Kas. St. Brd. Agric., 35 (1937) 116-217.
Blackman, F. F. Optima and limiting factors. Ann. Bot., 19 (1905) 281-295.
Blair, W. F. Faunal relationships and geographic distribution of mammals in Oklahoma. Amer. Midl. Nat., 22 (1939) 85-133.
Blair, W. F. and T. H. Hubbell. The biotic districts of Oklahoma. Amer. Midl. Nat., 20 (1938) 425-454.
Bodenheimer, F. S. Problems of animal ecology. Oxford University Press, 1938.
Brown, L. The distribution of the white-tailed jack rabbit (Lepus townsendii campanius Hollister) in Kansas. Trans. Kas. Acad. Sci., 43 (1940) 385-389.
Burt, W. H. Territoriality and home range concepts as applied to mammals. Jour. Mammalogy, 24 (1943) 346-351.

BIBLIOGRAPHY

Carpenter, J. R. Concepts and criteria for the recognition of communities. Jour. Ecol., 24 (1936) 285-289.
—, The biome. In, Theodore Just (Editor), Plant and animal communities, 75-91, Notre Dame, 1939.
Carter, F. L. A study of jackrabbit shifts in range in western Kansas. Trans. Kas. Acad. Sci., 42 (1939) 431-435.
Cary, M. A biological survey of Colorado. USDA No. Amer. Fauna, 33 (1911).
Chapman, R. N. Animal ecology. New York, 1931.
Craig, W. North Dakota life: plant, animal, and human. Amer. Geog. Soc. Bull., 40 (1908) 321-332; 401-415.
Cross, E. C. Periodic fluctuations in numbers of the red fox in Ontario. Jour. Mammalogy, 21 (1940) 294-306.
—, Color phases of the red fox (Vulpes fulva) in Ontario. Ibid., 22 (1941) 25-40.
Daubenmire, R. F. Merriam's life zones of North America. Quart. Rev. Biol., 13 (1938) 327-332.
Dexter, R. W. Ecology in recent biology texts. Ecology, 23 (1942) 377-378.
Dice, L. R. Distribution of land vertebrates of southeastern Washington. Berkeley, 1916.
—, Notes on the communities of the vertebrates of Riley county, Kansas, with special reference to amphibians, reptiles, and mammals. Ecology, 4 (1923) 40-53.
—, Life zones and mammalian distribution. Jour. Mammalogy, 4 (1923) 39-47.
— The relation of mammalian distribution to vegetation types. Sci. Mo., 33 (1931) 312-317.
—, The biotic provinces of North America. Ann Arbor, 1943.
Dice, L. R. and P. M. Blossom. Studies of mammalian ecology in southwestern North America, with special attention to the colors of desert mammals. Carnegie Inst., Wash., Pub. No. 485, 1937.
Evans, E. A. Jr. Carbon dioxide utilization in animal tissues. Science, 96 n.s. (1942) 25-29.
Formosov, A. N. Mammalia in the steppe biocensse. Ecology, 9 (1928) 449-460.
Garritson, M. S. The American bison. New York, 1938.
Gause, G. F. The struggle for existence. Baltimore, 1934.
—, The principles of biocoenology. Quart. Rev. Biol., 11 (1936) 320-336.
Godwin, H. Review of, Plant and animal communities. Edited by Theodore Just. New Pathologist, 39 (1940) 430-432.
Greulach, V. A. 'Photoperiodism' versus 'photoperiodicity'. Science, 101 n.s. (1945) 353-354.
Grinnell, J. The burrowing rodents of California as agents in soil formation. Jour. Mammalogy, 4 (1923) 137-149.
—, Review of the recent mammal fauna of California. Univ. Cal Pub. in Zool., 40 (1935) 71-234.
Hankinson, P. L. The vertebrate life of certain prairie and forest regions near Charleston, Illinois. Bull., Ill. St. Lab. Nat. Hist., 11 (1915) 281-303.

Hardy, R. The influence of types of soil upon the local distribution of some mammals in southwestern Utah. Ecol. Mono., 15 (1945) 71-108.
Haskell, E. F. Mathematical systematization of 'environment', 'organism', and 'habitat'. Ecology, 21 (1940) 1-16.
Hesse, R. Ecological animal geography; an authorized, rewritten edition based on the German edition of 1924. Prepared by W. C. Allee, and K. P. Schmidt. New York, 1937.
Hibbard, C. W. A revised checklist of Kansas mammals. Trans. Kas. Acad. Sci., 36 (1933) 230-249.
Hill, J. E. and C. W. Hibbard. Ecological differentiation between two harvest mice, (Ruthrodontomys) in western Kansas. Jour. Mammalogy, 24 (1943) 22-25.
Hollister, N. A systematic account of the prairie dog. No. Amer. Fauna, 40 (1916) 1-37.
Howell, A. J. Revision of the North American ground squirrels. No. Amer. Fauna, 56 (1938).
Jewell, M. E. Aquatic biology of the prairie. Ecology, 8 (1927) 289-299.
Just, T. (Editor) Plant and animal communities. Reprinted, Amer. Midl. Nat. Notre Dame, Indiana, 1939.
Kendeigh, S. C. A study of Merriam's temperature laws. Wilson Bull., 44 (1932) 129-143.
—,The rôle of environment in the life of birds. Ecol. Mono., 4 (1934) 301-417.
Lantz, D. Coyotes in their economic relations. USDA Div. Biol. Surv. Bull., 20 (1905).
Larson, F. The rôle of bison in maintaining the short grass plains. Ecology, 21 (1940) 113-121.
Longhurst, W. Observations on the ecology of the Gunnison prairie dog in Colorado. Jour. Mammalogy, 25 (1944) 24-37.
Marcovitch, S. Plant lice and light exposure. Science, 58 n.s. (1923) 537-538. (The first report of experimental work on the influence of photoperiodism on animals.)
Merriam, C. H. A memorial, by W. H. Osgood. Bibliography of C. H. Merriam, by H. W. Grinnell. Jour. Mammalogy, 24 (1943) 421-436; 436-458.
—,Result of a biological survey of the San Francisco mountain region and the desert of the Little Colorado, Arizona. USDA No. Amer. Fauna, 3 (1890).
—,The geographic distribution of life in North America, with special inference to the mammalia. Proc. Biol. Soc. Wash., 7 (1892) 1-64.
—,The geographic distribution of animals and plants in North America. USDA Yearbook, 1894, 203-214.
—,Laws of temperature control of the geographic distribution of terrestrial animals and plants. Natl. Geog. Mag., 6 (1894) 229-.
—,Life zones and crop zones of the United States. USDA Div. Biol. Surv. Bull., 10 (1898).
—,Prairie dogs of the great plains. USDA Yearbook, 1901, 257-270.
Murie, A. Ecology of the coyote in Yellowstone park. Fauna Series No. 4 (1940), Conserv. Bull., 4, Natl. Park Serv. Dept. Interior.

Murphy, R. C. Animal geography; a review. Geog. Rev., 28 (1938) 140-144.
Nelson, E. W. Rabbits of North America. USDA No. Amer. Fauna, 29 (1909).
Nice, M. M. The rôle of territoriality in bird life. Amer. Midl. Nat., 26 (1941) 441-487.
Noble, G. K. The rôle of dominance in the social life of birds. Auk. 56 (1939) 263-273.
Nuttall, G. H. F. The University of Cambridge symbiosis in animals and plants. Am. Nat., 57 (1923) 449-475.
O'Brien, B. Some biological effects of solar radiation. Rpt. Smithsonian Inst., 1943 (1944) 109-134.
Palmer, T. S. The jack rabbits. USDA Div. Biol. Surv. Bull., 8 (1896).
Park. O. Nocturnalism— The development of a problem. Ecol. Mono., 10 (1940) 485-536.
Pearse, A. S. Animal ecology. New York, 1926. Second edition, Ibid., 1939.
Phillips, J. F. V. The biotic community. Jour. Ecol., 19 (1931) 1-24.
—, Succession, development, the climax, and the complex organisms: an analysis of concepts. Jour. Ecol., 22 (1934) 554-571; 23 (1935) 210-246, 488-508. C. Gillman. East African vegetation types: Some notes on the dangers of ecological studies unchecked by co-ordination with the results of other sciences. Jour.Ecol., 24 (1936) 502-505. (This is a reply to Professor John Phillips's criticism of Gillman's Notes submitted in Ms discussed by Phillips in the article just cited.)
Pitelka, F. A. Distribution of birds in relation to major biotic communities. Amer. Midl. Nat., 25 (1941) 113-137.
Presnall, C. C. Wild life conservation as affected by American Indian and Caucasian concepts. Jour. Mammalogy, 24 (1943) 458-.
Rowan, W. On photoperiodism, reproductive periodicity, and the annual migrations of birds and certain fishes. Proc. Boston Soc. Nat. Hist., 38 (1926) 147-189.
Rübel, E. The replaceability of ecological factors and the law of the minimum. Ecology, 16 (1935) 336-341.
Ruthven, A. G. The faunal affinities of the prairie region of central North America. Amer. Nat., 42 (1908) 388-394.
—, The environmental factors in the distribution of animals. Geog. Rev., 10 (1920) 241-248.
Scott, T. G. Some food coactions of the northern plains red fox. Ecol. Mono., 13 (1943) 427-479.
Shackleford, M. W. Animal communities of an Illinois prairie. Ecology, 10 (1929) 126-154.
Shelford, V. E. Physiological animal geography. Jour. Morph., 22 (1911) 551-618.
—, Ecological succession. Biol. Bull., 21 (1911) 9-35, 127-151; 22 (1912) 1-38; 23 (1912) 59-99, 331-370.
—, Animal communities in temperate America, as illustrated by the Chicago region; a study in animal ecology. Geog. Soc. Chicago, 1913, Reprinted with corrections, 1937.

Shelford, —(continued)
—, Principles and problems of ecology as illustrated by animals. Jour. Ecol., 3 (1915) 1-23.
—, Terms and concepts in animal ecology. Ecology, 7 (1926) 389.
—, Naturalists' guide to the Americas. Prepared by a committee ... of the Ecological Society of America, Baltimore, 1926.
—, Laboratory and field ecology. Baltimore, 1929.
—, Some concepts of bioecology. Ecology, 12 (1931) 455-467.
—, Basic principles on the classification of communities and habitats and the use of terms. Ibid., 13 (1932) 105-120.
—, Life zones, modern ecology and the failure of temperature summing. Wilson Bull., 44 (1932) 144-157.
—, The smaller animals of the great plains. Science, 91 (1940) 167-.
—, Deciduous forest man in the grassland of North America. Proc. Eighth Amer. Sci. Cong. Wash., 1940, Auspices Govt. U.S.A., (12 Vol. Dept. St., 1943), Vol. IX History and Government, 203-206.
Shelford, V. E. and S. Eddy. Methods for the study of stream communities. Ecology, 10 (1929) 382-391.
Shelford, V. E. and S. Olson. Sere, climax, and influent animals with special reference to the transcontinental coniferous forest of North America. Ecology, 16 (1935) 375-402.
Smith, H. M. The amphibians of Kansas. Amer. Midl. Nat., 15 (1934) 377-528.
Soper, J. D. History, range, and home life of the northern bison. Ecol. Mono., 11 (1941) 347-412.
Sumner, F. B. Some biological problems of our southwestern deserts. Ecology, 6 (1925) 352-371.
Taylor, W. P. Significance of extreme intermittent conditions in distribution of species and management of natural resources with a restatement of Liebig's Law of Minimum. Ecology, 15 (1934) 374-379.
—, Significance of the biotic community in ecological studies. Quart. Rev. Biol., 10 (1935) 291-307.
—, What is ecology and what good is it? Ecology, 17 (1936) 333-346.
Vestal, A. G. As associational study of Illinois sand prairie. Bull. Ill. St. Lab. Nat. Hist., 10 (1913) 1-96.
—, Local distribution of grasshoppers in relation to plant associations. Biol. Bull., 25 (1913) 141-180.
Visher, S. S. The biogeography of the northern great plains. Geog. Rev., 2 (1916) 89-115.
Weese, A. O. The animal and its environment. In, G. E. Potter, Textbook of zoology, 681-695. St. Louis, 1938.
Young, S. P. and E. A. Goldman. The wolves of North America. Amer. Wildlife Inst. Washington, 1944.

Insects

Allee, W. C. Insect ecology. Ann. Ent. Soc. Amer., 20 (1927) 439-.
Hall, E. D. Problem of the range grasshopper. Jour. Econ. Ent., 30 (1937) 904-910.
Britton, W. E. The academic training of the entomologists in colleges and experiment stations of the United States. Jour. Econ. Entom., 8 (1915) 73-78.

Byars, L. F. An ecological study of ants of the plains region of Boulder county, Colorado. Univ. Colo. Studies, 24 (1941) 11.

Chapman, R. N. and associates. Studies in the ecology of sand dune insects. Ecology, 7 (1926) 416-426.

Cook, C. W. Insects and weather as they influence growth of cactus on the central great plains. Ecology, 23 (1942) 209-214.

Cook, W. C. The distribution of the pale western cutworm.... Ecology, 5 (1924) 60-69.

—,Some weather relations of the pale western cutworm.... Ibid., 7 (1926) 37-47.

—,Weather and the probability of outbreaks of the pale western cutworm. Mo. Weath. Rev., 56 (1928) 103-106.

—,A bioclimatic zonation for studying the economic distribution of injurious insects. Ecology, 10 (1929) 282-293.

Essig, E. O. A history of entomology. New York, 1931.

—,The value of insects to the California Indians. Sci. Mo., 38 (1934) 181-186.

Folsom, J. W. Entomology with special reference to its ecological aspects. Philadelphia, 1934.

Frost, S. W. General entomology. New York, 1942.

Hafen, L. R. and C. C. Rister. Western America. New York, 1941.

Hayes, W. P. Prairie insects. Ecology, 8 (1927) 238-250.

Hayes, W. P. and C. O. Johnston. The reaction of certain grasses to chinch bug attack. Jour. Agric. Res., 31 (1925) 575-583.

Headlee, T. J. and G. A. Dean. The mound building prairie ant, (Pogonomyrmex occidentalis Cresson). Kas. Expt. Sta. Bull.,154 (1908) 165-180.

Hebard, M. The orthoptera of South Dakota. Proc. Acad. Nat. Sci. Philadelphia, 77 (1925) 33-155.

—,The orthoptera of Montana. Ibid., 80 (1928) 211-306.

—,The orthoptera of Colorado. Ibid., 81 (1929) 303-425.

—,The orthoptera of Alberta. Ibid., 82 (1930) 377-403.

—,The orthoptera of Kansas. Ibid., 83 (1931) 119-227.

—,Notes on Montana orthoptera. Ibid., 84 (1932) 251-257.

—,Orthoptera of the upper Rio Grande valley and the adjacent mountains in northern New Mexico. Ibid., 87 (1935) 45-82.

Howard, L. O. The education of the entomologists in the service of the U. S. Dept. Agric. Jour. Econ. Entom., 7 (1914) 274-280.

—,The rise of applied entomology in the United States. Agric. Hist., 3 (1929) 131-139.

Isely, F. B. Seasonal succession, soil relations, numbers, and regional distribution of northeastern Texas acridians. Ecol. Mono., 7 (1937) 317-344.

—,Relations of acrididae to plants and soils. Ibid., 8 (1938) 551-.

LeConte, J. L. The coleoptera of Kansas and eastern New Mexico. Smithsonian Inst. Misc. Contrib. to Knowledge, 11 (1859).

Long, W. S. A checklist of Kansas birds. Trans. Kas. Acad. Sci., 43 (1940) 433-456.

McCullock, J. W. The Hessian fly in Kansas. Kas. Agric. Expt. Sta. Tech. Bull., 11 (1923).

Malin, J. C. An introduction to the history of the bluestem-pasture region of Kansas. Kas. Hist. Quart., 11 (1942) 3-28.

Malin, —(continued)
—, Winter wheat in the golden belt of Kansas. Lawrence, 1944.
Osborn, H. Meadow and pasture insects. Columbus, Ohio, 1939.
Parker, J. R. Grasshopper migrations and modifications. (Review of Uvarov, 1923). Ecology, 6 (1925) 458-459.
Parrot, P. J. The growth and organization of applied entomology in the United States. Jour. Econ. Entom., 7 (1914) 50-64.
Payne, N. M. The effect of environmental temperatures upon insect freezing points. Ecology, 7 (1926) 99-106.
Shelford, V. E. Suggestions as to the original habitat and distribution of various native insect pests. Jour. Econ. Entom., 8 (1915) 171-174.
—, Methods for the experimental study of the relation of insects to weather. Jour. Econ. Entom., 19 (1926) 251-260.
Shelford, V. E. and W. P. Flint. Populations of the chinch bug in the upper Mississippi valley from 1923 to 1940. Ecology, 24 (1943) 435-455.
Shotwell, R. Life histories and habits of some grasshoppers of economic importance on the great plains. USDA Tech. Bull., 774 (1941).
Smith, R. C. The neuroptera and mecoptera of Kansas. Bull. Brook. Ent. Soc., 20 (1925) 165-171.
—, Upsetting the balance of nature, with special reference to Kansas and the great plains. Science, 75 (1932) 649-654.
Sweetman, H. L. The biological control of insects. Ithica, 1936.
Turner, G. T. and D. F. Costello. Ecological aspects of the pricklypear problem in eastern Colorado and Wyoming. Ecology, 23 (1942) 419-426.
Uvarov, B. P. Locusts and grasshoppers. London: The Imperial Bureau of Entomology, 1928.
—, Insects and climate. Trans. Entom. Soc. London, Vol. 79 (1931) Part I.
Van Dyke, E. C. The distribution of insects in western North America. Ann. Entom. Sec. Amer., 12 (1919) 1-12.
Wardle, R. A. The principle of applied entomology. New York, 1929.
Webster, F. M. The diffusion of insects in North America. Psyche, 10 (1903) 47-58.
Weiss, H. B. The pioneer century of American entomology. New Brunswick, N. J., 1936.
Weiss, H. B. and G. M. Ziegler. Thomas Say, early American naturalist. Springfield, Ill., and Baltimore, 1931.
Wheeler, W. M. Ants, their structure, development and behavior. Columbia Univ. Biol. Series IX. New York, 1910. Reprint, 1926.
—, Essays in philosophical biology. Cambridge, 1939.
Whelan, D. B. The winter fauna of the bunch grass of eastern Kansas. Ecology, 8 (1927) 94-97.

Chapter 4, Climatology

Abbott, C. G. Solar variation and weather. Ann. Rpt. Smithsonian Inst. Washington, 1944. 119-153.

Antevs, E. Rainfall and tree growth in the Great Basin. Pub. 469, Special Pub. 21, Carnegie Inst. Washington, 1938.

Bair, R. A. Climatological measurements for use in the prediction of maize yield. Ecology, 23 (1942) 79-88.

Bowman, I. Our expanding and contracting 'desert.' Geog. Rev., 25 (1935) 43-61.

Briggs, L. J. and J. O. Belz. Dry farming in relation to rainfall and evaporation. USDA Bur. Plt. Ind. Bull., 188 (1910).

Blumenstock, G. Jr. Drought in the United States analyzed by means of the theory of probability. USDA Tech. Bull., 819 (1942) 1-63.

Brunt, D. Climatic cycles. Geog. Jour., 89 (1937) 214-230.

Cady, H. P. Is the rainfall in Kansas increasing? Coll. Kas. St. Hist. Soc., 12 (1911-12) 132-133.

Costellani, Sir A. Climate and acclimatization; Some notes and observations. London, 1938.

Chilcott, E. C. The relation between crop yields and precipitation in the great plains area. USDA Misc. Cir., 81 (1927).

—,The relation between crop yields and precipitation in the great plains area. Supplement 1, Crop rotations and tillage methods. Ibid., 81 (1931).

Clements, F. E. Climatic cycles and human populations in the great plains. Carnegie Inst., Washington, 1938. Supp. Pub. 43. Reprinted from Sci. Mo., 47 (1938) 193-210.

Crowe, P. R. An analysis of rainfall probability. Scottish Geogr. Mag., 49 (1933) 73-91.

—,The rainfall regime of the western plains. Geog. Rev., 26 (1936) 463-484.

Douglass, A. E. Climatic cycles and tree-growth; a study of the annual rings of trees in relation to climate and solar activity. Pub. 289, 3 Volumes, Carnegie Inst., Washington 1919-1936.

—,The secret of the southwest solved ... Natl. Geog. Mag., 66 (1929) 737-770.

—,The annual rings of trees. Tree growth and climatic cycles [by] A. E. Douglass. The language of tree rings [by] W. S. Glock. Suppl. Pub., 9, Carnegie Inst., Washington, 1934. Reprinted from Sci. Mo., 37 (1933) 481-495 and 38 (1934) 501-510.

Fassig, O. L. Fifty years of North American rainfall. Deviations from the average. Amer. Met. Soc. Bull., 13 (1932) 205-206.

Finch, V. C., G. T. Trewartha, M. H. Shearer, and F. L. Caudle. Elementary Meteorology. New York, 1942.

Flora, S. D. Is climate changing? 31st Bien. Rpt. Kas. St. Brd. Agric., 1937-38, 36: 30-33.

Glock, W. Principles and methods of tree-ring analysis. Pub. 486, Carnegie Inst. Washington, 1937.

—,Growth rings and climate. Bot. Rev., 7 (1941) 649-713.

Grunsky, C. E. The improbability of rainfall cycles. U. S. Mo. Weather Rev., 55 (1927) 66-68.

Guthrie, L. J. Record-breaking annual precipitation, 1846-1850. U. S. **Mo. Weather Rev.**, 66 (1938) 95-96.

Hainsworth, R. G., O. E. Baker, and A. P. Brodell. Seedtime and harvest today. USDA **Misc. Pub.**, 485 (1942).

Hanson, W. Cultural landscapes of the dissected drift plain in southwestern Nebraska. Dallas, Texas, 1938.

Hopkins, A. D. Bioclimatics, a science of life and climate relations. USDA **Misc. Pub.**, 280 (1938).

Hoyt, J. C. Droughts of 1930-1934. U. S. Geog. Surv. **Water-Supply Paper** 680 (1936) 106.

—, Drought of 1936 with discussion on the significance of drought in relation to climate. **Ibid.**, 820 (1938) 62.

Huntington, E. Tree growth and climatic interpretation. Paper in **Pub.**, 352, Carnegie Inst. Washington, 1925.

—, The climatic factor as illustrated in arid America. With contributions by Charles Schuchert, Andrew E. Douglass, and Charles J. Kullmer. Carnegie Inst. Washington, 1914, **Pub.** 192.

Jones, S. B. Classification of North American climates. **Econ. Geogr.**, 8 (1932) 205-208.

Jones, S. B. and R. Bellaire. The classification of Hawaiian climates; a comparison of the Köppen and Thornthwaite systems. **Geog. Rev.**, 27 (1937) 112-119.

Kendall, H. M. Notes on climatic boundaries in the eastern United States. **Geog. Rev.**, 25 (1935) 117-124.

Kincer, J. B. The climate of the great plains as a factor in their utilization. **Ann. Assn. Amer. Geogrs.**, 13 (1923) 67-80.

—, Is our climate changing. A study of long-time temperature trends. U. S. **Mo. Weather Rev.**, 61 (1933) 251-259.

Klages, K. H. W. Ecological crop geography. New York, 1942.

Lackey, E. E. Annual-variability rainfall maps of the great plains. **Geog. Rev.**, 27 (1937) 665-670.

Malin, J. C. Grassland, "treeless", and "sub-humid", **Geog. Rev.**, 37 (1947) 241-250.

Markham, S. F. Climate and the energy of nations. New York, (Revised) 1944.

Russell, R. J. Climates of California. Univ. Cal. **Pubs.** Geog., 2 (1926) 73-84.

—, Dry climates of the United States, I Climatic map. **Ibid.**, 5 (1931) 1-41.

—, Dry climates of the United States, II Frequency of dry and desert years, 1901-1920. **Ibid.**, 5 (1932) 245-274.

—, Climate years. **Geog. Rev.**, 24 (1934) 92-103.

—, Climatic change through the ages. USDA **Yearbook**, 1941, Climate and man.

—, Climates of Texas. **Ann. Assn. Amer. Geogrs.**, 35 (1945) 37-52.

Schulman, E. Some propositions in tree-ring analysis. **Ecology**, 22 (1941) 192-195.

Sears, P. B. Deserts on the march. Norman, Oklahoma, 1935.

Stetson, H. T. Solar radiation and atmosphere. **Ann. Rpt.** Smithsonian Inst., (1942) 151-172.

Setzer, J. A new formula for precipitation effectiveness. **Geog. Rev.**, (1946) 247-263.

Tannehill, I. R. Drought, its causes and effects. Princeton, 1947.
Thornthwaite, C. W. The climates of North America according to a new classification. Geog. Rev., 21 (1931) 633-655.
—, The climates of the earth. Ibid., 23 (1933) 433-440.
—, Atmospheric moisture in relation to ecological problems. Ecology, 21 (1940) 17-28.
—, Atlas of climatic types in the United States, 1900-1939. USDA Misc. Pub. No. 421 (1941).
Thornthwaite, C. W., C. F. Sharpe, and E. F. Dosch. Climate and accelerated erosion in the arid and semiarid southwest, with special reference to Polacca wash drainage basin, Arizona. USDA Tech. Bull., 808 (1942).
Trewartha, G. T. An introduction to weather and climate. New York, 1937. Second edition, 1943.
USDA Yearbook, 1941. Climate and man.
Van Royen, W. Prehistoric drought in the central great plains. Geog. Rev., 27 (1937) 637-650.
Visher, S. S. Laws of temperature. Ann. Amer. Geogrs., 13 (1923) 15-40.
—, The laws of wind and moisture. Ibid., 169-207.
—, Climatic effects of the proposed wooded shelter belt in the great plains. Ibid., 25 (1935) 63-73.
—, Regionalization of the United States on a precipitation basis. Ibid., 32 (1942) 355-370.
Yao, Shan Yu. The chronological and seasonal distribution of floods and droughts in Chinese history, 206 B.C.— A.D. 1911. An essential portion as a PhD thesis at the Univ. of Pennsylvania. Reprint from Harvard Jour. Asiatic Studies, Volume 7, Philadelphia, 1942.

Chapter 5, Geology and Geography

Geology

Adams, F. D. The birth and development of the geological sciences. Baltimore, 1938.
Bruce, E. L. Mineral deposits of the Canadian shield. Toronto, 1933.
Chamberlin, T. C. Biographical memoir. Natl. Acad. Sci., Memoirs, Vol. 17, (Sixth).
—, The origin of the earth. Chicago, 1916.
—, The two solar families; the sun's children. Chicago, 1929.
Chamberlin, T. C. and F. R. Moulton. Progress reports on nebular hypothesis research project, No. 2 (1903) and later. Carnegie Inst. Washington, 1903— Yearbook No. 3 (1904).
Chamberlin, T. C. and R. D. Salisbury. College textbook of geology. Rewritten and revised by R. T. Chamberlin and P. MacClintok. New York, 1909, 1927.
—, Geology, (3 Volumes). New York, 1904, 1906.
Dana, E. S. and C. Schuchert, etc. A century of science in America with special reference to the American Journal of Science, 1818-1918. New Haven, 1918.

Davis, W. M. and R. A. Daly. Geology and geography. Chapter 19 in Morison, Samuel Eliot (Editor). The development of Harvard University since the inauguration of President Eliot, 1869-1929. Cambridge, 1930.

Eddington, Sir A. The expanding universe. New York, and Cambridge, 1933.

Elias, M. K. (Editor) Symposium on loess, 1944. Amer. Jour. Sci., 243 (1945) 225-303.

Frye, J. C. The high plains surface in Kansas. Trans. Kas. Acad. Sci., 49 (1946) 71-86.

Geikie, Sir A. The founders of geology. London, 1897. Second, enlarged edition, 1905.

Gamow, G. Biography of the earth. New York, 1941.

Henderson, W. B. David Dale Owen: Pioneer geologist of the middle west. Indiana Hist. Coll., Volume 27. Ind. Hist. Bureau, Indianapolis, 1943.

Hubble, E. The problem of the expanding universe. Science, 95 n.s. (1942) 212-215.

—, The problem of the expanding universe. Ann. Rpt. Smithsonian Inst., 1942, 119-132. Reprinted from Sigma Xi Quart. Vol.30, No. 2, 1942. Included also in Sci. in Progress, Series III, 1942.

Jeans, J. H. The universe around us. New York, and Cambridge, Fourth edition, 1944.

Keyes, C. R. The genetic classification of geological phenomena. Jour. Geol., 6 (1898) 809-815.

Longwell, C. R., A. Knopf, R. F. Flint, C. Schuchert, and C. O. Dunbar. Outlines of geology. Second edition of combined works. New York, 1941.

Merrill, G. P. The first hundred years of American geology. New Haven, 1924.

National Research Council, Committee on physics of the earth. Natl. Res. Coun. Bull., 80 (1931) Natl. Acad. Sciences.

Nickels, J. M. Bibliography of North American geology, 1785-1918. Geol. Surv. Bull., Nos. 746, 747, 1918-1928; Bull. No. 823, Cf. E. M. Thom for continuation 1929-1939; Bull. No. 937 (1944).

Robertson, H. P. The expanding universe. Science in Progress, Second series, 1940.

Russell, H. N. The solar system and its origin. New York, 1935.

Schuchert, C. Stratigraphy of the eastern and central United States. New York, 1943.

See, T. J. J. Researches of the evolution of the stellar systems, 2 Volumes. Lynn, Massachusetts, 1896-1910.

Thom, E. M. Bibliography of North American geology, 1929-1939. USDA Int. Geol. Surv. Bull., 937 (1944).

Wagner, H. R. The plains and the Rockies; a bibliography of original narratives of travel and adventure, 1800-1865. San Francisco, 1921. Second edition. Revised and extended by C. L. Camp, San Francisco, 1937.

Warner, C. A. Texas oil and gas since 1543. Houston, Texas, 1940.

Warren, Lt. G. K. Memoir to accompany the map of the territory of the United States from the Missouri river to the Pacific ocean. Washington, 1859.

Warren, —(continued)
—, Map of the territory of the United States from the Mississippi river to the Pacific ocean ... to accompany the reports of the explorations for a railroad route.... Information to 1 May, 1857. Washington, 1859.
Willis, B. American geology, 1850-1900. Science, 96 (1942) 167-.
Wheeler, G. M. Report of surveys west of the 100 meridian, Part III, Volume I. Western exploring expeditions, 1857-1880 under governmental authority. Complete list given. This continued Warren's list. Washington, 1875-1889.

Geography

Association of American Geographers. A conference on regions, December, 1934. Ann. Assn. Amer. Geogrs., 25 (1935) 121-174.
—, Conventionalizing geographic investigations and presentation. Ann. Assn. Amer. Geogrs., 24 (1934) 77-122.
Atwood, W. W. The physiographic province of North America. Boston, 1940.
Barrows, H. H. Geography as human ecology. Ann. Assn. Amer. Geogrs. 13 (1923) 1-15.
Baulig, H. Amerique Septentrionale, Vol. 13 (2 parts) 1935, in Geographie Universelle (15 Volumes) published under the editorship of P. Vidal de la Blache, and L. Gallois. Paris, 1927-1935.
Bengston, N. A. and W. Van Royen. Fundamentals of Economic Geography. New York, 1935.
Boardman, P. Patrick Geddes, maker of the future. With an introduction by Lewis Mumford. New York, 1944.
Bowman, I. Forest physiography; physiography of the United States and principles of soils in relation to forestry. New York, 1911.
—, The new world. New York, 1921.
—, Geography in relation to the social sciences. Report of the Commission on the social studies, Amer. Hist. Assn., (Part 5) New York, 1934.
—, Geography in the creative experiment. Geog. Rev., 28 (1938) 1-19.
Brigham, A. P. The Association of American Geographers. Ann. Assn. Amer. Geogrs., 14 (1924) 109-116.
British Association for the Advancement of Science. Handbook of Canada, 1897. Toronto, 1897.
Brown, R. H. The American geographies of Jedidiah Morse. Ann. Assn. Amer. Geogrs., 31 (1941) 145-217.
Brunhes, J. Human geography, an attempt at a positive classification, principles and examples. Translated from the French. Chicago, and New York, 1920.
—, Human geography. In, H. E. Barnes (Editor) The history and prospects of the social sciences. New York, 1925, 55-105.
Bryan, K. The Columbian agriculture in the southwest, as conditioned by periods of alluviation. Ann. Assn. Amer. Geogrs., 31 (1941) 219-242.
—, In, M. K. Elias (Editor) Symposium on loess. Amer. Jour. Sci., 243 (1945) 225-303.

Bryan, P. W. Man's adaptation of nature: Studies of the cultural landscape. New York, 1933.
Cahnman, W. J. The concept of Raum and the theory of regionalism. Amer. Social Rev., 9 (1944) 455-462.
Capers, G. M. Jr. The biography of a river town: Memphis, its heroic age. Chapel Hill, North Carolina, 1939.
Carter, G. F. Plant geography and culture history in the American southwest. New York, 1945.
Case, E. C. and D. R. Bergsmark. College geography. New York, 1932.
Colby, C. C. Changing currents of geographic thought in America. Ann. Assn. Amer. Geogrs., 26 (1936) 1-37.
Cornish, G. A. Geography of commerce for Canadians. Toronto, 1933.
Cornish, V. The great capitals; an historical geography. London, and New York, 1923.
Davis, W. M. Geographical essays. Edited by D. W. Johnson, Boston, 1909.
—, The progress of geography in the United States. Ann. Assn. Amer. Geogrs., 14 (1924) 159-215.
—, A retrospect of geography. Ibid., 22 (1932) 211-230.
Dryer, C. R. A century of geographic education in the United States. Ann. Assn. Amer. Geogrs., 14 (1924) 117-149.
Earle, E. M. (Editor). Makers of modern strategy. Military thought from Michiavelli to Hitler. Princeton, 1943.
Elias, M. K. (Editor). Tertiary prairie grasses and other herbs from the high plains. Geol. Soc. Amer. Special Papers, 41 (1942).
—, Symposium on loess, 1944. Amer. Jour. Sci., 243 (1945) 225-303.
Fairgrieve, J. Geography and world power. Eighth and revised edition, New York, 1941.
Febvre, L. P. V. A geographical introduction to history. Translated from French. New York, 1925.
Fenneman, N. M. The circumference of geography. Ann. Assn. Amer. Geogrs., 9 (1919) 3-11.
—, Physiographic divisions of the United States. Ibid., 18 (1928) 261-363. Third revised edition, with maps.
—, Physiography of the western United States. New York, 1931.
Finch, V. C. Geographical science and social philosophy. Ann. Assn. Amer. Geogrs., 29 (1939) 1-28.
Finch, V. C. and G. T. Trewartha. Elements of geography. New York, 1936.
Fischer, E. The passing of the European age. A study of the transfer of western civilization and its renewal in other countries. Cambridge 1944.
Flanders, D. P. Geopolitics and American post-war policy. Pol. Sci. Quart., 60 (1945) 578-585.
Forsyth, W. D. The myth of open spaces. London, 1943.
Gilbert, E. W. The exploration of western America, 1800-1850. An historical geography. Cambridge, 1933.
Gottmann, J. Vaubon and modern geography. Geog. Rev., 34 (1944) 120-128.
Hack, John T. The physical basis of Hopi agriculture. (Report of the Peabody Museum Awatovi expedition) Harvard Univ., Peabody Museum Papers, 1942.

Hartshorne, R. Recent developments in political geography. Amer. Pol. Sci. Rev., 29 (1935) 785-804, 943-966.
—, The nature of geography. Ann. Assn. Amer. Geogrs., 29 (1939) 171-658. Second edition, 1946.
Huntington, E. Geography and natural selection. Ann. Assn. Amer. Geogrs., 14 (1924) 1-16.
—, The geography of human productivity. Ann. Assn. Amer. Geogrs., 33 (1943) 1-31.
—, Mainsprings of civilization. New York, 1945.
Huntington, E. and S. W. Cushing. Principles of human geography. Fourth edition. (Huntington Geography Series), New York, 1934.
Huntington, E., F. E. Williams, and S. Van Valkenburg. Economic and social geography. (Huntington Geography Series), New York, 1933.
Jacks, G. V. and R. O. Whyte. Vanishing lands; a world survey of soil erosion. New York, 1939.
James, P. E. An outline of geography. Boston, 1935.
Joerg, W. L. G. The geography of North America: A history of its regional exposition. Geog. Rev., 26 (1936) 640-663.
Johnson, C. W. Shifting desert economies. Jour. Geog., 45 (1946) 200-204.
Johnson, D. The geographic prospect. Ann. Assn. Amer. Geogrs., 19 (1929) 167-231.
Jones, C. F. Economic geography. New York, 1935.
Leith, C. K. World minerals and world policies; a factual study of minerals in their political and international relations. New York, 1931.
Leith, C. K., J. W. Furness, and C. Lewis. World minerals and world peace. Washington, Brookings Institute, 1943.
Lobeck, A. K. Geomorphology. An introduction to the study of landscapes. New York, 1939.
Mackinder, Sir H. J. Britain and the British seas. London, 1902. New York, 1914.
—, The geographical pivot of history. Geog. Jour., 23 (1904) 421. Reprinted by Andreas Dorpalen. The world of General Hanshofer; Geographers in action. New York, 1942.
—, Man-power as a measure of national and imperial strength. Natl. Rev., 45 (1905) 136-143.
—, Democratic ideals and reality. New York, 1919 1942.
—, The human habitat. Scottish Geog. Mag., 47 (1931) 321-335.
—, The round world and the winning of the peace. Foreign Affairs, 21 (1943) 595-605.
Malin, J. C. Mobility and history. Agric. Hist., 17 (1943) 177-.
—, Space and history. Ibid., 18 (1944) 65-74; 107-126.
Marsh, G. P. Man and nature; or physical geography as modified by human action. 1864.
—, The earth as modified by human action; a last revision of Man and nature. New York, 1874.
Mattern, J. Geopolitik: Doctrine of national self-sufficiency and empire. Baltimore, 1942.
Nichols, G. E., et. al. Organic adaptation to environment. New Haven, 1924.

Norris, G. W. Autobiography. (J. E. Lawrence, Editor). New York 1945.
Peattie, R. New college geography. Boston, 1932.
—, Look to the frontiers, a geography for the peace table. New York, 1944.
Pomfret, J. E. The geographic pattern of mankind. New York, 1935.
Powell, J. W. The physiography of the United States. New York, 1896.
Raisz, E. Atlas of global geography. New York, 1944.
Raup, H. M. Trends in the development of geographic botany. *Ann. Assn. Amer. Geogrs.*, 32 (1942) 319-354.
Renner, G. T. Human geography in the air age. A text for high school students. New York, 1942.
Sauer, C. O. Geography of the upper Illinois valley and history of development. *Ill. St. Geol. Survey*, 1916.
—, Recent developments in cultural geography. Chap. 4, in, E. C. Hayes (Editor) Recent developments in the social sciences. Philadelphia, 1927.
—, Foreword to historical geography. *Ann. Assn. Amer. Geogrs.*, 31 (1941) 1-24.
Smith, J. R. Grassland and farmland as factors in the cyclical development of Eurasian history. *Ann. Assn. Amer. Geogrs.*, 33 (1943) 135-162.
Spykman, N. J. America's strategy in world politics; The United States and the balance of power. New York, 1942.
Staples, Z. C. and G. M. York. Economic geography. Second edition. New York, 1934.
Taylor, E. G. R. Whither geography? A review of some recent geographical texts. *Geog. Rev.*, 27 (1937) 129-135.
Thomas, F. The environmental basis of society. New York, 1925.
Turner, F. J. The significance of the frontier in American history, (1893), in his volume, The frontier in American history, 1-38, (New York, 1920).
Van Valkenburg, S. Elements of political geography. New York, 1939.
Van Valkenburg, S. and E. Huntington. Europe. New York, 1935.
Vernadsky, W. I. The biosphere and the noosphere. *Amer. Scientist*, 33 (1945) 1-12.
Vidal de la Blache, P. Principles of human geography. New York, 1926.
Weigert, H. W. Generals and geographers; the twilight of geopolitics. New York, 1942.
—, Mackinder's Heartland. *Amer. Scholar*, 15 (1945-1946) 43-54.
Weigert, H. W. and V. Stefansson. Compass of the world. A Symposium on political geography. New York, 1944.
Whitaker, J. R. World view of destruction and conservation of natural resources. *Ann. Assn. Amer. Geogrs.*, 30 (1940) 143-162.
—, Sequence and equilibrium in destruction and conservation of natural resources. *Ibid.*, 31 (1941) 129-144.
Whitbeck, R. H. and V. C. Finch. Economic geography. Third edition. New York, 1935.
Whitbeck, R. H. and O. J. Thomas. The geographic factor: Its rôle in life and civilization. New York, 1932.

White, C. L. and E. J. Foscue. Regional geography of Anglo-America. New York, 1943.
White, C. L. and G. T. Renner. Geography: An introduction to human ecology. New York, 1936.
Whitney, J. D. The metallic wealth of the United States, described and compared with that of other countries. Philadelphia, 1854.
—, United States: Facts and figures illustrating the physical geography of the country and its material resources. Boston, 1889.
Whittlesey, D. The impress of effective central authority upon landscape. Ann. Assn Amer. Geogrs., 25 (1935) 85-97.
—, The earth and the state; a study of political geography. New York, 1939, 1944.
—, The horizon of geography. Ann. Assn. Amer. Geogrs., 35 (1945)1-.
Willey, G. R. Plant geography and culture history in the American southwest: A review. Geog. Rev., 36 (1946) 132-134.
Zon, R. and W. N. Sparhawk. Forest resources of the world. New York, 1923.

Chapter 6, Soil

Soil science

Atkinson, H. J. Soil colloids, I. Sci. Agric., 23 (1943) 273-286; with R. C. Turner, II. Separation by peptization. Soil Science, 57 (1944) 233-240; with R. C. Turner, and A. Leahey, III, Relationship to soil fertility. Ibid., 57 (1944) 243-246.
Baver, L. D. Soil physics. New York, 1940.
Bean, L. H. Crop yields and weather. USDA Misc. Pub. 471 (1942).
Billings, W. D. Quantitative correlations between vegetational changes and soil development. Ecology, 22 (1941) 448-456.
Clarke, G. R. The study of soil in the field. Review note, Jour. Ecol., 24 (1936) 508.
Emmett, H. E. G. and E. Ashby. Some observations of the relation between the hydrogen-ion concentration of the soil and plant distribution. Annals Bot., 48 (1934) 869-876.
Faulkner, E. H. Plowman's folly. Norman, Oklahoma, 1943.
—, A second look. Norman, Oklahoma, 1947.
Graham, E. R. Soil development and plant nutrition: I, Nutrient delivery to plants by the sand and silt separates. Proc. Soil Sci. Soc. Amer., 6 (1941) 259-261; II, Mineralogical and chemical composition of sand and silt separates in relation to the growth and chemical composition of soybeans. Soil Sci., 55 (1943) 265-273.
Gustafson, A. F. Soils and soil management. New York, 1941.
Hilgard, E. W. Soils, their formation, properties, composition, and relations to climate and plant growth in the humid and arid regions. New York, 1906.
Hough, A. F. Soils in a virgin hemlock-beech forest on the northern Allegheny Plateau. Soil Sci., 54 (1942) 335-341.
—, Soil factors and stand history in a virgin forest valley on the northern Allegheny Plateau. Ibid., 56 (1943) 19-28.

Jenny, H. Factors of soil formation: A system of quantitative pedology. New York, 1941.
Joffe, J. S. Pedology. Introduction by the late Curtis F. Marbut. New Brunswick, N. J. 1936.
Kardos, L. T. and C. C. Bowlsby. Chemical properties of some representative samples of certain great soil groups and their relation to genetic soil classification. Soil Sci., 52 (1941) 335-.
Kelley, W. P. Modern concepts of soil science. Soil. Sci., 62 (1946) 469-476.
Kellogg, C. E. Development and significance of the great soil groups of the United States. USDA Misc. Pub., 229, 1936.
—, Soil survey manual. Ibid., 274, 1937, 1938.
—, The soils that support us; an introduction to the study of soils and their use by men. New York, 1941.
Klages, K. H. W. Ecological crop geography. New York, 1942.
Krusekoff, H. H. Life and work of C. F. Marbut. Morgantown, 1942.
Lyon, T. L. and H. O. Buckman. The nature and properties of soils. Fourth edition. New York, 1943.
Magistad, O. C. Plant growth relations on saline and alkali soils. Bot. Rev., 11 (1945) 181-230.
Marbut, C. F. et. al. Soils of the United States. USDA Bur. Soils Bull., 96, Edition 1913. Revision No. 55 and No. 78.
Marbut, C. F. Soil classification report. Amer. Assn. Soil Survey Workers Ann. Rpt., 2 (1922) 24-33.
—, Soils of the great plains. Ann. Assn. Amer. Geogrs., 13 (1923) 41-66.
—, The great soil groups of the world and their development. Translated from German of Glinka, 1867-1927. Ann Arbor, Mich., 1927.
—, A scheme for soil classification. Proc. Intern. Congr. Soil Sci., 4 (1928) 1-31.
—, Soils of the United States (July 1935), Part III. Atlas Amer. Agric., 1936.
—, Introduction (dated Nov. 1934) to Joffe, Pedology. New Brunswick, N. J., 1936.
Marshall, C. E. Colloids in agriculture. London, 1935.
Millar, C. E. and L. M. Turk. Fundamentals of soil science. New York, 1943.
Prince, A. L., S. J. Toth, A. W. Blair, and F. E. Bear. Forty-year studies of nitrogen fertilizers. Soil Sci., 52 (1941) 247-262.
Retzer, J. L. and M. B. Russell. Differences in the aggregation of a prairie and gray brown podsol. Soil Sci., 52 (1941) 47-58.
Robinson, G. W. Soils: Their origin, constitution and classification. Second edition. London, 1936.
Russell. E. J. Soil conditions and plant growth. New York, 1937.
Russell, J. C. and W. McRuer. The relation of organic matter and nitrogen content to series and type in virgin grassland soils. Soil Sci., 24 (1927) 421-452.
Seifritz, W. The physiology of plants. New York, 1938.
Shantz, H. L. A memoir of Curtis Fletcher Marbut. Ann. Assn. Amer. Geogrs., 26 (1936) 113-123, with bibliography of Marbut's writings.
Shantz, H.L. and C.F. Marbut. The vegetation and soils of Africa. Amer. Geog. Soc., Research Series, 13. W.L.G. Joerg (Editor), New York 1923.

Shaw, C. F. Potent factors in soil formation. *Ecology*, 11 (1930) 239-245.
Shorey, E. C. The liming of soils. (Revised by O. Schreiner) USDA *Farmers' Bull.*, 1845 (1940).
Sigmond, A. A. J. de. The principles of soil science. Translated from Hungarian by A. B. Yolland; translation edited by G. V. Jacks. London, 1938.
Stewart, G. R. A study of soil changes associated with the transition from fertile hardwood forest land to pasture types of decreasing fertility. *Ecol. Mono.*, 3 (1933) 107-146.
Stone, M. H. Soil reaction in relation to the distribution of native plant species. *Ecology*, 25 (1944) 379-486.
Taylor, W. P. Some animal relations to soils. *Ecology*, 16 (1935) 127-136.
Throckmorton, R. J. and F. L. Duley. Soil fertility. Kas. Agric. Exp. Sta. *Bull.*, 260 (1932) 60.
Tyulin, A. T. The composition and structure of soil organo-mineral gels and soil fertility. *Soil Sci.*, 45 (1938) 343-357.
USDA *Yearbook*, 1938, Soils and men.
Whitney, M. Soil and civilization. A modern concept of the soil and the historical development of agriculture. New York, 1925.
Wilcox, E. V. Selenium *versus* General Custer. *Agric. Hist.*, 18 (1944) 105-106.
Wolfanger, L. A. Major soil groups and some of their geographic implications. *Amer. Geog. Rev.*, 19 (1929) 94-114.
—, Major soil divisions of the United States. New York, 1930.
Wynd, F. L. and J. R. Romig. Chemical characteristics of soils in the vicinity of Midland, Douglas county, Kansas. *Soil. Sci.*, 56 (1943) 135-142.

Soil microbiology

The best elementary introduction, probably, is Lyon & Buckman, *The nature and properties of soils*, (1943), Chapter 5, "The organisms of the soil." It contains excellently selected footnote citations to periodical and monographic literature. From that the reader can turn to Waksman and Starkey (1931) for a popular treatise, to Waksman (1945), or to Waksman (1932) for a comprehensive treatise, and E. J. Russell (1937). The most important single body of significant periodical literature is *Soil Science*, 1916— .

Bodily, H. L. The activity of microorganisms in the transformation of plant materials in soil under various conditions. *Soil Sci.*, 57 (1944) 341-350.
Booth, W. E. Algae as pioneers in plant succession and their importance in erosion control. *Ecology*, 22 (1941) 38-46.
Braun, P. J. Plant sociology, the study of plant communities. New York, 1932.
Buchanan, R. E. and E. I. Fulmer. Physiology and biochemistry of bacteria. 3 Volumes. Baltimore, 1928, 1930.

Burrill, T. J. Micro-organisms of soil and human welfare. Science, 20 n.s. (1904) 426-434.
Calkins, G. N. and F. M. Summers (Editors). Protozoa in biological research. New York, 1941.
Cutler, D. and L. Clump. Problems of soil microbiology. London, 1935.
Gates, G. E. The earthworm fauna of the United States. Science, 70 (1929) 266-267.
Gause, G. F. The principles of biocoenology. Quart. Rev. Biol., 11 (1936) 320-336.
—, Experimental populations of microscopic organisms. Ecology, 18 (1937) 173-179.
Greaves, J. E. and L. W. Jones. Inoculation of soil with azotobacter chroococcum. Soil Sci., 53 (1942) 229-232.
—, The survival of microorganisms in alkali soils. Soil Sci., 52 (1941) 359-364.
Hanna, W. J. and E. R. Purvis. Effect of borax and lime on activity of soil microorganisms in Norfolk fine sandy loam. Soil Sci., 52 (1941) 275-282.
Huberty, M. R. and A. R. C. Haas. The pH of soil as affected by soil moisture and other factors. Soil Sci., 49 (1940) 455-478.
Jacot, A. P. Soil structure and soil biology. Ecology, 17 (1936) 358-379.
—, The fauna of the soil. Quart. Rev. Biol., 15 (1940) 28-58.
Jansen, H. L. The fungus flora of the soil. Soil Sci., 31 (1931) 123-158.
Martin, J. P. and S. A. Waksman. Influence of microorganisms on soil aggregation and erosion, II. Soil Sci., 52 (1941) 381-394.
Myers, H. E. and T. M. McCalla. Changes in soil aggregation in relation to bacterial numbers, H-ion concentration and length of time soil was kept moist. Soil Sci., 51 (1941) 189-199.
Newman, A. S. and A. G. Norman. The activity of the microflora in various horizons of several soil types. Soil Sci. Soc. Amer. Proc., 6 (1941) 187-194.
—, The activity of subsurface soil populations. Soil Sci., 55 (1943) 377-391.
Pennak, R. W. Ecology of the microscopic metozoa inhabiting the sandy beaches of some Wisconsin lakes. Ecol. Mono., 10 (1940) 537-615.
Skinner, C. E. and F. Dravis. A quantitative determination of chitin destroying microorganisms in soil. Ecology, 18 (1937) 391-.
Skinner, E. C. and E. M. Mellem. Further experiments to determine the organisms responsible for decomposition of cellulose in soils. Ecology, 25 (1944) 360-365.
Smith, F. B. Contributions to the development of soil microbiology from the southeastern United States. Science, 96 n.s. (1942) 95-.
Stokes, J. L. The rôle of algae in the nitrogen cycle of the soil. Soil Sci., 49 (1940) 265-275.
Vandecaveye, S. C. and H. Katznelson. Microbial activities in soil: VI. Microbial numbers and nature of organic matter in various genetic soil types. Soil Sci., 50 (1940) 295-311.

Waksman, S. A. Microbiological analysis of soils as an index to soil fertility: V. Method for the study of nitrification. Soil Sci., 15 (1923) 241-260.
— Soil microbiology in 1924; an attempt at an analysis and at a synthesis. Soil Sci., 19 (1925) 201-249.
—, Principles of soil microbiology. Second edition. Baltimore, 1932.
—, Associative and antagonistic effects of microorganisms: I. Historical review of antagonistic relationships. Soil Sci., 43 (1937) 51-68; with J. W. Foster, II. Antagonistic effects of microorganisms grown on artificial substrates. Ibid., 69-76; with I. J. Hutchings, III. Association and antagonistic relationships in the decomposition of plant residues. Ibid., 77-92.
—, Antagonistic interrelationships among microorganisms. Chronica Botanica, 6 (No. 7, 1940) 145-148.
—, Antagonistic relations of micro-organisms. Bac. Reviews, 5 (1941) 231-291.
—, Three decades with soil fungi. Soil Sci., 58 (1944) 89-116.
—, Soil microbiology as a field of science. Science, 102 n.s. (1945) 339-344.
Waksman, S. A. and F. C. Gerretsen. Influence of temperature and moisture upon the nature and extent of decomposition of plant residues by microorganisms. Ecology, 12 (1931) 33-60.
Waksman, S. A., E. S. Horning, M. Welsch, and H. B. Woodruff. Distribution of antagonistic actinomycetes in nature. Soil Sci., 54 (1942) 281-296.
Waksman, S. A. and C. E. Skinner. Microorganisms concerned in the decomposition of cellulose in the soil. Jour. Bact., 12 (1926) 57-84.
Waksman, S. A. and R. L. Starkey. Partial sterilization of soil microbiological activities and soil fertility. Soil Sci., 16 (1923) 137-156, 247-268, 343-357.
Waksman, S. A. and H. B. Woodruff. Survival of bacteria added to soil and the resultant modification of soil population. Soil Sci., 50 (1940) 421-427.
—, The soil as a source of microorganisms antagonistic to disease producing bacteria. Jour. Bact., 40 (1940) 581-600.
—, Actinomyces antibioticus, a new soil organism antagonistic to pathenogenic and non-pathogenic bacteria. Ibid., 52 (1941) 231-
—, Selective antibiotic action of various substances of microbial origin. Ibid., 44 (1942) 373-384.
—, The occurrence of bacteriostatic and bactericidal substances in the soil. Soil Sci., 53 (1942) 233-239.
West, P. M. and A. G. Lockhead. The nutritional requirements of soil bacteria— a basis for determining the bacterial equilibrium of soils. Soil Sci., 50 (1940) 409-420.
White, D. P. Prairie soil as a medium for tree growth. Ecology, 22 (1941) 398-407.
Williams, O. B. A quantitative and qualitative determination of the bacterial flora of some representative virgin and cultivated Texas soils. Soil Sci., 19 (1925) 163-168.
Winogradsky, S. The direct method in soil microbiology and its application to the study of nitrogen fixation. Soil.Sci.,25(1928).

Chapters 8, 9. The grassland and early explorations.

Abert, Lt. J. W. Journal of Lt. J. W. Abert from Bent's Fort to St. Louis. Sen. ex. doc., 438, 29 Congress, 1 session, Pub. doc. 477.

Bieber, R. P. (Editor). Exploring southwestern trails, 1846-1854. Glendale, 1938.

Bryan, Lt. F. T. Report on a reconnaissance of a route from San Antonio, via Fredericksburg to El Paso. Sen. ex. doc., 64, 31 Congress, 1 session, Pub. doc., 562.

Cooke, P. St. G. Report of Lt. Col. P. St. George Cooke of his march from Santa Fé, New Mexico, to San Diego, upper California. Dated February 5, 1847. Folding map. H. ex. docs., 41, 30 Congress, 1 session, Pub. doc., 517.

—, The conquest of New Mexico and California. New York, 1878.

Cross, Major C. A report in the form of a journal, to the Quartermaster General of the march of the regiment of mounted riflemen to Oregon, from May 10 to October 5, 1849. Sen. ex. doc., 1, 32 Congress, 2 session, Pub. doc., 587.

Emory, Brig. Gen. W. H. Notes of a military reconnaissance from Fort Leavenworth, in Missouri, to San Diego, in California, including part of the Arkansas, Del Norte, and Gila rivers, 1846, 1847. H. ex. doc., 41, 30 Congress, 1 session, Pub. doc., 517.

Foreman, G. Marcy and the Gold Seekers: the Journal of Captain R. B. Marcy, with an account of the gold rush over the southern route, by Grant Foreman. Norman, Oklahoma, 1939.

French, Capt. S. G. Report in relation to the road opened between San Antonio, Texas, and El Paso del Norte, June, 1849. Sen. ex. doc., 64, 31 Congress, 1 session, Pub. doc., 562.

Hasse, A. R. Reports of explorations printed in the documents of the United States government. Washington, 1899.

Marcy, Capt. R. B. Report of exploration and survey of route from Fort Smith, Arkansas, to Santa Fé, New Mexico, made in 1849. Folding map. H. ex. doc., 45, 31 Congress, 1 session, Pub. doc., 577.

—, Explorations of the Red river of Louisiana in the year 1852.... Sen. ex. doc., 54, 32 Congress, 2 session, Pub. doc., 666. Printed as both a Senate and House document of the 33 Congress, 1 session.

—, Adventure on the Red river, report on the exploration of the headwaters of the Red river by Captain Randolph B. Marcy, and Captain G. B. McClellan, edited and annotated by Grant Foreman. Norman, Oklahoma, 1937.

—, Message of the President to the United States communicating ... a copy of the report and maps of Captain Marcy of his explorations of the Big Wichita and the headwaters of the Brazos rivers, 1854. Sen. ex. doc., 60, 34 Congress, 1 session, Pub. doc., 821

Michler, Lt. N. Report of a reconnaissance of the country between Corpus Christi, and the military post on the Leona ... 1849. Sen ex. doc., 64, 31 Congress, 1 session, Pub. doc., 562.

—, Report on the examination of the route from the upper valley of the south branch of the Red river to the Rio Pecos, 1849. Sen. ex. doc., 64, 31 Congress, 1 session, Pub. doc., 562.

Nevins, A. Fremont, pathmaker of the west. New York, 1939.
Nicollet, I. N. Report intended to illustrate a map of the hydrographical basin of the upper Mississippi ... Sen. ex. doc., 237, 26 Congress, 2 session, Pub. doc., 380.
Parker, W. B. Notes taken during the expedition through unexplored Texas, in the summer and fall of 1854. Philadelphia, 1856.
Simpson, Lt. J. H. Report and map [4 maps] of the route from Fort Smith, Arkansas, to Santa Fé, New Mexico, 1849. Sen. ex. doc., 12, 31 Congress, 1 session, Pub. doc., 554.
—, Report of exploration and survey of route from Fort Smith, Arkansas, to Santa Fé, New Mexico, made in 1849. H. ex. doc., 45, 31 Congress, 1 session, Pub. doc., 577.
Stone, I. Immortal wife [Jessie Frémont]. New York, 1944.
Torrey, J. Plantae Frémontianae. Smithsonian Inst., Pub. 46 (1953).
Whiting, Lt. W. H. C. The report of Lt. W. H. C. Whiting's reconnaissance of the western frontier of Texas. Sen. ex. doc., 64, 31 Congress, 1 session, Pub. doc., 562.
Wislizenus, Dr. F. A journey to the Rocky Mountains in the year 1839. St. Louis, 1912.
—, Memoir of a tour to northern Mexico..., with a scientific appendix and 3 maps. Sen. Misc. doc., 26, 30 Congress, 1 session, Pub. doc., 511.

Chapter 10, Equilibrium

Use the bibliographies in earlier chapters relating to the respective subjects, plus the following:

Aiton, A. S. Coronado's muster roll. Amer. Hist. Rev., 44 (1939).
Bennett, H. H. Soil conservation. New York, 1939.
Bowman, I. Limits of land settlement; a report on present day possibilities. New York, 1938.
Bunce, A. C. Economics of soil conservation. Ames, Iowa, 1942.
Chapline, W. R. (in charge). History of western range research. Agric. Hist., 18 (1944) 127-143.
Cornelius, D. R. Establishment of some true prairie species following reseeding. Ecology, 27 (1946) 1-12.
Costello, D. F. and G. F. Turner. Judging condition and utilization of shortgrass ranges. USDA Farmers' Bull., 1949 (1944).
Fenneman, N. M. Cyclic and non-cyclic aspects of erosion. Science, 83 (1936) 87-94.
Forsling, C. L. and E. V. Storm. The utilization of browse forage as summer range for cattle in southwestern Utah. USDA Circular 62, 1929.
Haley, J. E. E. Charles Goodnight, cowman and plainsman. Boston, and New York, 1936.
Hall, E. R. Control of predatory animals. Hearing before the Committee on Agriculture, House of Representatives, 61 Congress, 2 session, on H. R. 9599. By Mr. Leavitt, April 29, 30, and May 1, 1930. Serial O, E. R. Hall testimony, Pp. 56-66.
—, Predatory mammal destruction. Jour. Mammalogy, 2 (1930) 362-372.

McCoy, I. Journal of Isaac McCoy for the exploring expedition of 1830. Edited by Lela Barnes. Kas. Hist. Quart., 5 (1936) 339-.
Malin, J. C. Indian policy and westward expansion. Humanistic Studies, Vol. II, No. 3, Univ. Kas. Press, Lawrence, Kansas, 1921.
—, An introduction to the history of the bluestem-pasture region of Kansas. Kas. Hist. Quart., 11 (1942) 3-28.
—, Winter wheat in the golden belt of Kansas. Univ. Kas. Press, Lawrence, Kansas, 1944.
—, Dust storms, 1850-1900. Kas. Hist. Quart., 14 (1946) 129-144, 265-296, 391-413.
Meyerhoff, H. A. Floods and dust storms. Science, 83 (1936) 622.
National Resources Board. Supplementary report of the Land Planning Committee, Part V. Soil erosion (with colored folding map). Washington, 1935.
Nicollet, I. N. Report intended to illustrate a map of the hydrographical basin of the upper Mississippi.... Washington, Pub. docs. 380, 464.
Palmer, W. J. Report of surveys across the continent, in 1867-68, on the thirty-fifth and thirty-second parallels, for a route extending the Kansas Pacific railway to the Pacific ocean at San Francisco, and San Diego. By Gen. Wm. J. Palmer, December 1, 1868. Philadelphia, 1869.
Person, H. S. Little waters ... Their use and relations to the land. Washington, 1935- 1936.
Rodgers, A. D. III. John Torrey, a story of North American botany. Princeton, 1942.
Sears, P. B. Deserts on the march. Norman, Oklahoma, 1935.
—, Floods and dust storms. Science, 83 (1936) 9.
Seltzer, F. M. Archeological perspective in the northern Mississippi valley. Smithsonian Misc. Coll., 100 (1940) 253-290.
Shelford, V. E. Deciduous forest man and the grassland fauna. Science, 100 (1944) 135-140, 160-162.
Shantz, H. L. The relation of plant ecology to human welfare. Ecol. Mono., 10 (1940) 311-342.
Shreve, F. The problems of the desert. Sci. Mo., 38 (1934) 199-209.
Stefansson, V. The friendly Arctic; a study of five years in the polar regions. New York, 1922.
Steward, J. H. Native cultures of the intermontane (Great Basin) area. Smithsonian Misc. Coll., 100 (1940) 445-502.
Stewart, G, W. P. Cottam, and S. S. Hutchings. Influence of unrestricted grazing on northern salt desert plant associations in western Utah. Jour. Agric. Res., 60 (1940) 289-316.
Strong, W. D. From history to prehistory in the northern great plains. Smithsonian Misc. Coll., 100 (1940) 353-394.
Talbot, M. W. Indicators of southwestern range conditions. USDA Farmers' Bull., 1782 (1937).
Thompson, W. R. Biological control and the theories of the interactions of populations. Parasitology, 31 (1939) 299-388.
Throckmorton, R. I. and L. L. Compton. Erosion by wind in Kansas. Rpt. Kas. St. Brd. Agric. Doc., 1937, Vol. 56, 224A, Topeka, 1938.
Van Royan, W. Prehistoric droughts in the central great plains. Geog. Rev., 27 (1937) 637-650.

Warren, G. K. Preliminary report of explorations in Nebraska and
 Dakota. *Amer. Jour. Sci.*, 77 (1859) 378-380.
Webb, W. P. The great plains. Boston, 1931.
Wedel, W. R. Culture sequence in the central great plains. *Smithsonian Misc. Coll.*, 100 (1940) 291-352.
—, Environment and native subsistence economies in the central great plains. *Smithsonian Misc. Coll.*, 101 (1941) No. 3, 29 pages and 5 plates.
—, Prehistory and environment in the central great plains. *Trans. Kas. Acad. Sci.*, 50 (1947) 1-18.
Weigert, H. and V. Stefansson (Editors). Compass of the world. New York, 1944.

Chapters 11, and 12.

Science and social theory; history and the methodology of science

See the bibliography to *Essays on Historiography*, especially Chapter 3. There is much of interest also in such periodicals as *Philosophy of Science* (1934—), *The Journal of the history of ideas* (1940—), and *ETC: A review of general semantics* (1943—).

Adams, C. C. The relation of general ecology to human ecology. *Ecology*, 16 (1935) 316-335.
—, A note for social minded ecologists and geographers. *Ibid.*, 19 (1938) 500-502.
—, Selected references on the relation of science to modern life. *102 Ann. Rpt. N. Y. St. Museum, St. Mus. Bull.*, 322 (1940) 79-96.
—, Symposium on relation of ecology to human welfare— the human situation. Ann. meeting Ecol. Soc. Amer., at Columbus, Ohio, 1939. C. C. Adams, chairman. *Ecol. Mono.*, 10 (1940) 307-374.
Alihan, M. A. Social ecology. New York, 1938.
Allee, W. C. Animal aggregations. Chicago, 1931.
—, Animal life and social growth. Baltimore, 1932.
—, The social life of animals. New York, 1938.
—, Group organization among vertebrates. *Science*, 95 n.s. (1942) 289-293.
—, Regeneration, development and genotype. *Ibid.*, 95 n.s. (1942) 364-369.
—, Where angels fear to tread; a contribution from general sociology to human ethics. *Science*, 97 n.s. (1943) 517-525.
Bain, R. The field and problems of biological sociology. Chapter 3, In, L. L. Bernard's The fields and methods of sociology. New York, 1934.
Baitsell, G. A. (Editor). Science in progress. (National Sigma Xi lectures) 3 series, 1938, 1940, and 1942.
Bernard, L. L. and J. Bernard. Origins of American sociology. New York, 1943.
Bowerman, W. G. Residential mortgage loans and sunspot numbers. *Popular Astronomy*, 52 (1944) 123-127.
—, Sunspots and the weather. *Ibid.*, 52 (1944) 159-169, 228-239.

Brozek, J. and A. Keys. General aspects of interdisciplinary research in experimental human biology. Science, 100 (1944) 507-12.
Burgess, E. W. The urban community. Proc. Amer. Sociol. Soc. Chicago, 1925.
Child, C. M. The physiological foundations of behavior. New York, 1924.
Coolidge, W. D. The rôle of science institutions in our civilization. Science, 96 n.s. (1942) 411-417.
Conklin, E. G. Science and ethics. Science, 86 n.s. (1937) 595-603.
Crowther, J. G. The social relations of science. New York, 1941.
Fairchild, H. P. Dictionary of sociology. New York, 1944.
Garner, J. W. Introduction to political science. New York, 1910.
Gerard, R. W. Higher levels of integration. Science, 95 n.s. (1942) 309-313.
Gerard, R. W. and A. E. Emerson. Extrapolation from the biological to the social. Science, 101 n.s. (1945) 582-585.
Harding, T. S. The degradation of science. New York, 1931.
Herskovits, M. J. The myth of the negro past. New York, 1941.
Hofstadter, R. Social Darwinism in American thought, 1860-1915. Philadelphia, 1944.
Hoagland, H. Adventure in biological engineering. Science, 100 n.s. (1944) 63-67.
Hollingshead, A. B. Human ecology. In, R. E. Park (Editor) An outline of the principles of sociology, Part II, 63-168. New York, 1939.
Howells, W. Mankind so far. Amer. Museum Nat. Hist. Sci. Series. New York, 1944.
Hughes, E. C. The ecological aspect of institutions. Amer. Social Rev., 1 (1936) 180-189.
Lehman, H. C. Man's most creative years: then and now. Science 98 (1943) 393-399.
Lewis, J. H. The biology of the negro. Chicago, 1942.
Loeb, L. The biological basis of individuality. Springfield, Ill., and Baltimore, 1945.
Lotka, A. J. Element of physical biology. Baltimore, 1925.
—, Length of life; a study of the life table. New York, 1936.
Lovejoy, A. O. The great chain of being: A study of the history of an idea. Cambridge, 1936.
Lynd, R. S. Knowledge for what? The place of social science in American culture. Princeton, 1939.
MacIver, R. M. History and social causation. In, The tasks of economic history. December 1943, a supplement to Jour. Econ. Hist., 135-145.
McKenzie, R. D. The ecological approach to the study of the human community, Chapter III, In, R. E. Parks, E. W. Burgess, and R. D. McKenzie (Editors), The city, with bibliography by Louis Wirth. Chicago, 1925.
—, The scope of human ecology. In, E. W. Burgess (Editor), The urban community. Chicago, 1926.
—, Movement and the ability to live. Proc. Inst. Internatl. Relations, Riverside, California, 1926.
—, The concept of dominance and world organization. Amer. Jour.

McKenzie, —(continued)
 Sociology, 33 (1927-28) 28-42.
—, Ecological succession in the Puget Sound region. Pubs. Amer. Social Soc., 23 (1929) 60-80.
—, The metropolitan community. New York, 1933.
—, The field and problems of demography, human geography, and human ecology. In, L. L. Bernard, Fields and methods of sociology. New York 1933.
—, Readings in human ecology. Ann Arbor, Michigan, 1934.
Mather, K. F. Man's physical environment and man's behavior. Sigma Xi Quart., 29 (1941) 130-142.
Mellon, M. G. Science, scientists, and society. Science, 97 n.s. (1943) 361-368.
Nordenskiöld, E. The history of biology. Original Swedish edition in 3 Volumes, 1920-24. New York, 1928, 1929, 1932.
Novikoff, A. B. The concept of integrative levels and biology. Science, 101 (1945) 209-215.
—, Continuity and discontinuity in evolution. Ibid., 102 (1945) 405-406.
Park, R. E. The city: Suggestions for the investigation of human behavior in city environment. Amer. Jour. Sociol., 20 (1915) 577-612.
—, Urbanization as measured by newspaper circulation. Ibid., 35 (1929) 60-79.
—, Human ecology. Ibid., 42 (1936) 1-15.
—, Symbiosis and socialization: A frame of reference for the study of society. Ibid., 45 (1939) 1-25.
—, An outline for the principles of sociology. New York, 1939.
Park, R. E. and E. W. Burgess. Introduction to the science of sociology. Chicago, 1921.
Park, R. E., E. W. Burgess, and R. D. McKenzie. The city. Chicago, 1925.
Pearl, R. Introduction to medical biometry and statistics. Philadelphia, and London, 1940.
Queen, S. A. and L. F. Thomas. The city. A study of urbanism in the United States. New York, 1939.
Redfield, R. (Editor). Levels of integration in biological and social systems. Biol. Symposia, Vol. 8, 1942.
Reuter, E. B. Racial theory. Amer. Jour. Sociol., 50 (1945) 452-461, at 458.
Schilpp, P. A. American neglect of a philosophy of culture. Phil. Rev., 35 (1926) 434-446.
Sigerist, H. E. The history of science in postwar education. Science, 100 n.s. (1944) 415-420.
Smith, T. V. The American philosophy of equality. Chicago, 1927.
Sigma Xi, Society of, Rensselaer Chapter. A revaluation of our civilization; a forum on civilization. The Rensselaer Chapter, The Society of Sigma Xi, Albany, New York. The Rensselaer Polytechnic Institute, Engineering and Science Series, No. 57, 1944.
Thompson, W. S. Population studies. Amer. Jour. Sociol., 50 (1945) 436-442.
—, Population problems. Three editions, 1930, 1935, 1942, New York, 1942.

Thompson, W. S. and P. K. Whelpton. Population trends in the United States. New York, 1933.
Weldon, T. D. States and morals. New York, 1947.
Wirth, L. Human ecology. Amer. Jour. Sociol., 50 (1945) 483-488.
Woodger, J. H. Biological principles. London, 1929.
—, The concept of organism and the relation between embryology and genetics. Quart. Rev. Biol., 5 (1930) 1-22, 438-463.
Woelfel, N. Molders of the American mind. A critical review of the social attitudes of seventeen leaders in American education. New York, 1933.

Genetics

Bartlett, H. H. et al. The concept of the genus. A symposium. Torrey Bot. Club Bull., 67 (1940) 349-389.
Clausen, J., D. D. Keck, and W. M. Hiesey. The concept of species based on experiment. Amer. Jour. Bot., 26 (1939) 103-106.
—, Experimental studies on the nature of species. I, Effect of varied environment on western North American plants. Pub. 520 Carnegie Inst., Washington, 1940.
Conklin, E. G. Jean Baptiste-Pierre Antoine de Monet, Chevalier de Lamarck. Genetics, 29 (1944) i-iv.
Dobzhansky, T. Genetics and the origin of species. Second edition. New York, 1941.
Fisher, R. A. The genetical theory of natural selection. Oxford 1930.
Goldschmidt, R. Physiological genetics. New York, 1938.
—, The material basis of evaluation. New Haven, 1940.
Haldane, J.B.S. The causes of evolution. London, 1932.
—, The Marxist philosophy and the sciences. London, 1938.
—, New paths in genetics. New York, 1942.
Huxley, J. S. (Editor). The new systematics. Oxford, 1940.
—, Evolution, the modern synthesis. New York, 1942.
Mayr, E. Speciation phenomena in birds. Amer. Naturalist, 74 (1940) 249-278.
—, Systematics and the origin of species. From the viewpoint of the Zoologist. New York, 1942.
Morgan, T. H. Evolution and genetics. Princeton, 1925.
—, The theory of the gene. New Haven, 1926.
—, The scientific basis of evolution. London, 1932.
Sandow, A. Social factors in the origin of Darwinism. Quart. Rev. Biol., 13 (1938) 315-326.
Waddington, C. H. An introduction to modern genetics. London, 1939.

Philosophy of organism

Agar, W. E. Whitehead's philosophy of organism. An introduction for biologists. Quart. Rev. Biol., 11 (1936) 16-34.
Bews, J. W. Human ecology. Introduction by Jan C. Smuts. London, 1935.
Blyth, J. W. Whitehead's theory of knowledge. Brown University Series, Vol. 7. Princeton, 1941.

Bowman, A. A. A sacramental universe, being a study in the metaphysics of experience. Princeton, 1939.
Burns, C. D. Review: Jan C. Smuts, Holism and evolution. Int. Jour. Ethics, 37 (1926-1927) 314.
Conger, G. P. Review: Jan C. Smuts, Holism and evolution. Jour. Phil., 24 (1927) 136-137.
Crafforn, F. S. Jan Smuts. A biography. Garden City, 1943.
Dunham, A. M. Animism and materialism in Whitehead's organic philosophy. Jour. Phil., 29 (1932) 31-47.
Emmet, D. M. Whitehead's philosophy of organism. London, 1932.
Frank, L. K. Time perspective. Jour. Social Phil., 4 (1939) 293-.
Hulett, J. E. Jr. The person's time perspective and the social rôle. Social Forces, 23 (1944) 155-159.
Kraus, R. Our master: The life of Jan Christian Smuts. New York, 1944.
Lillie, R. S. General biology and the philosophy of organism. Chicago, 1945.
Mellin, S. G. General Smuts. London, 1936.
Murphey, A. E. Review: Whitehead, Process and reality. Int. Jour. Ethics, 40 (1929-1930) 433-435.
Smart, H. R. Review: A. N. Whitehead, Symbolism; its meaning and effect. Phil. Rev., 37 (1928) 388-389.
Smuts, General, the Rt. Hon. J. C. Holism and evolution. London, 1926.
Swabey, W. C. Review: Whitehead, Science and the modern world. Phil. Rev., 35 (1936) 272-279.
Torrey, T. W. Organisms in time. Quart. Rev. Biol. 14 (1939) 275-.
Whitehead, A. N. Religion in the making. New York, 1926.
—, Science and the modern world. New York, 1926.
—, Symbolism; its meaning and effects. New York, 1927.
—, Process and reality. New York, 1929.
—, Adventures of ideas. New York, 1933.
Wirman, H. N. Review: Alfred North Whitehead. Religion in the making. Int. Jour. Ethics, 37 (1927) 312-313.

Chapter 13. Pioneering toward grassland regionalism

Albion, R. G. The communications revolution. Amer. Hist. Rev., 37 (1932) 718-720.
Alway, F. J. and E. S. Bishop. Nitrogen content of the humus of arid soils. Jour. Agric. Res., 5 (1916) 909-916.
Aughey, S. Improvement of western pasture-land. Science, 1 (1883) 335.
Ball, C. R. and A. H. Leidigh. Milo as a dry-land grain crop. Farmers' Bull., 322 (1908).
Clapham, J. H. The economic development of France and Germany, 1815-1914. First edition, 1921. Second edition, Cambridge, 1923.
—, An economic history of modern Britain. 3 Volumes. Cambridge, 1926-1938.
Chilcott, E. C. Some soil problems for practical farmers. USDA Yearbook, 1903, 441-452.
—, Dry-land farming in the great plains area. Ibid., 1907, 451-468. Washington, 1908.

Chilcott,—(continued)
—, A study of cultivation methods and crop rotation for the great plains area. USDA Bur. Plant Indst. Bull., 187 (1910).
—, Some misconceptions concerning dry-farming. USDA Yearbook, 1911, 247-256, Washington, 1912.
Chilcott, E. C. and J. S. Cole. Growing winter wheat on the great plains. USDA Farmers' Bull., 895 (1917).
Chilcott, E. C., J. S. Cole, and W. W. Burr. Spring wheat in the great plains area; relation of cultural methods to production. USDA Bull., 214 (1915).
—, Oats in the great plains area; relation of cultural methods to production. USDA Bull., 218 (1915).
—, Corn in the great plains area; relation of cultural methods to production. Ibid. Bull., 219 (1915).
—, Barley in the great plains area; relation of cultural methods to production. Ibid., 222 (1915).
—, Crop production in the great plains area; relation of cultural methods to yields. Ibid., 268 (1915).
Chilcott, E. C., J. S. Cole, and J. B. Kuska. Winter wheat in the great plains area; relation of cultural methods to production. USDA Bull., 595 (1917).
Chilcott, E. C., W. D. Griggs, and C. A. Burmeister. Corn, milo, and kafir in the southern great plains area; relation of cultural methods to production. USDA Bull., 242 (1915).
Failyer, G. H. Management of soils to conserve moisture, with special reference to semiarid conditions. Farmers' Bull., 266 (1906).
Fireman, P. Russian soil investigations. Exp. Sta. Rec., 12 (1900-1901) Nos. 8 and 9, 704-712; 807-818.
Free, E. E. and J. M. Westgate. The control of blowing soils. USDA Farmers' Bull., 421 (1910).
Gilpin, W. Physical geography of our continent. Natl. Intell., October 13, 15, 22, and December 3, 1857.
—, The agricultural capabilities of the great plains. U. S. Pat. Office Rpt. Agric., 1857, 294-296. Condensed from Natl. Intell., October 13, 1857.
—, The central gold regions. The grain, pastoral, and gold regions of North America. With some new views of its physical geography; and observations on the Pacific railroad. Philadelphia, and St. Louis, 1860.
—, Notes on Colorado; and its inscription in physical geography of the North American continent. London, 1870.
—, The mission of the North American people, geographical, social, and political. Philadelphia, 1873. Second revised edition, 1874.
—, The cosmopolitan railway, compacting and fusing together all the world's continents. San Francisco, 1890.
Goodykoontz, C. B. The wild west, or Beadle's dime novels. In, H. O. Brayer (Editor) 1946.
—, The westerners 1945 brank book. Denver, 1946.
Grace, O. J. The effect of different times of plowing small-grain stubble in eastern Colorado. USDA Bull., 253 (1915).
Haystead, L. If the prospect pleases: The west the guidebooks never mention. Norman, Oklahoma, 1945.

Hilgard, E. W. The agriculture and soils of California. Rpt. USDA 1878, 476-507.
—, Report on cotton production in the United States.... In, U. S. Tenth Census, (1880) Vols. 5, and 6, Washington, 1884, Pub. doc. serial Nos. 2129-2152.
—, California Agric. Exp. Sta., Summary of its work in Rpt., of Office of Exp. Stas. In, Rpt., of Sec. Agric., 1889, 496-501.
—, A report on the relation of soil to climate. USDA Weather Bur. Bull., 3 (1892) 59.
—, Steppes, deserts, and alkali lands. Pop. Sci. Mo., 48 (1896) 602-616.
—, Some physical and chemical peculiarities of arid soils. Proc. Soc. Prom. Agric. Sci., 1898, 70-76. Abs. In, Exp. Sta. Rec., 11 (1899-1900) 717.
—, The causes of the development of ancient civilization in arid countries. North Amer. Rev., 175 (1902) 309-315.
—, Soil management. Science, 20 n.s. (1904) 605-608.
—, Some peculiarities of rock weathering and soil formation in the arid and humid regions. Amer. Jour. Sci., 4 ser. Vol. 171, 21 (1906) 261-269. Abst. E.S.R., 18 (1906-1907) 12.
—, Soils, their formation, properties, composition, and relations to climate and plant growth in the humid and arid regions. New York, 1906.
Hilgard, E. W., T. C. Jones, and R. W. Furnas. Climatic and agricultural features, and agricultural practices and needs of the arid regions of the Pacific slope, with notes on Arizona and New Mexico. USDA Rpt. 20, 1882.
—, Report on the climatic and agricultural features and the agricultural practices and needs of the arid regions of the Pacific slope. Under direction of U.S. Commissioner of Agric., 1882.
Hilgard, E. W.
 Tributes to Hilgard and evaluation of his work:
 In Memoriam: Eugene Waldemar Hilgard. Berkeley, Cal., 1916. Exp. Sta. Rec., 35 (1916) 595.
 Editorials (Unsigned, E. W. Allen, Editor) in Experiment Station Record.
 Introduction to Dr. Peter Fireman, "Russian soil investigations," Exp. Sta. Rec., 12 (1900-1901) 701-703.
 Abstract and review of soils. Ibid., 18 (1906-1907) 315-316.
 On occasion of Hilgard's retirement. Ibid., 21 (1909) 1-4.
 In memory of Hilgard. Ibid., 34 (1916) 301-303.
 Review: E. W. Hilgard, Soils. Amer. Jour. Sci., 172 (1906) 468-
 Review: E. W. Hilgard, Soils. Science, 24 n.s. (1906) 681-684.
 Loughridge, The scientific work of E. W. Hilgard. Ibid., 43 n.s. (1916) 450-453.
 G. P. Merrill, Eugene Waldemar Hilgard. Dict. Amer. Biol., 9 (1932) 22-23.
 E. J. Russell, The first international agrogeological congress. [Budapest, 1909], Nature (London), 84 (1910) 157-158.
 E. A. Smith, Memorial of Eugene Waldemar Hilgard. Bull., Geol Soc. Amer., 28 (1917) 40-54.
 Hilgard, and R. H. Loughridge, Bibliography. Ibid., 54-67.

Holmes, G. K. Practices in crop rotation. USDA Yearbook, (1902) 519-532.
Hulbert, A. B. Soil, its influence on the history of the United States. New Haven, 1930.
Jefferson, M. The civilizing rails. Econ. Geog., 4 (1928) 217-231.
Johnson, W. D. The high plains and their utilization. 21 Ann. Rpt. U. S. Geog. Surv. Part IV (1899-1900) 609-741. 22 Ann. Rpt... Part IV (1900-1901), (1902) 635-669.
Loughridge, R. H. Humus in California soils. Cal.Sta. Bull., 242 (1914) 49-92.
Powell, J. W. Report on the lands of the arid region of the United States, with a more detailed account of the lands of Utah. Washington, Govt. Print. Off., 1878.
—, Institutions for the arid lands. Century, 18 n.s. (1890) 111-116.
—, Physiographic regions of the United States. New York, 1895.
Means, T. H. Reclamation of alkali lands in Egypt, as adapted to similar work in the United States. USDA Bur. Soils Bull., 21 (1903).
Richardson, A. D. Beyond the Mississippi from the great river to the great ocean, life and adventure on the prairies, mountains, and Pacific coast ... 1857-1867. Hartford, Conn., 1867.
Shaler, N. S. Improvement of the native pasture-lands of the far west. Science, 1 (1883) 186-187.
—, Physiography of North America. In, Justin Winsor (Editor) Narrative and critical history of North America. (8 Volumes) Vol.4 (1884) Chapter 1, Introduction.
—, A first book of geology. Boston, 1884.
—, Kentucky: a pioneer commonwealth. (American Commonwealth Series, H. E. Scudder, Editor), First edition, 1885. Boston, 1896.
—, Introduction to Richard A. F. Penrose Jr., Nature and origin of deposits of phosphate of lime. U. S. Geol. Surv. Bull., 46 (1888).
—, Animal agency in soil working. Pop. Sci. Mo., 32 (1888) 484-487.
—, Chance or design. Andover Rev., (1889) 1-17.
—, Aspects of the earth. A popular account of some familiar geological phenomena. New York, 1889.
—, The story of our continent. Boston, 1891.
—, Nature and man in America. New York, 1891.
—, The origin and nature of soils. U. S. Geol. Surv., 12 Rpt. Part 1, (1892) 219-345.
—, The United States of America; a study of the American commonwealth, its natural resources, people, industries, manufactures, commerce, and its work in literature, science, education, and self-government. 3 Volumes, New York, 1894. Edited by Shaler.
—, Domesticated animals, their relation to man and to his advancement in civilization. New York, 1895.
—, American highways; a popular account of their conditions, and of the means by which they may be bettered. New York, 1896.
—, The economic aspects of soil erosion. Natl. Geog. Mag., 7 (1896) 328-338, 368-377.
—, Outlines of the earth's history; a popular study in physiography. New York, 1898.

—, The neighbor; the natural history of human contacts. Boston and New York, 1904.
—, The individual: a study of life and death. New York, 1900.
—, The citizen: a study of the individual and the government. New York, 1904.
—, Man and the earth. New York, 1905.
—, The autobiography of Nathaniel Southgate Shaler, with a supplementary memoir by his wife, [Mrs. Sophia Penn Page Shaler.] Boston, and New York, 1909.
   John E. Wolff, Memoir of Nathaniel Southgate Shaler. Bull., Geol. Soc. Amer., 18 (1907) 592-609. Bibliography of N. S. S. 599-609.
Shantz, H. L. Natural vegetation as an indicator of the capabilities of land for crop production in the great plains. USDA Bur. Plt. Indst. Bull., 20 (1911).
Smythe, W. E. Conquest of arid America. Revised edition, New York, 1905.
Wells, D. A. Recent economic changes. New York, 1889.
Whitney, M. Soils in their relation to crop production. USDA Yearbook, 1894 (1895) 129-164.
—, Reasons for cultivating the soil. Ibid., 1895 (1896) 123-130.
—, Climatology and soils [of cotton]. USDA Off. Exp. Sta. Bull., 33 (1896) 143-169.
—, Soil moisture: a record of the amount of water contained in soils during the crop season of 1896. USDA Div. Soils, Bull., 9 (1897).
—, Some interesting soil problems. USDA Yearbook, 1897 (1898) 429-.
—, Soil investigations in the United States. Ibid., 1899 (1901) 335-.
—, Purpose of soil survey. Ibid., 1901 (1902) 117-132.
—, Soil fertility. USDA Farmers' Bull., 257 (1906).
—, Soils of the United States, based upon the work of the Bureau of Soils, to January 1, 1908. USDA Bur. Soils, Bull., 55 (1909).
—, A study of crop yields and soil composition in relation to soil productivity. USDA Div. Soils Bull., 57 (1909).
—, Fertilizers for wheat soils. Ibid. Bull., 66 (1910).
—, Soil and civilization: a modern concept of soil and the historical development of agriculture. New York, 1925.
Whitney, M. and F. K. Cameron. Chemistry of the soil as related to crop production. USDA Bur. Soils Bull., 22 (1903).
—, Investigations in soil fertility. Ibid. Bull., 23 (1904).

Chapters 14-20
Historical studies dealing with the occupation of the grassland of North America

Alvord, C. W. Centennial history of Illinois. 5 Volumes. Springfield, 1920.
Anderson, E. S. The beet sugar industry of Nebraska as a response to geographical environment. Econ. Geog., 1 (1925) 373-386.
—, The potato industry in Nebraska. Ibid., 6 (1930) 37-53.
Anderson, R. H. Grain drills through thirty-nine centuries. Agric. Hist., 10 (1936) 157-205.

Bailey, R. W. and A. R Craft. Contour-trenches control floods and erosion on range lands. Federal Security Agency, CCC Forestry Pub. 4 (1937).

Baker, O. E. A graphic summary of American agriculture based largely on the census. USDA Yearbook, 1921.

—, The same, with data for 1930. USDA Misc. Pub., 105. 1922, 1931.

—, The Agriculture of the great plains region. Ann. Assn. Amer. Geogrs., 13 (1923) 109-167.

—, The potential supply of wheat. Econ. Geog., 1 (1925) 15-53.

—, Agricultural regions of North America: Map. Part I. The basis of classification, Econ. Geog., 2 (1926) 459-494. II. The south. Ibid., 3 (1927) 50-57; III. The middle country. Ibid., 309-340; IV. The corn belt. Ibid., 447-466; V. The hay and dairying belt. Ibid., 4 (1928) 44-74; VI. The spring wheat region. Ibid., 399-434; VII. The middle Atlantic trucking region. Ibid., 5 (1929) 36-69; VIII. Pacific sub-tropical crops region. Ibid., 6 (1930) 166-190, 278-308; IX. The north Pacific hay and pasture region. Ibid., 7 (1931) 109-153; X. The grazing and irrigated crops region. Ibid., 7 (1931) 325-364, 8 (1932) 325-377; XI. The Columbia Plateau wheat region. Ibid., 9 (1933) 167-198.

Baver, L. D. Soil physics. New York, 1940.

Bieber, R. P. (Editor). Introduction to Joseph McCoy, Historic sketches of the cattle trade of the west and southwest. Glendale, California, 1939.

Bogardus, J. F. The great basin. Econ. Geog., 6 (1930) 321-337.

Boggess, A. C. The settlement of Illinois, 1778-1830. Chicago, 1908.

Bonner, J. C. The genesis of agricultural reform in the cotton belt. Jour. South. Hist., 9 (1943) 475-500.

—, The plantation overseer and southern nationalism as revealed in the career of Garland D. Harmon. Agric. Hist., 19 (1945) 1-11.

Boynton, C. B. and T. B. Mason. Journey through Kansas.... Cincinnati, 1855.

Branch, E. D. Westward. The romance of the American frontier. New York, 1930.

Briggs, H. E. Frontiers of the northwest. A history of the upper Missouri valley. New York, 1940.

Browne, W. A. Agriculture in the Llano Estacado. Econ. Geog., 13 (1937) 155-174.

Burlingame, M. G. The Montana frontier. Helena, Montana, 1942.

Call, L. E. Safeguards against drouth; storing surplus feed. Kansas Agricultural Convention, 1937. Report of the Kas. St. Brd. Agric., for Quarter ending March 31, 1937, 53-61 at 56.

Carlston, C. W. Notes on the early history of water-well drilling in the United States. Econ. Geol., 38 (1943) 119-136.

Cassady, J. T., et al. Revegetating semi-desert range lands in the southwest. Fed. Sec. Agency, CCC Forestry Pub. 8 (1940).

Caughey, J. W. A criticism of the critique of Webb's, The great plains. Miss. Val. Hist. Rev., 27 (1940) 442-444.

Chambers, W. I. Lower Rio Grande valley of Texas. Econ. Geog., 6 (1930) 364-373.

Chapin, F. S. Cultural change. New York, 1928.

Chepil, W. S. Dynamics of wind erosion: I. Nature of movement of soil by wind. II. Initiation of soil movement. Soil Science, 60 (October, November 1945) 305-320; 397-411.
Cooke, P. St. G. Report of Lt. Col. P. St. George Cooke of his march from Santa Fé, New Mexico to San Diego, upper California. Dated February 5, 1847. Folding map. H. ex. docs. 41, 30 Congress, 1 session, Pub. doc., 517.
—, The conquest of New Mexico and California. New York, 1878.
Dana, R. H. Jr. Two years before the mast. First Pub. 1840. Many editions since.
Danhof, C. H. The fencing problem in the eighteen fifties. Agric. Hist., 18 (1944) 168-186.
Davidson, F. A. Relation of taurine cattle to climate. Econ. Geog., 3 (1927) 466-485.
Davis, J. S. Standards and content of living. Amer. Econ. Rev., 35 (1945) 1-15.
—, What kind of an agriculture does America want? Proc., Conference, agricultural economists, Fed. Res. Bank of Minneapolis, Oct. 1945.
Deering, F. USDA, Manager of American agriculture. Norman, 1945.
Dick, E. The sod-house frontier, 1854-1890. New York, 1937.
Edwards, A. D. Influence of drought and depression on a rural community; a case study in Haskell county, Kansas. Farm Security Ad. Soc. Res. Report 7, Washington, 1939.
Evans, H. K. (Edited by R. H. Burns) Sheep trailing from Oregon to Wyoming. Miss. Val. Hist. Rev., 28 (1942) 581-592.
Faulkner, E. H. Plowman's folly. Norman, Oklahoma, 1943.
Finch, V. C. and O. E. Baker. Geography of the world's agriculture. Washington, 1917.
Finnell, H. H. Problem area groups of land in the southern great plains. USDA Soil Conserv. Serv., Washington, 1939.
Fletcher, R. S. Organization of the range cattle business in eastern Montana. Montana St. Col. Agric. Exp. Sta. Bull., 265, (1932).
Foster, E. E. Type-of-farming areas of the United States, 1930. U. S. Bur. Census in Coop. Bur. Agric. Econ. Accompanying text by E. E. Foster. Washington, 1933.
Freeman, O. W. and H. H. Martin. The Pacific Northwest. A regional, human, and economic survey of resources and development. New York, 1942.
Gates, P. W. The Illinois Central Railroad and its colonization work. Harvard Econ. Studies, Vol. 42. Cambridge, 1934.
—, Recent land policies of the federal government. In, Certain aspects of land problems and government land policies. (Part VII of the Supplementary report of the land planning committee of the National Resources Board). Washington, 1935.
Gibson, J. S. Agriculture of the southern high plains. Econ. Geog., 8 (1932) 245-261.
Goodrich, C., et al. Migration and economic opportunity. Philadelphia, 1936.
Gordon, C. Report on cattle, sheep, swine.... Tenth Census of the United States (1880) Vol. 3 (1884).
Gras, N. S. B. The development of metropolitan economy in Europe and America. Amer. Hist. Rev., 27 (1922) 695-708.

Gras,—(continued)
—, An introduction to economic history. New York, 1922.
—, A history of agriculture in Europe and America. New York, 1925.
Gray, L. C. History of agriculture in the southern United States to 1860, 2 Volumes. Carnegie Inst., Washington, 1933.
Great Plains Committee. The future of the great plains. U. S. Govt. Print. Off., 1936.
Hafen, L. R. and C. C. Rister. Western America. New York, 1941.
Hartshorne, R. A new map of the dairy area of the United States. Econ. Geog., 11 (1935) 347-355.
—, Agricultural land in proportion to agricultural population in the United States. Geog. Rev., 29 (1939) 488-492.
Haven, C. T. and F. A. Belden. A history of the Colt revolver and the other arms made by Colt's Patent Fire Arms Manufacturing Company, 1836-1940. New York, 1940.
Haystead, L. 'Faulknerizing' soil. Fortune, 31 (1945) 170,172,175.
—, Centralizing might save family farm. Ibid., 31 (1945) 158,160.
—, Not all tenants are Jeeter Lester. Ibid., 32 (1945) 190, 192.
Hayter, E. W. Barbed wire fencing— a prairie invention; its rise and influence in the western states. Agric. Hist., 13 (1939) 189-207.
—, The western farmers and the drivewell patent. Ibid., 16 (1942) 16-28.
Hazelton, J. M. History and handbook of Hereford cattle and Hereford bull index. Kansas City, Missouri Hereford Journal Company, 1925.
—, A history of linebred Anxiety 4th. Herefords of straight Gudgell and Simpson breeding. Kansas City, Missouri Associated Breeders of Anxiety 4th Herefords, 1939.
Henry, R. S. Letter to Isabel (Bowler) Paterson. In, The God of the machine. (New York, 1943) 222-223 footnote.
Henry, T. C. Address, First annual county fair, Dickinson county, Kansas, at Abilene. Abilene Chronicle, November 10, 1870.
Hesseltine, W. B. Regions, classes and sections in American history. Jour. Land and Pub. Utility Econ., 20 (1944) 35-44.
Jarchow, M. E. Farm machinery in frontier Minnesota. Minn. Mag. Hist., 23 (1942) 316-327.
Kansas Agricultural Experiment Station, Bulletins, No.
   248 (1929), S. C. Salmon and R. I. Throckmorton. Wheat production in Kansas.
   249 (1930), J. W. Zahnley. Soybean production in Kansas.
   250 (1930), T. B. Stinson and H. H. Laude. A report of the Tribune branch Agricultural Experiment Station.
   251 (1930), J. A. Hodges, F. E. Elliott, and W. E. Grimes. Types of farming in Kansas.
   253 (1931), A. E. Aldous and J. W. Zahnley. Tame pastures in Kansas.
   261 (1932), A. D. Weber and W. E. Connell. Wheat as a fattening feed for cattle.
   262 (1932), R. L. von Trebra and F. A. Wagner. Tillage practices for south-western Kansas.

Kansas Agric. Exp. Sta. Bull.—(continued)
- 264 (1933) R. F Cox and W. E. Connell. Lamb feeding experiments with Atlas Sorgo.
- 265 (1933) H. H. Laude and A. F. Swanson. Sorghum production in Kansas.
- 266 (1934) A. F. Swanson and H. H. Laude. Varieties of sorghum in Kansas.
- 268 (1934) L. F. Payne. The comparative nutritive value of sorghum grain, corn, and wheat as poultry feeds.
- 271 (1935) A. F. Swanson. Pasturing winter wheat in Kansas.
- 272 (1935) A. E. Aldous. Management of permanent pastures.
- 273 (1936) A. L. Hallstead and O. R. Mathews. Soil moisture and winter wheat with suggestions on abandonment.
- 275 (1937) H. E. Reed. Sheep production in Kansas.
- 277 (1938) C. E. Aubel. Swine production in Kansas.
- 280 (1938) A. F. Swanson and H. H. Laude. Barley production in Kansas.
- 287 (1939) R. F. Cox. Feeding range lambs in Kansas.
- 289 (1940) R. L. Larmour. A comparison of hard red winter and hard red spring wheats.
- 291 (1940) K. L. Anderson. Deferred grazing of bluestem pastures.
- 292 (1941) R. W. Jugenheimer, A. L. Clapp, and H. D. Hollembeak. Kansas corn tests.
- 293 (1941) R. I. Throckmorton and H. E. Myers. Summer fallow in Kansas.
- 294 (1941) R. J. Doll. Planning the farm business in the bluestem belt of Kansas.
- 302 (1942) A. B. Cardwell and S. D. Flora. The climate of Kansas.
- 303 (1942) H. A. Hackley and H. Howe. Farm tenure law in Kansas.
- 304 (1942) A. F. Swanson and H. H. Laude. Sorghums for Kansas.
- 305 (1942) W. H. Pine. Area analysis and agricultural adjustments in Nemaha county, Kansas

Kansas Agricultural Experiment Station, Technical Bulletins, No.
- 13 (1924) S. O. Salmon. Seeding small grain in furrows.
- 18 (1925) M. C. Sewell and L. E. Call. Tillage investigations relating to wheat production.
- 47 (1939) E. C. Miller. Physiological study of the winter wheat plant at different stages of its development.

Karpinas, B. D. The length of time required for the stabilization of a population. Amer. Jour. Soc., 41 (1936) 504-513.
Klages, K. H. Ecological crop geography. New York, 1942.
Kollmorgen, W. M. and R. W. Harrison. The search for the rural community. Agric. Hist., 20 (1946) 1-8.
Kraenzel, C. F. New frontiers of the great plains. Jour. Farm Econ., 24 (1943) 571-589.
Kroeber, A. L. Cultural and natural areas of native North America. Berkeley, 1939.
Lee, S. C. The theory of the agricultural ladder. Agric. Hist., 21 (1947) 53-61.
Lewton, F. L. Notes on the old plows in the United States National Museum. Agric. Hist., 17 (1943) 62-64.

Malin, J. C. Indian policy and westward expansion. Univ. Kas. Humanistic Studies, 2 (1921) 1-108.
—, The Kinsley boom of the late eighties. Kas. Hist. Quart., 4 (February, May, 1935) 23-49, 164-187.
—, The turnover of farm population in Kansas. Ibid., 4 (1935) 339-.
—, Adaptation of the agricultural system to sub-humid environment. Illustrated by the Wayne Farmers' Club of Edwards county, Kansas. Agric. Hist., 10 (1936) 118-141.
—, J. A. Walker's early history of Edwards county. Kas. Hist. Quart., 9 (1940) 259-284.
—, Agricultural adaptation to the plains. Dict. Amer. Hist., (5 Volumes, New York, 1940) 1: 22-24.
—, Bluestem pastures. Ibid., (5 Volumes, New York, 1940) 1:205-206.
—, Dry-farming. Ibid., 2:171.
—, Local historical studies and population. In, Caroline Ware (Editor), The cultural approach to history. (New York, 1940) 300-.
—, An introduction to the history of the bluestem pastures. Kas. Hist. Quart., 11 (1942) 3-28.
—, John Brown and the legend of fifty-six. Philadelphia, 1942.
—, Mobility and history: Reflections on agricultural policies of the United States in relation to a mechanized world. Agric. Hist., 17 (1943) 177-191.
—, Space and history: Reflections on the closed-space doctrines of Turner and Mackinder and the challenge of those ideas by the air age. Ibid., 18 (1944) 65-74, 107-126. See Geog. Rev., 35 (1945) Geographical Record, 338-339.
—, Winter wheat in the golden belt of Kansas. Lawrence, Kansas,1944.
Reviews: Winter wheat in the golden belt of Kansas.
    Edwards, Agric. Hist., 19 (1945) 20.
    Jarchow, Amer. Hist. Rev., 50 (1945) 372-373.
    Ekblaw, Econ. Geog., 20 (1944) 153.
    Ableiter, Geog. Rev., 35 (1945) 688.
    Dunbar, Jour. Econ. Hist., 5 (1945) 109-110.
    Hodges, Jour. Farm Econ., 26 (1944) 814-816.
    Garland, Jour. Geog., 44 (1945) 298-299.
    Renne, Jour. Land and Pub. Utility Econ., 21 (1945) 84-86.
    Farrell, The Kansas Industrialist, Kansas State College, Manhattan, October 19, 1944.
    Enlow, Land Policy Rev., 8 (1945) 35.
    Jarchow, Minnesota History, 25 (1944) 375-376.
    Schmidt, Miss. Val. Hist. Rev., 31 (Dec. 1944, Mar. 1945) 461-463; 641-644; 32 (1945) 305.
    Edwards, Pac. Hist. Rev., 14 (1945) 225-226.
    Kraenzel, Rural Sociology, 9 (1944) 403-404.
    Edwards, Social Studies, 36 (1945) 231.
—, Dust storms 1850-1900. Kas. Hist. Quart., 14 (May, August, November, 1946) 129-144, 265-296, 391-413.
Miller, R. C. The background of populism in Kansas. Miss. Val. Hist. Rev., 11 (1925) 469-489.
Moore, E. S. Possibilism. In, H. P. Fairchild (Editor), Dictionary of Sociology. New York, 1944.

Oliphant, J. O. Winter losses of cattle in the Oregon country, 1847-1890. Wash. Hist. Quart., 23 (1932) 3-17.
—, The cattle trade from the northwest of Montana. Agric. Hist., 6 (1932) 69-83.
—, The eastward movement of cattle from the Oregon country. Agric. Hist., 20 (1946) 19-43.
Olson, L. Vergil and conservation. Agric. Hist., 18 (1944) 153-.
Osgood, E. S. The day of the cattleman. Minneapolis, 1929.
Owsley, F. W.
    In the American Historical Review, 52 (July, 1947) 845-849, Frank. W. Owsley described the methodology he developed with his students in studying land tenure in certain southern states. The use of census material bears many resemblances to that presented in chapters 16 - 18. His methodology and the resulting studies are excellent.
Park, R. E. Urbanization as measured by newspaper circulation. Amer. Jour. Sociol., 35 (1929) 60-79.
Paxson, F. L. When the west is gone. New York, 1930.
Peacock, C. T. et al. Textbook of methods and implements. The Plowman's Association of Colorado, 1938.
Peterson, E. Forward to the land. Norman, Oklahoma, 1942.
Poggi, E. M. The prairie province of Illinois; a study of human adjustment to the natural environment. Urbana, Illinois, 1934.
President's Committee. Report on farm tenancy. Washington, 1937.
Puri, A. N. Physical characteristics of soils: V. The capillary tube hypotheses of soil moisture. Soil Sci., 48 (1939) 505-520.
Rockefeller Foundation. Proceedings, Conference on the great plains area, April 17-18, 1942, New York City, under the auspices of the Rockefeller Foundation. New York, mimeographed.
—, Conference on the northern plains. June 25-27, 1942. Lincoln, Nebraska. Transcript of discussions held under the auspices of the Rockefeller Foundation. New York, mimeographed.
Rose, J. K. Corn yield and climate in the corn belt. Geog. Rev., 26 (1936) 88-102.
Salter, L. A. Jr. Farm prosperity and agricultural policy. Jour. Pol. Econ., 51 (1943) 13-22.
—, Land tenure in process. A study of farm ownership and tenancy in a Lafayette county (Wisconsin) township. Wisc. Agric. Exp. Sta. Research Bull., 146 (1943).
Sanders, A. H. The story of the Herefords. Chicago, 1914.
—, Shorthorn cattle. Chicago, 1918.
—, A history of Aberdeen-Angus cattle.... Chicago, 1928.
Sauer, C. O. A geographic sketch of early man in America. Geog. Rev., 34 (1944) 528-573.
Shannon, F. A. Appraisal of, The great plains: a study in institutions and environment, by Walter Prescott Webb. Soc. Sci. Res. Coun., Critique of research in the social sciences: III (1940).
—, The farmers' last frontier, 1865-1900. New York, 1945.
Schell, H. S. Drought and agriculture in eastern South Dakota in the eighteen nineties. Agric. Hist., 5 (1931) 162-180.
Shryock, R. H. American historiography: a critical analysis and a program. Proc. Amer. Phil. Soc., 87 (1943) 35-46.

Shryock. —(continued)
—, The need for studies in the history of American science. Isis, 35 (1944) 10-13.

Sigerist, H. E. The history of science in postwar education. Science, 100 (1944) 415-420.

Smith, W. O. The Sharps rifle. Its history, development and operation. New York, 1943.

Stuart, G. Forty years on the frontier, as seen in the Journal and reminiscence of. 2 Volumes. Edited by P. C. Phillips, Cleveland, 1925.

Sviatlovsky, E. E. and W. C. Eells. The centrographical method and regional analysis. Geog. Rev., 27 (1937) 240-254.

Tanner, A. M. Communications on barbed wire patents. Scien. Amer., 67 (1892) 313; 68 (1893) 283. (The present author wishes to acknowledge his indebtedness to Miss M. Klema for calling attention to these French barbed wire patents).

Thornthwaite, C. W. Asstd. by H. I. Slentz. Internal migration in the United States. Philadelphia, 1934.

Throckmorton, R. I. and L. L. Compton. Soil erosion in Kansas and methods of control. In, Soil erosion by wind. Report of Kas. St. Brd. Agric. for Quarter ending December 31, 1937, 7-44 at 8.

Towne, C. W. and E. N. Wentworth. Shepherds empire. Norman, 1945.

Turner, R. E. The industrial city: center of cultural change. In, The cultural approach to history, Caroline Ware (Editor) 228-242.

USDA Farmers' Bulletins, No.

436 (1911) C. W. Warburton. Winter oats for the south. [Illustrated (Figure 3) a southern open furrow drill].

1094 (1922) J. S. Call and A. L. Hallstead. Methods of winter wheat production at the Fort Hays branch station.

1163 (1920) O. R. Mathews. Dry farming in western South Dakota.

1173 (1923) D. E. Stephens, M. A. McCall, and A. F. Bracken. Experiments in wheat production on dry lands of the western United States.

1175 (1923) J. B. Sieglinger. Grain-sorghum experiments at the Woodward field station in Oklahoma.

1361 (Issued Oct. 1923, Revised, 1924). V. V. Parr. Brahman (zebu) cattle.

1520 (1927, 1939) W. J. Morse and J. L. Carter. Soybeans: culture and varieties.

1545 (1927) B. Hunter. Dry farming methods and practices in wheat-growing in the Columbia and Snake river basins.

1690 (1932) W. Ashley and A. H. Glaves. Plowing with moldboard plows.

1761 (1936) W. M. Hurst and W. R. Humphries. Harvesting with combines.

1764 (1936) J. H. Martin, J. S. Cole, and A. T. Semple. Growing and feeding grain sorghums.

1771 (1937) E. C. Chilcott. Preventing soil blowing in the southern plains.

1776 (1937, 1938) W. V. Kell and G. F. Brown. Strip cropping for soil conservation.

BIBLIOGRAPHY

1779 (1937) W. H. Black. Beef cattle breeds for beef and for beef and milk.
1789 (1938) C. I. Hamilton. Terracing for soil and water conservation.
1797 (1938) J. S. Cole. Implements and methods to control soil blowing on the northern great plains.
1806 (1938) K. S. Quisenberry and J. A. Clark. Hard red winter wheat varieties.
1812 (1939) M. M. Hoover. Native and adopted grasses for conservation of soil and moisture in the great plains and western states.
1816 (1938) C. K. Shedd and E. V. Collins. Mechanizing the corn harvest.
1823 (1939) G. Steward. Reseeding range lands of the intermountain region.
1833 (1939) G. K. Rule. Crops against the wind on the southern great plains.
1840 (1939, 1944) R. Y. Bailey. Kudzu for erosion control in the southwest.
1844 (1940) J. H. Martin and J. C. Stephens. The culture and use of sorghums for forage.
1848 (1940) G. K. Rule and R. W. Netterstrom. Soil defense in the Pacific southwest.
1852 (1940, 1942) G. K. Rule. Toward soil security on the northern great plains.
1902 (1942) J. A. Clark. Varieties of spring wheat for the north central states.
1913 (1942) E. L. Flory and C. G. Marshall. Regrassing for soil protection in the southwest.
1917 (1942, 1944) L. S. Carter and G. R. McDole. Stubble-mulch farming for soil defense.
1923 (1943) R. McKee and J. L. Stephens. Kudzu as a farm crop.
1924 (1943) L. R. Short. Reseeding to increase the yield of Montana range lands.

United States Forest Service. The western range. Sen. doc., 199, 74 Congress, 2 session, 1936.
Van Hise, C. R. The conservation of natural resources in the United States. New York, 1910.
Warren, J. A. Agriculture in the central part of the semi-arid portion of the great plains. USDA Bur. Plt. Ind. Bull., 215 (1911).
Warren, L. Lincoln's parentage and childhood. New York, 1926.
White, L. The insular integrity of industry in the Salt Lake oasis. Econ. Geog., 1 (1925) 206-235.
—, Transhumance in the sheep industry. Ibid., 2 (1926) 414-425.
Widtsoe, J. A. Dry farming. New York, 1913.
Wentworth, E. N. Eastward sheep drives from California and Oregon. Miss. Val. Hist. Rev., 28 (1942) 507-538.
Wissler, C. The American Indian. New York, First edition 1917. Second edition 1922. Third edition, 1938.
Wolters, G. F. A socio-economic analysis of four rural parishes in Nemaha county, Kansas. Washington, 1938.

Zink, N. E. Dry-farming regions in Utah. Econ. Geog., 15 (1939) 421-431.

Comparative agricultural geography

Bergsmark, D. R. Economic geography of Asia. New York, 1935.
Case, E. C. The pastoral and agricultural industries of Kenya colony and protectorate. Econ. Geog., 6 (1930) 243-256.
—, Agriculture and commerce of Uganda. Ibid., 352-363.
Cressy, G. B. Foundations of Chinese life. Econ. Geog., 15 (1939) 95-104.
Fleur, A. and E. J. Foscue. Agricultural production in China. Econ. Geog., 3 (1927) 297-308.
Jonasson, O. Agricultural regions of Europe. Econ. Geog., 1 (1925) 277-314.
Jones, C. F. Agricultural regions of South America. Econ. Geog., 4 (1928) 1-30. (Large folding map, colored); 159-186; 267-294; 5 (1929) 108-140; 277-307; 390-421; 6 (1930) 1-36.
Madigan, C. T. The Australian sand-ridge deserts. Geog. Rev., 26 (1936) 205-227.
Marbut, C. F. Agriculture in the United States and Russia. A comparative study of natural conditions. Geog. Rev., 21 (1931) 598-.
Merriam, G. P. The regional geography of Australia. Econ. Geog., 2 (1926) 86-107.
Moyer, R. T. Agricultural soils in a loess region of north China. Geog. Rev., 26 (1936) 414-425.
—, Agricultural practices in semi-arid north China. Scientific Mo., 55 (1942) 301-316.
Shantz, H. L. Agricultural regions of Africa. Econ. Geog., 16 (1940) 1-47, 121-161, 341-389; 17 (1941) 217-249, 353-379; 18 (1942) 229-246, 343-362; 19 (1943) 217-269.
Shaw, E. B. Swine industry of China. Econ. Geog., 14 (1938) 381-.
Simkins, E. The coast plains of south India. Econ. Geog., 9 (1933) 19-50, 136-159.
Taylor, G. Agricultural regions of Australia. Econ. Geog., 6 (1930) 109-134, 213-242.
—, The distribution of pasture in Australia. Geog. Rev., 27 (1937) 291-294.
Thompson, W. R. Moisture and farming in south Africa. South Africa: Central News Agency, Ltd., 1936.
Thorp, J. Geography and the soils of China. Published by the National Geological Survey of China, in cooperation with the Institute of Geology of the National Academy of Peiping, and with the support of the China Foundation for the Promotion of Education and Culture. Nanking, 1936.
Tulaikov, N. M. Agriculture in the dry region of U. S. S. R. Econ. Geog., 6 (1930) 54-80.
Van Valkenburg, S. Agricultural regions of Asia. Econ. Geog., 7 (1931) 217-237; 8 (1932) 109-133; 9 (1933) 1-18, 109-135.
Walker, E. H. The plants of China and their usefulness to man. Ann. Rpt. Smithsonian Inst., Washington, 1943.

Whittlesey, D. Major agricultural regions of the earth. *Ann*. Assn. Amer. Geogrs., 26 (1936) 199-240. The bibliographies incorporated into this monograph are invaluable.

Chapter 21. An open system

Abbot, C. G. Solar radiation as a source of power. *Ann*. *Rpt*. Smithsonian Inst., Washington, 1943, 99-107.

Armstrong, E. F. The sea as a storehouse. *Ann*. *Rpt*. Smithsonian Inst., Washington, 1943, 135-149.

Bigelow, H. B. Oceanography: Its scope, problems, and economic importance. Boston, and New York, 1931.

Bowman, I. Commanding our wealth. *Science*, 100 (1944) 229-241.

—, The new geography. A radio talk, July 15, 1945, sponsored by the United States Rubber Company. Printed copy, 1945.

Cook, O. F. Natural rubber. *Ann*. *Rpt*. Smithsonian Inst., 1943, 363-411.

Entrican, A. R. Our soft wood complex. *Jour*. *Forestry*, 27 (1929) 768-781.

Fowler, G. H. Science of the sea. An elementary handbook of practical oceanography for travellers, sailors, and yachtmen. Prepared by the Challenger Society for the promotion of the study of oceanography. London, 1912.

Frolich, Dr. P. K. Petroleum, past, present, and future. *Science*, 98 n.s. (Nov., Dec., 1943) 457-463; 484-487.

Guyot, A. H. The earth and man. Lectures on comparative physical geography in its relation to the history of mankind. Translation by C. C. Felton. Boston, 1860.

Hottel, H. C. Artificial converters of solar energy. *Ann*. *Rpt*. Smithsonian Inst., Washington 1941. (With bibliography).

Ilin, M. Men and mountains. Man's victory over nature. Philadelphia, 1935.

Jeans, Sir J. The universe around us. New York, First to Fourth editions, 1929,- 1944.

Lawrence, E. O. The new frontiers in the atom. *Ann*. *Rpt*. Smithsonian Inst., Washington, 1941.

Mackinder, Sir H. J. The round world and the winning of the peace. *Foreign Affairs*, 21 (1943) 595-605.

Marshak, I. I. New Russia's primer; the story of the five-year plan. By, M. Ilin (pseud.) Translated from Russian by G. S. Counts ... and N. P. Lodge. Boston, and New York, 1931.

—, Men and mountains. Man's victory over nature. Translated from Russian by B. Kinkead. Philadelphia, and London, 1935.

Maury, M. F. The physical geography of the sea. Sixth edition, 1856. New York, and London, 1859.

Manchester, H. Super-wood has arrived. *American Forests*, (1944). Condensed in *Readers Digest*, 45 (1944) 43-44.

Mather, K. F. The future of man as an inhabitant of the earth. *Ann*. *Rpt*. Smithsonian Inst., Washington, 1940, 215-229. Reprint from *Sigma Xi Quart*., 28 (1940) with changes, 3-14, 20.

—, The resources of the continent. *Science*, 96 n.s. (1942) 125-127.

Murray, Sir J. The ocean. A general account of the science of the sea. New York, and London, (n.d.)

Reclus, E. The earth. A descriptive history of the phenomena of the life of the globe. Translated by B. B. Woodward. Edited by H. Woodward. London, and New York, 1871.

—, The ocean, atmosphere, and life. Being a second series of a descriptive history of the life of the globe. New York, 1873.

Robbins, W. J. The importance of plants. Science, 100 (1944) 440-43.

Stefansson, V. The North American Arctic. In, Weigert & Stefansson (Editors), Compass of the World. An earlier version of this article in Foreign Affairs, (1939), a later version (1945), in, The arctic in fact and fable. Headline Series No. 51, Foreign policy assn., 1945.

—, Arctic manual. New York, 1944.

Sverdrup, H. U. The unity of the sciences of the sea. Sigma Xi Quart., 28 (1940) 105-115.

Thoreau, H. D. A week on the Concord and Merrimack rivers, 1849. Reprinted in, The writings of Henry David Thoreau (11 Vols.) Vol. I (1893) 400-401. (The original writing of this book was 1839. The first printing, in 1849).

Vernadsky, W. I. The biosphere and the noösphere. Amer. Sci., 33 (1945) 1-12.

Weyl, H. The open world; three lectures on the metaphysical implications of science. Translated with assistance of L. Huffman. New Haven, 1932.

BIBLIOGRAPHY OF MAPS

In mapping regionalism it is important to consider the several different possible bases for defining boundaries. No region can be given definite boundaries, and the approximate limits, according to the several standards, do not necessarily agree. It will be found however, that the characteristic regional nuclei of some such areas coincide fairly closely and it is this basic fact that is significant rather than exact outer boundary lines.

Physiographic Maps. Chronologically listed the principal physiographic maps that present the physical features of the grassland, or a part of it, either separately or as a part of the United States or of North America, are Powell (1895); Johnson (1899-1900); Chilcott (1907); Bowman (1911); Marschner (1915) in Section 1. Atlas of American Agriculture (1936); Lobeck (1931, 1939, p. 624); Fenneman (1928, 1931); and Atwood (1940, 1941) with a map of landforms of the United States by Erwin Raisz, 1939.

Climate Maps. Four major publications of the United States government are particularly important for climatic maps; The Atlas of American Agriculture; USDA Yearbook, 1941, Climate and Man; Thornthwaite, Atlas of climate types in the United States, 1941; and Van Dersal, 1938, Native woody plants of the United States, containing a map of climate provinces and plant growth.

Other publications of outstanding importance are Livingston and Shreve (1921), and the several presentations of new types of climate methods published in the Geographical Review (New York); Thornthwaite (1931), Russell (1934), Kendall (1935), Crowe (1936), and Lackey (1937).

Vegetational Maps. Among governmental publications on vegetation, the most important for present purposes is the Atlas of American Agriculture (1936), the vegetation section being prepared by Shantz and Zon (1928). The double-page natural vegetation map (1923) is colored and is the best generalized map available, especially for forest distribution. For the eastern United States, the forests are indicated in five zones from north to south. The grassland is differentiated as to the main divisions of distribution, following the Clements contention that there are only two divisions, but which Shantz calls tall and short grasses (omitting the mixed grass region). In a black and white map, p. 6, the five main grassland divisions of the United States are shown. Van Dersal (1938), Figure 1, opposite p. 16, presents plant growth regions (Mulford). The USDA Yearbook, 1938, Soils and Men p. 839, contains a Shantz map of vegetation and the Yearbook, 1941, Climate and Man p. 107, world distribution of vegetation; p. 450, "Grasslands of the United States, showing the dominant types of grasses in each region as determined by climate"; and p. 451, "Sections of the United States where native short grasses and prairie grasses are well adapted and are of primary importance." Hitchcock, Manual of _grasses_ (1935), gives small distribution maps for each of the more important grass species.

In Livingston and Shreve (1921) Part I, "Distribution of vegetation in the United States", pp. 29-91, contains a series of maps devoted to several species of plants, Plates 2-32 inclusive. For present purposes Plates 10 and 11 devoted to grass species distribution, buffalo grass (Plate 10) and two gramas, buffalo and June grass (Plate 11) are important. The book contains also a large folding map, in color, of the life zones of the United States according to Merriam.

Shantz (1923) p. 83, gave a full page of "Vegetation of the Great Plains." Transeau (1935) has a folding map of the prairie peninsula with particular respect to its forest boundary, and a series of distribution maps of trees. These should be compared with Livingston's and Shreve's (1921) tree distribution maps. The Aikman (1935) and Carpenter (1940) maps of the central North American grassland present the three-fold east-west division of the grass associations into tall, mixed, and short grasses. They are the best maps drawn on that basis, in contrast with the Clements and Weaver (1929, 1938), Clements and Shelford (1939), and the Shantz (several examples cited) drawn on the two fold division, although Shantz constructed his division differently. The best North American map of the desert is Shreve (1942).

Biotic Unities. Maps drawn to show biotic unities coordinated with geographical area (biotic communities, areas, regions, biomes, provinces), are designed to combine plant and animal data. The principal examples are Clements and Shelford (1939), Weese (1938), and Carpenter (1940), who use the name Biome, a Clements term (1916). Dice's (1943) biotic provinces were defined on a different basis and are presented in a large folding map. Pitelka (1941) p. 114, has a map of "Major biotic communities of North America...." and nine maps, or map groups, showing distribution of bird species.

Soil Maps. Soil mapping was dominated almost completely by the government and therefore soil maps are to be found in governmental publications. Of the Marbut maps (1931) in the Atlas of American Agriculture (1936), Plate 4, is "Distribution of parent materials of soils, classified according to process of accumulation"; Plate 2, "Distribution of the great soil groups (soil provinces)"; and Plate 6, "Distribution of soils without normal profile." All of these maps are in color. The first of the Kellogg series is in USDA Misc. Pub., 229 (1936) and next USDA Yearbook 1938, Soils and Men, has large folding maps in color in the pocket. These maps indicate modifications of the Marbut scheme of classification. In the USDA Yearbook 1941, Climate and Man, are small black and white maps of the United States and of the world, pp. 111, 276-277, and 278.

In non-governmental publications, Marbut (1923) gave a map of the "Soils of the Great Plains," and Kellogg (1941) included several maps in his book The Soils that support us. Adaptations of the maps from governmental publications appear in the standard soil textbooks Joffee (1936), Jenny (1941), Millar and Turk (1943), and Lyttleton and Buckman (1943). Joffee (1936) has a map of soil zones of the world after Afanasiev, and one of Siberia after Nikiforov.

Agricultural Geography. The bibliography for Chapter 15 contains several titles under the name of O. E. Baker for 1921 and later giving agricultural maps, with crop and livestock distribution. Bowman (1935) and Hartshorne (1939) give additional maps, and still another series appears in USDA Yearbook, 1941, Climate and Man. The crop distribution maps should be compared with the natural vegetation maps to study the extent of correlations between the native and the agricultural vegetation; both the similarities and the differences. Books of particular interest in this connection are K. H. Klages, Ecological crop geography (New York, 1942), J. A. Hodges, et al., Types of farming areas in Kansas (1930), and E. F. Foster's descriptive text, on the types of farming areas map of the United States (Bureau of the Census and Bureau of Agricultural Economics, Washington, 1933).

Rural Cultural Areas. A. R. Mangus, Rural cultural regions of the United States (1940); both text and a large folding map prepared in the USDA Bureau of Agricultural Economics. This map must be regarded only as an interesting experimental project, the results being of doubtful validity.

Historical Atlases; The historical atlases commonly used by historians are unsatisfactory, on the whole, for a study of the grassland and for the kind of material that is essential to any effective correlation of the sciences and history. Some of the maps, however, are useful, but discrimination must be exercised. Every worker in this field will find it necessary to go to the sources for nearly every important map, and to study each map maker's reasons for his particular conclusions.

Paullin, C. O. (Editor), Atlas of the historical geography of the United States, (New York, and Washington, 1932). In this atlas the physical divisions map is based on Fenneman (1928), and the soil regions map on Marbut. The vegetation map followed Livingston and Shreve (1921), and is totally inadequate to the purpose of a study of the grassland as of the 1940s. The climate maps were taken mostly from the Atlas of American Agriculture, and the year of publication, 1932, antedates the new approaches to the study of climate.

Lord, C. H. and E. H. Lord, Historical Atlas of the United States, (New York, 1944, 1945). This is the most accessible inexpensive atlas and contains some maps usable for the present purpose, but there is not sufficient explanation for identification of the basis of most of these maps. This is important in a field which is so new and where there is so much disagreement.

Adams, J. T. (Editor in chief), Atlas of American History, (New York, 1943), accompanies the Dictionary of American History. This atlas is practically useless for the natural science background of regionalism. It has a limited amount of usable material of the conventional kind for dealing with the west.

APPENDIX

Grassland map studies

The Prairie Peninsula. The mapping of the tall grass prairie involves many interesting problems and possibilities, and the following combinations of maps seems to possess an exceptional interest. Thornthwaite (1941), Plate 3 shows normal (12 month) climate of the United States and Plate 4 the normal crop season (March through August) climate of the United States. By superimposing Plate 4 on Plate 3, the overlap of the humid portion of the crop season map on the subhumid portion of the year-round map defines closely the strictly tall grass prairie; in Kansas and Nebraska, the western limit falls between 96° and 97°. This overlap suggests that it is the dry winter, with its associated complex of factors, that is critical to the difference between grass and forest in this area.

In Transeau's monograph, The Prairie Peninsula (1935), he offered polar charts of annual precipitation patterns which emphasize year-round distribution of moisture as characteristic of the forest areas, and growing season distribution, with dry winters, as applying to the grassland (Figure 8, p. 428, and Figures 9-11, inserts between pp. 428 and 429). These seem to contribute confirmation to the conclusions suggested by the combination of Thornthwaite maps, Plates 3, and 4. Other interesting material is to be derived from Thornthwaite's (1941) and from Russell's (1932) year-climate frequency maps.

Kendall's (1935) year-climate variability composite map shows the right-angle intersection of the east-west zone of variability with the north-south zone of variability as falling in Iowa and Illinois. Transeau (1935) recognized the significance of Kendall's map to the Prairie Peninsula problem and reproduced it as Figure 7, p. 427 of his study. If this Kendall map is superimposed upon the Thornthwaite maps, the unique characteristics of the area become cumulatively convincing. Russell's (1934) nuclear and transitional climates concept also fit into this pattern.

The unique characteristics of the prairie soils are recognized by pedologists and no substantial area of equivalent soils seems to exist elsewhere. Marbut's solum division line lies in the midst of the maze of lines in Kendall's north-south variability zone, and the prairie soils, with a pH approximately neutral, lie between the acid pedalfer and the alkaline pedocal soils.

By such a combination of maps the prairie peninsula of tall grass becomes clearly defined cartographically in terms of the factors of variability and the range of extremes. It is peculiarly a mid-latitude vegetation, the north-south variation being restricted by the intervention of physical features which limit the temperature and light factors characteristic of low and high latitudes. To the northward are the Great Lakes which present a large water surface, and to the south the Ozark mountains and plateaux with high elevations.

Chapter Twenty-two

INTRODUCTION TO THE ADDENDA

Eleven years have passed since this book, <u>The Grassland of North America: Prolegomena to its History</u>, was written, although publication did not come until 1947. In the decade much has happened that would justify a rewriting of the book, but that is not practicable. Too much of the program to which it and the <u>Essays on Historiography</u> (1946) were indeed a Prolegomena remains to be done. Some recognition seems fitting, however, of the extent and direction of change. First, important developments, factual and theoretical, have occurred in the several sciences surveyed. Second, the author has continued to learn, has done some fresh thinking, and is prepared to go much farther in revision of traditional views than was practicable at the earlier date. But in no case is it necessary to repudiate the former major commitments.

The three articles reprinted here with the permission of the Board of Editors of the <u>Scientific Monthly</u> appeared in that journal May 1950, January 1952, and April 1953. The first of them is a condensation of a paper prepared for a symposium, "The Orientation of Ecology", arranged by Dr. Charles C. Adams for the annual meeting of the Ecological Society of America, September 1948. The third article is based on a paper presented before a joint session of AAAS Section G, the Ecological Society of America, and the Grassland Research Foundation in a symposium on "The Western Range", which was held during the 1952 annual meeting of the American Association for the Advancement of Science in St. Louis. The second of these articles proved the most explosive. The author was "admonished" both officially and privately, but persists in his heresy wholly unrepentant.

The three articles are indicative of several shifts in emphasis and some new departures. These should be coordinated, however, with the books that followed the Prolegomena: — Grassland Historical Studies: Natural Resources Utilization in a background of science and technology, Volume one, Geology and Geography (1950); The Nebraska Question, 1852-1854, (1954); On the Nature of History, (1954); The Contriving Brain and the Skillful Hand, (1955). Besides a bibliographical function, this introduction affords some reorientation and sharpening of perspective in relation to some of the more significant of the literature of the decade.

The subject matter reviewed

The subjects that merit special mention in this introduction are the interrelated themes of Pleistocene geology, American anthropology, meteorology, ecology, and soil science.

Pleistocene geology

The new conceptions about Pleistocene geological history in the middle latitudes of the grassland is best introduced by the monograph of John C. Frye and A. Byron Leonard, Pleistocene Geology of Kansas, (Lawrence, State Geological Survey of Kansas, Bulletin 99, [1952]). The four maps pp. 194-195 show successive stages in the historical evolution of the drainage pattern. Another illuminating approach is found in G. E. Condra and E. C. Reed, "The geological section of Nebraska", and "Correlation of the Pleistocene deposits of Nebraska", (Lincoln, Nebraska Geological Survey Bulletin 14 [1943] and 15 [1948]). They make clear, even to those with little experience in geology, the complexity of the erosion and deposition of materials. They are vivid reminders that the W. D. Johnson monograph on "The High Plains and their Utilization", (21 and 22 Annual Report of the United States Geological Survey, 1901, 1902) was a misleading oversimplification. The intensified research initiated during the second quarter of the twentieth century is yielding other important revisions of geological knowledge about the Pleistocene, its loess deposits and its buried soils in Kansas and other states of the Trans-Mississippi West, the work having been carried farther in Nebraska than elsewhere.

American anthropology

The appearance of man during Pleistocene time made anthropology and Pleistocene geology inseperable. This fact is strikingly presented in American archaeological work by the discovery during the late 1920s of remains from Folsom man in northeastern New Mexico and the consequent revolution in knowledge about human occupancy of the American continents. Instead of a maximum 3,000 years of man's presence, man of the Folsom cultural level may have occupied not only North America but the grassland as long ago as 10,000 years, and more primitive man a still longer time. As late as 1934 when the third edition of Clark Wissler's Indians of the Plains was published, he was not ready to accept outright the new evidence that the Great Plains had been long occupied by man of cultures earlier than the American Indians.

Only since World War II have the fuller implications of the new archaeological recoveries begun to be appreciated. The flooding of large acreages of primitive village sites by Missouri Basin dams has put beyond the possibility of recovery for study much of the material for the desired reconstruction of anthropological history. This was not a "book burning" about which some people became greatly excited, but it was complete destruction of the unread record-books of human history. Emergency salvage operations saved a small sampling of these books but not enough. Possibly the most comprehensive general guide to the new anthropology in which the American scene has its setting is to be found in Anthropology Today (1953), and An Appraisal of Anthropology today (1953), edited by Sol Tax, both products of a Symposium sponsored by the Wenner-Gren Foundation for Anthropological Research.

In the field of anthropogeography the provocative book by Carl Sauer, Agricultural origins and dispersals (1952), is important. The present writer's views on this book have been expressed in a review in Agricultural History, 28 (1954) 34-35, and On the Nature of History, Chapter 4, section 6.

Meteorology

Since World War II Meteorology has received a major revision. The phenomenon of Jet Streams was a wholly new discovery. Their nature, behavior, and effects are yet to be adquately understood, but their impact has already necessitated a reassessment of the whole field. The most comprehensive attempt is the Compendium of Meteorology (American Meteorological Society, 1951).

The new ecology

Among ecologists, the schools of ecology in the tradition of Clements and Cowles have been called in some quarters the Classical Ecologists to distinguish them from the newer thinking about ecology. Although the textbooks reflect little of the newer thought, the systems of the Classical Ecologists have been discarded more generally and extensively than is apparent on the surface. A masterly reassessment of the status of American ecology at midcentury is badly needed—something that would place in sharp and effective contrast the differences between the older and the newer modes of thinking—differences that are not merely shifts in emphasis, but very nearly if not quite of an absolute character.

Not given much recognition in the formative days of American ecology, H. A. Gleason now looms as probably the most original and significant of the early twentieth century Founding Fathers, with emphasis upon his individualistic concept of the plant association. Even earlier was E. W. Hilgard, before the time when the term ecology had been coined for the purposes to which it is applied in the twentieth century.

Although no single over-all statement of the new ecology can be cited, several book reviews and a few articles point up significant aspects. (Frank E.

Egler, "A commentary on American plant ecology, based on the textbooks of 1947-1949", Ecology 33 [1951] 673-649; "Weaver's Prairie", Ecology, 37 [January 1956] 208-209; Hugh M. Raup, Review: John E. Weaver, "North American Prairie", Quarterly Review of Biology, 30 [June 1955] 156-157. See also the work of Hugh M. Raup and associates at Harvard Forest, Petersham, Massachusetts, especially John C. Goodlett, "Vegetation adjacent to the border of the Wisconsin drift in Potter county, Pennsylvania", Harvard Forest Bulletin 25 [1954]. Attention is called also to the research papers of John T. Curtis and associates of the University of Wisconsin Department of Botany.)

The role of fire

Probably the most controversial aspect of the new knowledge about Pleistocene geology, anthropology, meteorology, and ecology, relates to the role of fire in influencing life—fire originating from natural causes and from acts of man, whether or not intentional. The classical Ecologists placed emphasis in the beginning almost exclusively upon climate as the primary factor in producing "climax" formations. Later this extreme position gave ground to other factors including fire. At the other extreme were men of quite different backgrounds, such as Carl Sauer and Omer Stewart, who questioned the existence of a climate-formed grassland. Man's use of fire was made primary as both men questioned also grass fires from natural causes. The present writer has offered what appears to be positive proof of lightning-caused grass fires. (See Malin, "The grassland of North America: Its occupancy and the challenge of continuous reappraisals", a contribution to a Symposium sponsored by the Wenner-Gren Foundation for Anthropological Research, June 1955, on Man's role in changing the Face of the Earth, [Chicago, University of Chicago Press, 1956]).

In rejecting both extremes of the fire and climate themes, a splitting of the difference is not acceptable. The problem is not merely one of degree, but of basic thinking about vegetation in all of its relations. Doctrinaire theoretical concepts block completely any realistic and independent thought about

what was and is actually present in any landscape. Each
and every place and time is unique and change is continuous, irreversible, and indeterminate. No vegetational cover is made up of simple stands of grass or
trees, but always of a complex and ever varying combination of a wide range of properties. Besides the
grasses proper, every grassland contains a large proportion of other plants, several types of which may be
designated as woody. To be sure that is no new discovery, but a reading of the literature from the camps
of extremists provides striking evidence that at both
ends of the spectrum of opinion those simple facts have
become peculiarly obscured in order to make a case.
Any resolving of the controversies about fire and woody
growth in a grassed area must turn in part upon an
agreed definition of "woody" growth and an acceptance
of the fact of the presence in any case of "woody"
plants that are not grass, or trees, or shrubs in a
conventional sense. Yet such a landscape may present
grass as its predominant property. In the North American grassland fluctuations in climate are associated
with important latitudinal and longitudinal variation
from year to year and from one climate extreme to another, because, besides climate in all its facets,
altitude, topography, geological structure, soil,
ground water, etc., all have important bearings on the
life expectancy of the different plants with their
different modes of propagation and survival. In transitional areas between clear predominance of trees or o
grass, the extent of fluctuations may be expected to be
pronounced. In such areas fire may be critical, but
whether or not fire promotes or retards woody growth,
such as trees and related shrubs, or grasses in relatio
to each other depends upon species, the season, the nature or the fire, and other factors. Nothing could be
more erroneous than a generalization that always fire
retards trees or promotes the invasion of grass at the
expense of trees, or the reverse.

Soil

The recurrence of dust storms during the mid-1950s was a reminder of those of the 1930s and of the fact that soil scientists had not made much contribution to the role of such phenomena in relation to soil formation. New facts made available by Pleistocene geology and archaeology have reinforced the assertions already made in the original edition of this book. Studies of atmospheric dust are in progress, but little has been done to illuminate the insistent problem, quantitative and theoretical, of additions to soil at the top derived from dustfall. Although the views presented in the article "Man, State of Nature, and Climax..," were denounced by certain private and governmental soil conservation enthusiasts, they were commended privately by competent soil scientists. "The Plow that broke the Plains" and other propaganda of the same vintage of the 1930s were so effective in building up false notions that a recognition of the fact that dust storms possess a constructive role can scarcely get a hearing. On occasion, destruction and creation are but the two sides of the same process, recombinations of what already exists. No one would question the importance to soil formation of loessial materials of the past. Why ignore dustfalls of the present? In this context, the time factor in the soil-formation equation is of an order of magnitude that is significant to current soil maintenance. Largely the agricultural soils of the world are formed out of transported materials, carried by water, or wind, or both, rather than being residual soils formed in place. The water and wind erosion add materials at the top that enter into horizon A as a factor in soil renewal. Regardless of the vicissitudes of political manipulation, floods and dust storms are constructive as well as destructive in the a-moral, endless transformations of nature.

A Finished or Open World

Eighteenth century rationalism constructed the concept of a Finished World, subject to the supposed harmonies of a State of Nature. In this Finished World Man had violated Nature—"civilized" man as distinguished from the "Noble Savage". As Carl Becker pointed out in his Heavenly City of the Eighteenth Century Philosophers this rationalism that thought of itself as scientific and sophisticated had really done nothing more philosophically than to transform the Christian tradition of the Garden of Eden, the Fall of Man, and the Redemption through Divine intervention. The single real difference was in the mode of redemption, which eighteenth century Philosophes thought men could do for themselves—the doctrine of Progress, or the unlimited perfectibility of man. The mission of a reformed mankind was to restore harmonious relations with Nature— the Finished World.

Followed to its logical conclusion, the evolutionary world of Spencer, Darwin, and Huxley of the late nineteenth century, with man as the end-product, would have discarded the whole finished world concept along with the doctrine of Progress as a mode of redemption of man. What happened, however, was not logical. American eclecticism seized upon a part of the new system without abandoning the old. In consequence, American social and scientific philosophy of the twentieth century was dominated by the eighteenth century Finished World and the depravity of Man, only slightly modified at some points by evolutionism. Political and social reform, and the concept of Progressivism and Liberalism reflect this background. The conservation movement, ecology, and soil science, as scientific disciplines, were founded upon eighteenth century concepts. Only a very few thinkers, if any, have achieved a clear break-through to thinking at a new order of magnitude about the uniqueness of history and the endless transformations within an open system in which man is not necessarily terminal. The concepts of an end-product and of an open system are irreconcilable and mutually exclusive.

INTRODUCTION TO THE ADDENDA

<u>Bibliographical Note</u>.

Waldo R. Wedel has reviewed the fire and vegetation problem in a paper, "The Central North American Grassland: Man-made or Natural", which is scheduled for publication in a volume of essays under the auspices of the Pan-American Union, Washington, D. C.

The best statement of the new point of view in ecology is a paper by Hugh M. Raup, "Vegetational adjustment to the instability of the site", presented at Edinburgh, Scotland, at the Sixth Technical Meeting, June 20-28, 1956, of the International Union for the Protection of Nature. The paper was mimeographed for the membership and is to be published in the <u>Proceedings</u> of the International Union for the Protection of Nature, (Brussels, 31 rue Vautier, Belgium). Dr. Raup supplied the present author with a copy of the mimeographed version.

Chapter Twenty-three

ECOLOGY AND HISTORY

THE present discussion of the relations of ecology and history is organized around five points: A statement of premises incident to collaboration of the several disciplines across the boundary lines; some implications of a general ecological interpretation of history based upon variations in options made available by cultural changes; an ecological re-examination of the history of the United States, with emphasis upon methods; a review of the grassland problem as a concrete example to illustrate the meaning of adaptation; and, finally, a consideration of the grassland in relation to mechanically powered-minerals culture which provided the background for its occupation after about 1850.

The premises upon which the discussion rests include the dependence of society upon nature and natural forces, not upon a conquest of nature; the unity of knowledge; and the existence of a body of methodology sufficiently common to all disciplines to afford a working basis for collaboration in matters of interdisciplinary nature. Both history and ecology may be defined as the study of organisms in all their relations, living together, the differences between plant, animal, and human ecology or history being primarily a matter of emphasis. Therefore, all forms of single- or limited-factor interpretations are rejected as fragmentation of knowledge, with its resultant distortion of facts.

At any level, cultures afford man the opportunity to exercise options in ordering his relations with environment with himself, and as the culture changes the range of the options shifts. Since ecology has become a recognized discipline, it is appropriate to re-examine history with special reference to ecological relations and the significance of these shifting options; the Mediterranean-centered culture yielded to Western Europe, and then to an Atlantic-centered pivot of world power. Included in this selective competition is the transit of culture to the world outside Europe, and then, since about the middle of the nineteenth century the invasion of the grassland of North America and elsewhere, together with the changing options under the influence of a mechanized-power culture. This article focuses attention primarily upon the meaning of ecological adaptation as applied to the United States.

The Turner Frontier Hypothesis and its Closed-Space Corollary

The validity of the Frederick Jackson Turner frontier interpretation of history has been challenged, and upon sound grounds within the traditional methods of writing history. But before leaving the problem of the frontier hypothesis, whether American or new world discovery and expansion of Europe, attention is called to another body of associated ideas, which may be challenged effectively by an ecological methodology applied to history. The term "geographical discovery" is, after all, a subjective term, purely relative in character, whereas scientific method presumes a definition of terminology in a form that is both objective and operational. The idea of discovery applied only to Europeans, not to the aborigines of the lands "discovered." The same is true of the concept of the "frontier"—the frontier of what? Obviously, the frontier of the culture of modern Europe. Again, what is the "new land" about which Turner and Halford J. Mackinder and others wrote so insistently? To whom was it new? Certainly not to the American Indian, to the aborigines of Australia, or of Africa, nor to the Chinese. Europe and European-peopled America considered their culture a superior, or master, culture, and themselves perhaps a master race, and thought of "discovery" and of "frontier of new land" in a subjective and egocentric sense. For the most part, the invading culture refused to recognize that the displaced cultures possessed any values, or that the peoples concerned possessed any rights which should be respected. Newness, however, implies the operation of the factor of time, subjective and relative, and time is a ruthless leveler of persons and things. Given a sufficient lapse of time, the tables might be turned upon these invaders by another invading culture.

A scientifically conceived ecological methodology applied to human history would emphasize ecological competition of two or more cultures for dominance in given earth areas, which could be made as objective and exacting as when applied to plants and animals—competition of cultures of differing degrees of complexity, the ecologically stronger invading the earth area occupied by another. But such a methodology would avoid value judgments and would recognize instead differences and similarities. The land involved would not be new land, but land that has been exploited for unknown generations, and that, in consequence of cultural invasion, became subject to a mode of exploitation different from that under a previous culture, and in that sense the cultural techniques might be called new to the land of the displaced culture. The land was not new; only the introduced technology of exploitation was new to the land. In further development of this line of thought, newness is involved because the more complex invading culture possessed technological tools and skills which made available different or wider ranges of options as applied to the exploitation of the area, bringing into the flow of utilization existent resources that were latent under the displaced culture. That point deserves special emphasis. The earth possessed all known, and yet to be known, resources, but they were available as natural resources only to a culture that was technically capable of utilizing them. There can be no such thing as the exhaustion of the natural resources of any area of the earth unless positive proof can be adduced that no possible technological "discovery" can ever bring to the horizon of utilization any remaining property of the area. An attempt to prove such an exhaustion is meaningless, because there is no possibility of implementing such a test. Historical experience points to an indeterminate release to man of such "new resources" as he becomes technologically capable of their utilization. At one stroke, such a concept renders the Turner-Mackinder doctrine of "closed space" meaningless, and correspondingly destroys the basis of the argument of the "closed space corollary to the Turner frontier hypothesis," which holds that a welfare state—a regimented social order—must be instituted to serve as a substitute for the "closed frontier" in order to preserve American democracy and opportunity.

Ecological Re-examination of the Data of History: The Occupation of the United States by Europeans

There are several other possible methods of approach to an ecological re-examination of the history of the United States. Four are indicated here. One fruitful method would be to trace the occupation of the area by Northern Europeans, allowing for a contrasting occupation of the Middle Americas by Southern Europeans. These Northern Europeans were the product of a forest culture in a maritime climate. For present purposes of stating the problem in terms of ecological data, the issues may be brought out effectively by asking a series of questions, but without now attempting answers. What ecological concepts or impedimenta did these Europeans bring to America as a part of their cultural heritage? What did they expect to find? What did they find that they expected, and what did they find otherwise than expected? How did they react to what they found? To what extent did they recognize differences? How did they make adjustments to the differences they recognized, that they did not recognize, or were slow to evaluate? How did that adjustment, or lack of it, or lag, affect their cultural concepts and behavior?

As the occupation of the interior of North America advanced, the population, accustomed to a forest-maritime environment, found it necessary to readjust to a continental climate and to a grassland. The series of questions formulated to apply to the transit of culture across the Atlantic applies equally well to the advance into the interior, but where the emphasis as between Europe and the American seaboard was upon similarities, the emphasis as between the maritime seaboard and the continental interior would be placed more sharply upon differences, especially west of the forest-prairie boundary. In both areas there would be substantial emphasis upon differences, however, as between the southern and northern latitudes, which involve temperature differences and short- and long-day photoperiodism.

A second possible approach to this ecological reexamination would be the history of the exploration of North America by naturalists: the geologist, the botanist, and the zoologist. The story of the geographical exploration of the continent has been told frequently, but scientific exploration came relatively late, and the story has not been told adequately, if at all, for any field of science. Such historical writing as has been done in connection with science emphasized the most obvious aspect: collection of material, its classification and naming —the making of a classified catalogue. An ecological examination of this process would supplement and extend the record of what has thus far been primarily taxonomic.

A variant of the foregoing approach would be found in a series of biographical studies of leading naturalists who participated in the scientific cataloguing of North America. This would be done with a view to reconstructing their ecological outlook and reactions to the work upon which they were engaged.

The construction of ecological traverses of America, or of regions, is another fruitful procedure. The idea of the traverse is borrowed from the geographer, but the materials, procedure, and objectives would be ecological. The primary focus of ecological traverses would be historical. They would be run for several successive time periods and in several directions, especially east and west and north and south, through areas of different topography, altitudes, and rainfall patterns. The materials would be written documents descriptive of ecological facts which would be evalued systematically. Such ecological traverses of chosen historical time periods could then be compared with current ecological traverses of identical routes based upon field studies of ecological survivals. I have presented sample studies for traverses, as of about 1849, in my book *The Grassland of North America*. These were undertaken experimentally, and the details of techniques and methodology need to be perfected as experience is gained in practice.

How much has man modified the ecological setting of history in America? No certain scientific answer can be given because the necessary historical-scientific investigations of an ecological character have not been made. The samples of ecological history completed indicate less fundamental change than is usually assumed by conservation propagandists. At present, answers must, perforce, be tentative, and largely a matter of personal opinion. This is peculiarly unfortunate when public attention is being bombarded by propaganda to authorize gigantic programs dealing with natural resources.

The Grassland Problem: The Meaning of Adaptation

When forest man met the prairie in the Ohio Valley and Great Lakes regions as the American frontier of white settlement moved westward, he was puzzled about the fact that these earth areas were covered with grass. A voluminous literature grew up dealing with the question of why the prairies were treeless. Although there may be exceptions, forest man tended to avoid settlement in the open prairies. If he took prairie land, it was contiguous to forest land, which he made the base of his farm establishment. Several reasons have been assigned for the retarded settlement of the prairies, some of which are inconsistent and apply obviously to different kinds of grassland: wetness, difficulty in plowing grass sod, low degree of fertility, prevalence of disease, lack of water, of wood, of navigation, and of protection against the hazards of climate. Because the problem has not been adequately defined, there is much contradiction, misunderstanding, and some controversy. Sometimes trees and sometimes grass occupied soil inferior for the standard agricultural crops of the area. It is possible, however, to trace the major steps in the substantial reversal of forest man's concepts about the soils of the grassland, but soil scientists generally have not yet arrived at an adequate understanding of them. A long list of other problems needs examination and reorientation in order to clarify history as well as to afford perspective on the present, both the successful adjustments and the numerous ecological blunders, and to define the meaning of adaptation.

The British breeds of beef cattle were introduced into the plains. One of these breeds, the Hereford, possessed latent characteristics that were released so conspicuously by the new environment as to enable it to dominate the region from Mexico to Canada. The beef and dairy breeds of Brahman cattle from Asiatic India possessed desirable characteristics for the hot, dry, Southern plains. Among field crops, the competitive experimentation with wheat varieties proved that the hard spring and winter wheats from Russia possessed adaptive capacities; likewise, the sorghums from Africa and Asia, mostly imported by way of the higher-rainfall East, established themselves in favor in various parts of the grassland; and, in addition, alfalfa from the Mediterranean area, by way of Chile, was introduced into the economy of California and the plains. It is important to emphasize that all the animals and plants involved in these so-called adaptations to the plains were introduced without biological change from their original environments to the new, and that in the original environments they possessed all the qualities that they demonstrated under the new conditions. In their original environment, they possessed in some, and possibly in most, cases characteristics that were relatively unnoticed or latent, but which became conspicuous or even decisive under the options of the plains environment. The process was one of selective experiment with materials already in existence elsewhere, and no one understood the biological mechanism involved. Much is still not understood. Only in the second quarter of the twentieth century did the second, or creative, phase of biological adjustment emerge in effective form, based upon the genetics of mutation and hybridization and the correlation of breeding and agronomic programs with the principles of developmental physiology. This phase has only begun, its potential is unknown, but for the first time the ecologist and the historian gain a new insight into the meaning of adaptation, the mechanism of biological behavior of the past, and one which opens the door to future adaptive breeding programs with chosen objectives.

Soil physics, tillage methods, and the tools with which to operate a farm were the product of the forest-maritime environment. Here again concepts stood in the way, as well as the practices and the tools designed to meet traditional conditions. But there was much in this cultural accumulation, latent or relatively so, that possessed values that became significant and important in the grassland: in plowing technology, the steel plow, the lister, and disc machinery; methods of handling the soil to conserve moisture and to retard wind erosion; harvesting machinery, particularly the header; and the beginnings of horse-powered agriculture.

This approach to history makes no commitments about the validity of the concepts of civilization, about its rise and fall, and about progress. What are they, anyway, but philosophical speculations? What is envisioned here is an indeterminate, selective, ecological competition of people occupying earth areas under the changing range of options afforded by cultures of differing degrees of complexity.

Chapter Twenty-four

MAN, THE STATE OF NATURE, AND CLIMAX: AS ILLUSTRATED BY SOME PROBLEMS OF THE NORTH AMERICAN GRASSLAND

THE state of nature as it is commonly accepted is nonexistent. When man appeared upon the scene he destroyed such a state, because he possessed the unique capacity to act with a purpose. No matter how primitive, he introduced the factor of planning, and the element of choice. The length of time man has occupied the North American continent has been variously estimated. Prior to about 1930, the maximum was set at about 3,000 years. After that approximate date, the Folsom man discoveries led a large segment of anthropologists to reach back 40,000–50,000 years. The new carbon-14 test, applied at the mid-twentieth century, affords a new measuring device, the application of which, not yet complete, is constructing a new calendar, with 10,000 years as the probable maximum and Folsom man possibly near half that figure.*

The contention of this paper is that the botanist, zoologist, and the soil scientist cannot expect any real success in many aspects of their own disciplines, in an ecological sense, until some tangible achievement is made in dealing with the problem of man's influence. Such a necessary recognition of the role of man affords a common ground between them and the geographer and the historian. First, the nature of the problem of man, as an ecological factor, must be discussed and defined, in order to appreciate the full range of the ramifica-

* Since this was written a further revision, involving a drastic reorientation, has been proposed by G. F. Carter (*Sci. Monthly,* 73, 297 [1951]). The effect of his proposal is only to add emphasis to what is said here.

tions involved. Once the significance of the problem is fully recognized, then, possibly, methodology can be formulated and techniques and tools devised that may achieve some exactitude of measurement, and thereby place the study upon a quantitative basis. If success attends such efforts, then the accumulation of necessary data will follow. All that can be attempted here is to undertake the first step of definition and discussion of the nature and scope of the problem as applied to a particular area of concentration, and some of its implications to all the disciplines concerned.

The true state of nature existed only prior to the emergence of man. The disciplines most concerned in the study of that time are geology and paleontology in their several phases. If the study of the state of nature as modified by aboriginal man is the center of interest, then to the two disciplines already mentioned must be added archaeology and anthropology, plant and animal ecology, geography, and soil and other sciences. It is history based upon human remains, or documents, prior to the invention of writing. If the focus of study is upon the occupance of the continent since European man arrived in 1492, the conventional conception of history derived from written documents is added to all the others. Of the three stages, the first deals with the true state of nature, the second with aboriginal man's modification, and the third with European man's modification. The first two involve a dual culture conflict, the last two a triple culture conflict. How can either regime of human occupance be determined and measured?

The speculative or philosophical status of most of the approaches to the problem of the North American grasslands under man's influence may be illustrated by a brief reference to the work of a number of distinguished specialists. Carl Sauer, a geographer, took an extreme view of aboriginal man's influence upon nature, attributing the grassland and the extinction of certain Pleistocene animals to man's use of fire.[1] His views carry great weight west of the Rocky Mountains. C. W. Thornthwaite also took an extreme view of culture control of environment, applied more particularly to European man in America. As the extensive erosion in the West was attributed by him to the influence of man, he held that the remedy lay in the same hands, especially by means of government planning. Because of his being in government service, his influence was widely disseminated by high-pressure official propaganda.[2] The position of Kirk Bryan was near the opposite extreme from Sauer and Thornthwaite, holding to the dominance of impersonal natural forces effecting climate changes sufficient to shift the balance of equilibrium on the side of erosion or introduce such instability that man's role was only a minor factor in the larger complex setting.[3] Under this interpretation, remedial measures by either private or governmental action were of limited benefit. Ellsworth Huntington occupied the extreme position also in maintaining climate determinism.[4]

In the field of plant ecology, F. E. Clements formulated a doctrinaire concept of ecological succession, climax, and disclimax, the last phase being charged particularly to the work of European man.[5] This theoretical construction had little realistic relation to such historical evidence as was available.

In the field of soil theory, C. F. Marbut and his associates and successors in the USDA adapted the highly doctrinaire Russian soil theory to the United States: The idea of profile and maturity as soil climax.[6] This stereotype also possesses doubtful value, execpt in a few limited spots, in understanding the problems of the grassland, where the soils are largely derived from transported materials, and disturbance is the most conspicuous characteristic of the environment. There is little room for the idealized view of the conservationist literature, that soil was derived from the disintegration of underlying rock as rapidly as stabilized erosion in a state of nature removed it at the top—a state of mythical equilibrium—until that state was destroyed by modern man. H. H. Bennett declared that the Missouri and Mississippi rivers ran clear during this "state of nature."[7]

George F. Carter offered the suggestion, relative to California aborigines and grasses, that the gathering of edible seeds aided in the dissemination and reproduction of some species, creating an artificial climax. The passing of the Indians and the invasion of Mediterranean rye were approximately coincident in time, and the elimination of the native grasses may have been the result of a cause-effect relationship.[8] O. F. Cook made the suggestion, revived by Higbee, that the dominance of the chicle forest in Maya areas of Central America reflects the deliberate protection of that species for its edible nuts.[9] Likewise, Carter and Higbee make a similar suggestion for the dominance of the oak-hickory-chestnut deciduous hardwood forest in the eastern mid-latitudes of the United States.[10]

As stated, all these interpretations are, in the scientific sense, unproved. They are in the nature of speculation, hypotheses, theories, philosophical assumptions—or whatever term most appropriately designates propositions that are not subject to scientific proof in the broad, general form in which the question is stated. Of course, many scientific data have been brought to bear upon them, and the scholarly standing of the men mentioned is such that much of their work stands the test, within limits. Is it possible, then, to formulate the essential questions in meaningful forms subject to proof?

The Problem of the Great Plains

Less broad than the propositions just stated is the problem of the so-called Great Plains portion of the North American grassland and immediately related areas. W. D. Johnson offered the first major interpretation of that area: that the Great Plains were the relatively undisturbed remnant of the apron of debris that had been washed down from the Rocky Mountains in the process of weathering. No rivers had cut through the Llano Estacado, but he viewed the mid-latitude section as having been dissected by such rivers as the Canadian, Arkansas, Smoky Hill, Republican, and Platte. Within the past two decades, however, the Johnson theory has had to be discarded as too simple. Geologists dealing with Nebraska and Kansas have arrived at the conclusion that the drainage system of these mid-latitudes has changed materially, possibly more than once, during the Pleistocene, and that the material has been eroded and redeposited accordingly. In other words, during Pleistocene time, the area was in a state of continuous disturbance under the influence of impersonal forces.[11]

Late in Pleistocene time man appeared, possibly 5,000–8,000 years ago, in the mid-latitudes (a more exact dating may emerge from the carbon-14 method of measuring time). From that point, the purposive influence of man was added to the impersonal forces contributing to continuous disturbance in the area. Archaeological work and its interpretation by anthropologists have only begun in the grass country east of the Rocky Mountains, and in some areas only sketchy preliminary surveys have been made for some subdivisions. Since Folsom and Yuma man, several successive cultures have occupied the country, the extent, density, and persistence of which are only beginning to be understood. Archaeological evidence is lacking as yet to construct satisfactory maps of distributions and historical sequences of occupance. Sites have been located upon which successive cultures have been recorded in archaeological remains, one above the other, and separated by accumulations of wind-blown material.[12] These records are unmistakable evidence of continuous disturbance, both impersonal and human, upon the same site. Possibly a reasonably accurate dating, or calendar, of these occupations and of intervening abandonment may be constructed, and upon a sufficient number of sites, with a distribution wide enough to be really meaningful.

By narrowing down step by step in space and time, particular sites come into perspective, and the bridge can be built from the purely archaeological documentation to the written, the intervening transition involving both. The expedition of Coronado into the Great Plains in 1541 is such an instance, although for other spots dated events may be associated as accurately or more so. The route of Coronado and his terminus in Quivira have long been matters of dispute, but the most recent studies brought to bear upon the problem, utilizing both archaeological and written evidence, have agreed upon the Rice-McPherson County, Kansas, area as the most probable site of Quivira.[13] As these counties were settled by Americans during the 1870s, the period between Coronado and American settlement was about 330 years. The archaeological estimates of time indicate that the same culture that Coronado met persisted until the early eighteenth century, allowing a time span of about 150 years between the two occupations.

Since American contacts were established in the area, intermittent disturbances of considerable proportions have been conspicuous. For instance, dust storms were an integral element in the physical and ecological history of the country within the period of written record, and during the years that the only direct and tangible human influence was that of the nineteenth-century Plains Indian occupants. The missionary-surveyor Isaac McCoy described a typical dust storm, in what is now east-central Kansas, in the fall of 1830.[14] The printed record of dust storms, with particular reference to Kansas, has been compiled by the present author for 1850–1900.[15] These accounts do not describe the particular sites involved in the Quivira remains, but those village sites are within the area covered and were subject to the same general conditions that affected the region as a whole.

Movement of Soil

The definition and delimitation of the problem under discussion may be facilitated by an approach from another direction, followed by the necessary narrowing of space and time to pin-point the essential issue at stake. Taking the Mississippi Valley drainage basin as a whole, the records of the U. S. Geological Survey indicate that no greater quantity of water or silt passed Baton Rouge into the Gulf of Mexico in 1951 than when records began, upwards of a century earlier. The high water mark of the Missouri River flood of 1951 did not exceed the high water mark of the flood of 1844 at Boonville, Missouri. Such facts run counter to the assumptions of most conservation propaganda, but they are vital in evaluating the influence of human action upon the area as a whole as well as upon any particular spot. In spite of great damage to man's improvements in the valley, especially urban development in flood bottoms (which never should have been made in the first place), the great flood of 1951 in the Kansas Valley was of great benefit, by and large, from the standpoint of agricultural resources as reflected in improved productivity of bottom land. Serious erosion occurred in some uplands, and the gains must be balanced against the losses for the valleys as wholes. For the Mississippi Valley as a whole, or for the Kansas Valley as a whole, natural resources in terms of water erosion meant, primarily, only a movement of soil from one place to another within the valley, and not a dead loss.

The problem of wind erosion must be approached in similar fashion. Sand and dust are moved by the wind from place to place, but the net loss is negligible, and probably not greater than formerly—that is, prior to occupation by Americans within the century. According to the archaeological evidence, the thickness of the wind-blown material that interlayers successive aboriginal village sites would indicate greater dust storms by far than occurred during the decade of the 1930s. The movement of dust by air currents is continuous, but the meteorological service had not kept records and had not devised instruments and techniques for recording the phenomena. During the still part of the night, dry dust fall is usually most conspicuous. A partly enclosed patio floor of the author's residence has afforded opportunity to verify dust fall; red, yellow, or black dust collected in measurable quantities repeatedly during the night, the color giving some indication of the probable place of origin. Repeated checking with the local weather observer and with published reports of near-by weather stations on such occasions has not revealed any record by them of such occurrences. Likewise, during the summer of 1951, when rain fell nearly every day, measurable quantities of dust fell with the rain. Water erosion moves soil from a higher to a lower position topographically, but airborne dust is deposited everywhere. The question for which there is no present answer is, How much? And neither is there an answer in exact measurements to the question, What is the net gain or loss from combined movements of soil materials? A further question should be entered in the record: Does this soil lose in its capacity to produce vegetation by being moved from place to place either by water or by wind? If the answer is always in the affirmative, then why is land in flood bottoms often greatly benefited by floods? If it is not lost in that case, then the burden of proof would seem to lie against the claim that it is always lost.

Aboriginal Cultures

It is a truism that archaeologists deal with excavations; one of their chief badges of office is a spade. But if all erosion carried soil to the ocean, they would have no remains of human culture to uncover. What are the sources of the cover, or overburden, under which the evidences of human occupation are buried? First is the accumulated matter derived from vegetation; second, if the site is on a slope, water erosion would wash material from higher points; third, dust fall derived from wind erosion would accumulate on high as well as on low spots, both dry and wet deposits.

One of the first men to publish descriptions of the Rice-McPherson County sites was John August Udden, a young Swedish teacher (later to become a distinguished geologist) at the Academy of Bethany College, Lindsborg, Kansas. In 1881, when he was twenty-two years of age, his students called his attention to the Paint Creek archaeological remains, and in his spare time he worked over a period of seven years at the site. His published report did not appear until 1900. He described two types of mounds, which he interpreted, respectively, as burial and dwelling sites. The burial mounds were "usually built on high bluffs or on upland hills." They were covered with earth and vegetation, and within them was "a pile or layer or rocks" under which were human remains. The mounds he called dwelling sites "do not occupy any conspicuously high places, but usually lie on or near some flat and fertile lowlands as on the border of an alluvial plain." They were really not mounds, he explained, "but merely flat surface of the ground where dwellings of an earlier race have once been standing. They would never be noticed, were it not for the relics of household art, chase, and warfare scattered about the place."[16] Wedel[13] denied that they were building sites. His verdict was that they were village refuse dumps, but for present purposes the original use is not at issue.

There is no exact dating either of the time of first or last occupance of the Paint Creek site, or of its duration. The accepted minimum of time elapsed since aboriginal abandonment and Udden's observations is about 150 years. The significant part of Udden's description is that "they would not be noticed, were it not for the relics." This would seem to imply that so far as soil and grass composition were concerned there was no apparent difference between the low, flat mounds and the intervening prairie. Upon digging through the mounds, Udden noted that upon reaching the prairie level, the ground was harder. Having lived in Kansas during a dry, dust-blown period of the later eighties, and as a geologist being professionally interested in wind-blown materials,[17] he attributed the mounds to dust accumulation around the houses. Even if we accept Wedel's interpretation of origin, the dust-and-sand-accumulation thesis would apply equally well. Outwardly, at least, the marks of men (scars?), representing possibly 200 or 300 years of occupance, had been erased within 150 years, except for some unevenness of the surface. The present author checked one of these sites with some care at two different times during the summer of 1951 and could find no evidence of surface variations in soil or vegetation that could be attributed to aboriginal occupation. The same aboriginal culture covered at least five or six central Kansas counties and extended down the east side of the Arkansas River valley at least as far south as Arkansas City, where some excavating has been done. These people lived by agriculture, cultivating advanced varieties of corn, beans, and melons, they dug native roots from the prairie soil, and they hunted wild game. Yet this area was always described by the first white settlers as virgin prairie, and by conservation enthusiasts as undisturbed by abuse at the hands of man until white Americans broke the sod and upset the perfect equilibrium of the state of nature.

One of the overriding facts that has been kept in the foreground throughout this paper is the virtually continuous character of disturbance to soil and to vegetation during the whole Pleistocene and Recent time. The disturbance was the consequence of impersonal forces alone, physical and animal, prior to the advent of man, and to both impersonal and human purposive forces since. Without going to the extremes of Sauer, we may say that aboriginal man, through occupance over long periods of time, through agricultural pursuits, use of fire, killing of selected game, utilization of selected plants, and digging in the earth for root plants, certainly destroyed vegetation altogether in some spots of considerable size again and again, and interrupted succession of vegetation over wide areas where he did not destroy it. His selected utilization of animals and plants undoubtedly exercised an important influence upon the relative dominance not only of those particular species, but upon the whole range of interrelations of equilibrium throughout the grassland as well as in particular spots. Even though the nature of these influences have not been, and probably cannot be, reconstructed with any degree of completeness, the fact of such human influence cannot be denied or ignored. Even the nineteenth-century Plains Indians exercised a similar influence, in spite of the fact that their culture was substantially different from the culture of the people found by Coronado. In spite of such a record of destruction and interruption of succession, there existed a wide range in succession stages in the grassland, associated with a variety of conditions resulting from disturbances derived from natural forces of water, wind, and animals, and from planned action of man. The fact should be clear also, that certain kinds of damage done to nature by the action of man are more easily repaired than is usually recognized—and especially than is alleged by conservation enthusiasts, whether private or official. In other words, there are influences of human action upon nature that are quickly, and largely, if not fully, reversible.

The Quivira sites are only one instance of many that might be studied in a similar manner throughout the grassland. The archaeological materials that are accumulating for interpretation are so numerous and so new that one hesitates to generalize about the possibilities of correlating the archaeological evidence with the written documents. The point that is certain is that the reappraisal of the role of man must modify subtantially the prevailing views of the history of the grasslands.

The Soil Question

What corollaries or conclusions are to be derived from these facts? What other aspects of the study of the grassland must be re-examined in a fundamental fashion in the perspective of such a historical approach to the region? What is soil? How was it formed? And what span of time was involved? What of agriculture in the grassland under the long-term utilization essential to the support of the existing human culture? What of the proposed valley development of natural resources and conservation planning that has for some time occupied the spotlight of political controversy? Certainly some fundamental reconsiderations are in order. Why the haste to flood many of the sites of prior human culture before adequate investigations and interpretations of human occupance can be made? Why flood them at all, period? Once archaeological evidence is destroyed, whatever enlightenment might have been derived from it regarding the experience of man is irretrievably lost, and through the blunders of jockeying for current political advantage. Instead, it should be preserved in the interest of finding the most tenable solutions.

The limits of this article preclude any attempt to discuss the three major positions. Subjects two and three are offered only as suggestions, and the first, the soil question, is discussed only briefly and primarily in the spirit of inquiry to direct thought into fresh channels.

How can the idealized type of theoretical climax, either of soil or of vegetation, have culminated under the conditions of continuous disturbance established by history? The calendar of events does not allow the time necessary, 100,000 or even 10,000 years to produce five inches of topsoil as claimed by conservationists, or to complete succession and climax of vegetation for the area as a whole. Difference was more characteristic than uniformity. If this interpretation of facts is in error, then the concept of climax must be redefined, especially in its time requirement, both as respects soil maturity (climax) and vegetational climax. Furthermore, in that case, the implication that vegetational succession and climax are dependent upon progressive soil changes in the direction of maturity needs restatement or abandonment.

Does soil age have any relation to fertility or productivity? And, if so, what is meant by soil age? Is soil anything more than a medium, useful but not essential, to the growth of vegetation? How shall soil age be measured? To this last question there are three possible and quite different answers: First, by length of time in place sufficient to produce a well-defined profile according to the Russian-American school of pedology, with horizons A, B, C, D. Second, the length of time since initial weathering from rock, including the history of the weathered material as it may have been removed from place to place by action of water and wind. Does the weathered material gain or lose properties by such change of place? Third, the length of time involved in successive geological time periods—successive cycles of sedimentation, consolidation, and erosion down to the present. This last point is illustrated and emphasized sharply by inferences that the writer draws from studies of the selenium problem and of trace elements. In undertaking to explain the high concentrations of selenium in the soil at some points in South Dakota and Wyoming, Trelease and Beath advance the view that the selenium in the Cretaceous rocks was probably derived from the magma, or molten rock, exposed in Rocky Mountain formation; that Tertiary formations were derived from reworked Cretaceous strata, and thereby selenium became more highly concentrated; and that modern concentrations are a further development of the process, aided by certain plants that serve as selenium accumulators.[18] To what extent is selenium only an unusual example of the problem that is involved in all soil evolution, a problem that must be recognized in any attempt to explain the presence or absence of particular properties in any soil?

An evolutionary approach to soil theory is posed from another direction by the deficiency of some soils in trace elements. This problem appears to arise in connection with (but not exclusively) soils derived predominantly from shales and sandstones rather than from marine limestone deposits. After World War II, Kansas established an experimental substation in the southeastern part of the state to study the problem of such soils, of crop production upon them, and of feeding animals upon those crops. As with the selenium problem, which represents the presence of undesirable elements, the absence of desirable elements introduces questions of human nutrition.

In the low rainfall areas of the grassland, the soils are derived largely from transported materials, largely wind-blown, and other parts from water-worked materials modified by the wind; the same is true of the bottom lands of the valleys in the higher rainfall areas. This fact is emphasized for Kansas by R. I. Throckmorton's soil map, constructed before the subject was confused by later theories of pedology.[19] In some limited areas of low rainfall in the west, and in larger areas in the higher rainfall eastern portion of the grassland, the soil has been formed in place, and is the residual product of soil formation that seems to fit approximately into the diagrammatic formula of the Russian-American school of Marbut and his successors. Between the soils that shift rapidly under the influence of wind and water and the predominantly residual-type soils are all possible variants, with repeated disturbances under the influence of wind, water, and man, both aboriginal and contemporary, prominent in their history. Man's influence upon the total situation may easily be exaggerated; likewise, the influence of aboriginal man may be ignored under the influence of egocentric twentieth-century man. Both these perversions of history and science have occurred, and most of the literature available is vitiated by one or the other—or both—these errors.

In an area as varied as the grassland of North America, each spot must be evaluated in its own right. On soils derived from transported materials, which are of substantial thickness—and these occupy much of the low rainfall grassland—the topsoil may be eroded by wind or water down to plow depth, not merely once but several times, without impairing the productivity of the land. E. W. Hilgard, who did the most fundamental thinking about soils that has been done in America, pointed this out for arid soils of California in 1892.[20] Farmers in the grassland in question have learned this unimpaired capacity from long experience, and have compelled reluctant scientists to recognize it.

In the study of soil, why should there be exaggerated emphasis upon soil maturity in terms of profile? Has profile, per se, any necessary significance, and if so, under what limited circumstances of place and associated situation? An evolutionary or historical approach to soil formation reaches back in time beyond the range of such oversimplified stereotypes.

Obviously, the North American grassland has supported a succession of human cultures over a long period of time. You say the present one is different? Yes, but so were each of the others different from what had gone before. Beware of the egocentric present-mindedness of the dominant thought of the mid-twentieth century! Beware of its arrogance and intolerance! Not only of its arrogance, and intolerance of all contemporaries who differ with it, but of its attempt to escape from history by ignoring or ridiculing the past. Time is a ruthless judge and, in the end, Time has the last word.

A large part of American research funds and energy is expended upon technological research of a short-term character to achieve functional ends, often primarily for political advantage. In the past, the basic thinking of the modern world has been done almost exclusively by Europeans.

Two world wars have destroyed Europe physically, but what is critical is that Europe offers no recognizable evidence that it has the vitality to revive its capacity to think creatively. And no Marshall Plan, or any other artificial stimulus, can perform that miracle. Neither is there any present recognizable evidence that either the United States or Russia has the capacity to assume the responsibility for the fundamental thinking they have helped to destroy. Government funds, foundation funds, university funds, poured out to any Johnny-come-lately for "quickie research," cannot meet the challenge. Fundamental results cannot be produced in a summer, a year, or two years, with progress reports semiannually, or annually, to be reviewed by administrative authority to determine whether results justify renewal of support for another term. Creative thinking may require the sacrifice of a lifetime, and expectations for the requisite basic thinking must be geared to the cumulative power of generations of untrammeled effort, a situation now virtually nonexistent. The intrinsic problems of grasslands everywhere await a more comprehensive body of thought of this order of potentiality and action.

References

1. Sauer, C. *Geogr. Rev.*, **34**, 529 (1944).
2. Thornthwaite, C. W., Sharpe, C. F. S., and Dasch, E. F. *Climate and Accelerated Erosion in the Arid and Semi-Arid Southwest, with Special Reference to the Polacca Wash Drainage Basin, Arizona.* USDA Tech. Bull. 808. Washington, D. C.: GPO (May 1942); Happ, S. C., *J. Morphol.*, **5**, 338 (1942).
3. Bryan, K. *Ann. Assoc. Am. Geogr.*, **31**, 219 (1941); ———. *Science*, **62**, 338 (1925); ———. *New Mexico Quart.*, **10**, 227 (1940); Whittlesey, D. *Ann. Assoc. Am. Geogr.*, **41**, 88 (1951).
4. Huntington, E. *Civilization and Climate.* New Haven, Conn.: Yale Univ. Press (1915); ———. *Mainsprings of Civilization.* New York: Wiley (1945).
5. Clements, F. E. *Plant Succession: An Analysis of the Development of Vegetation.* Washington, D. C.: Carnegie Institution, Pub. No. 242; Clements, F. E., and Shelford, V. E. *Bio-ecology.* New York: Wiley (1939).
6. Marbut, C. F. *Soils of the United States.* Part III,

Atlas of American Agriculture. Washington, D. C.: GPO (1935).
7. BENNETT, H. H. *Soil Conservation.* New York: McGraw-Hill, 1-3 (1939).
8. CARTER, G. F. *Sci. Monthly,* **70**, 73 (1950).
9. COOK, O. F. *Annual Report, Smithsonian Institution.* Washington, D. C.: GPO, 481-97 (1903); HIGBEE, E. *Geogr. Rev.,* **38**, 457 (1948).
10. HIGBEE, E. *Op. cit.;* CARTER, G. F. *Op. cit.*
11. JOHNSON, W. D. *Twenty-first Annual Report, U. S. Geological Survey,* Part IV. Washington, D. C.: GPO, 609-741 (1899-1900); *Twenty-second Annual Report, U. S. Geological Survey,* Part IV. Washington, D. C.: GPO, 635-69 (1900-1901); FLINT, R. F. *Glacial Geology and the Pleistocene.* New York: Wiley (1947); CONDRA, G. E., and REED, E. C. *The Geological Section of Nebraska* (2nd ed.). Lincoln: Nebraska Geol. Survey, Bull. 14 (1943); ———. *Correlation of the Pleistocene Deposits of Nebraska.* Lincoln: Nebraska Geological Survey, Bull. 15 (1948); FRYE, J. C. *Trans. Kansas Acad. Science,* **49**, 71 (1946); COLBERT, E. H., et al. *Bull. Geol. Soc. Am.,* **59**, 541 (1948); FLINT, R. F., et al. *ibid.,* **60**, (9), (1949).
12. WEDEL, W. R. *Trans. Kansas Acad. Science,* **50**, 16 (1947).
13. ———. *Explorations and Field-Work of the Smithsonian Institution in 1940.* Washington, D. C.: GPO, 71 (1941); ———. Smithsonian Inst. Misc. Coll. **101**, (7), (1942); ———. *Trans. Kansas Acad. Science,* **50**, 16 (1947); BOLTON, H. E. *Coronado, Knight of Pueblos and Plains.* Albuquerque, N. Mex.: Univ. of New Mexico Press (1949).
14. BARNES, L. *Kansas Histor. Quart.,* **5**, 364 (1936).
15. MALIN, J. C. *Ibid.,* **14**, 129, 265, 391 (1946).
16. UDDEN, J. A. *An Old Indian Village.* Rock Island, Ill.: Augustana Library, Pub. No. 2, 10 (1900).
17. ———. *The Mechanical Composition of Wind Deposits.* Rock Island, Ill.: Augustana Library, Pub. No. 1 (1898); ———. *The Cyclonic Distribution of Rainfall.* Rock Island, Ill.: Augustana Library, Pub. No. 4 (1905). (Udden's later career was identified primarily with Texas and included the study of oil and gas geology.)
18. TRELEASE, S. F., and BEATH, O. A. *Selenium: Its Geological Occurrence and its Biological Effects in Relation to Botany, Chemistry, Agriculture, Nutrition, and Medicine.* New York: Authors (1949).
19. THROCKMORTON, R. I. *Twenty-eighth Biennial Report of the Kansas State Board of Agriculture,* **33**, 91, 100-101 (1931-32).
20. HILGARD, E. W. *A Report on the Relations of Soil and Climate.* USDA Weather Bureau Bull. 3. Washington, D. C.: GPO, 19 (1892).

Chapter Twenty-five

SOIL, ANIMAL, AND PLANT RELATIONS OF THE GRASSLAND, HISTORICALLY RECONSIDERED

I. The Problem of Method and Point of View

WHEN a historian appears on a scientific program, it may be appropriate to ask some questions. What is science? What is history? The answer to these questions is not necessarily difficult. In dealing with the field sciences, as distinguished from the laboratory sciences, history and science may be only different facets of the same thing. Nowhere is this fact more relevant than when applied to any consideration of natural resources.

But, again, a question. What is a natural resource? The answer is that the properties of the earth become natural resources only as they involve man and are utilized by him. Without the intervention of man, although particular properties are actually present, they are latent, or unrealized. A natural resource is not determined by the properties of the earth, per se, but by the qualities of the mind of men. The first requisite of a natural resource is an idea. There are no known limits, therefore, to the multiplication of natural resources of the earth, and exhaustion of them is impossible, except, or unless, the capacities of man are exhausted—the capacities through which the latent properties of the earth are discovered and thus become properties new to man and available to his use as natural resources.[1] The record of the process by which the potential of the earth has been made actual is within the province of history, and to the study of it the historical method should apply, regardless of whether the intellectual enterprise is undertaken by the historian or by the scientist. In this context, history provides the background and prepares the setting, but at that point science may take over. Obviously, the boundary lines that have become traditional between the several accepted intellectual disciplines are artificial, and as they were adopted originally only for the purpose of making the intellect more effective through specialization, they are justified only so long as they accomplish, rather than hinder, that purpose. To achieve the goal of making the mind most effective, intellectual enterprise must possess both perspective and depth, and to sacrifice either defeats the full realization of the other. History is of particular importance in establishing perspective.

When the problem of history and historical method are introduced, the time has come to distinguish two schools of thought on the subject. First, the concept of objective history strives to reconstruct historical actuality as completely as possible, without respect to any possible use to which it might be put. Second, the subjective relativist functional notion about history holds that the only excuse for its practice is to make it useful for some present purpose. To accomplish this functional goal a selection from the whole corpus of historical actuality is made, utilizing only the so-called usable part. The great difficulty with this method is that the results are almost certain to be predetermined by the frame of reference adopted before the so-called historical investigation was begun, only those things being found, or considered applicable, that fitted the preconceived hypothesis. The requirement of usefulness more often than not defeats itself. On the contrary, the first method described, the pursuit of knowledge as intellectual enterprise, without respect to usefulness, offers more probability of turning up something useful, although not necessarily anything that was in the mind of the investigator when he began.

II. Illustration: Roe, *North American Buffalo*

In order to reduce the problem to something tangible, the book by Frank Gilbert Roe, *The North American Buffalo: A Critical Study of the Species in its Wild State* (1951), is taken as a remarkable example of excellent history that is at the same time essential to the scientist and to the historian in more ways than the book reviewers have thus far recognized. All reviews of the book that have come to the notice of the present writer have been quite favorable, but in spite of that fact the impression conveyed is primarily that it is just a good book that should be read sometime. Certainly, without the intent of the reviewers, that is almost equivalent to damning it with faint praise; because Roe's *North American Buffalo* is a book that is of such outstanding importance that anyone interested in the grassland of North America, whether he be historian or scientist, should read it immediately, the whole of it; not only read it once, but reread it and digest the contents thoroughly. This book should be read not only for what is actually said, but it should be studied in all its implications in order to search out all the possible ramifications. Studied with thoroughness, it should become the springboard for a wide variety of investigations in several disciplines not even contemplated by its author.

Frank G. Roe is a resident of Canada, and he is not a professional historian. He became interested in the buffalo problem as a by-product of a study of the earliest roads in old England and arrived at the unorthodox conclusion that they "were probably *not* originally wild animal tracks; nor were the earliest human (Indian) trails of this continent [North America] buffalo tracks." The contradictory historical evidence relating to the buffalo led him into a fifteen-year study of the buffalo, with special reference to the part of the North American continent, north of approximately 40° north latitude. He devoted only limited attention to the country south of the Republican River. Another self-imposed limitation was to exclude scientific considerations, but no scientist should be misled into assuming that the book has no value for scientists. The facts are quite otherwise. Roe's statement of reasons is fundamental:

> In dealing with an animal now extinct as a free world species in its most characteristic native habitat, the first task is to ascertain and classify the historical evidence: and not until this has been done can biological investigation proceed with much profit.

The evidence collected appears to support the conclusion that some variation did exist within the buffalo species. In relation to the long-accepted tradition that the buffalo made general annual migrations from the south to the north and return, Roe has demonstrated conclusively that no such general movement occurred, especially north of the Republican River country, or about the 40th parallel. He concedes that some such movement occurred south of that boundary, but, as a part of the self-imposed limitation of the geographical scope of his study, he did not undertake to survey southern literature intensively.

Roe's contention, that the buffalo movements were primarily random wanderings, appears to be fully demonstrated. The Indians who were largely dependent upon the buffalo followed these random wanderings as best they could. Thus the numbers of buffalo that might occupy a particular spot could be enormous, and damage to vegetation disastrous, but the incidence was not continuous. There is still an opportunity for other investigators to study thoroughly the problem of overgrazing, drought, and dust storms under aboriginal culture.

Although Roe did not go into the problem of the consequences entailed by his conclusions, it is in this context that we learn that surface erosion by wind and water was present and upon occasion severe under aboriginal conditions—recurring drought, fires, overgrazing, and trampling by animals, especially buffalo, as well as the wearing of innumerable paths to watering places. Dust storms upon a large scale were not caused by the "plow that broke the plains."[2]

Roe has demonstrated that the migrations of the buffalo were primarily random. But there are other aspects of his treatment of buffalo movements that are not so satisfactory. He challenged the notion that buffalo changed their direction of movement deliberately on account of encountering sparse grazing, and that they sought out more productive pastures over considerable distances. Yet he accepted the view that buffalo moved between the plains and the Rocky Mountains in midlatitudes, and between the rough wooded areas and the plains in the north country. Probably he is correct in the sense that he ruled out in the first case the notion of buffalo capacity to make choices on the basis of memory or instinct akin to rationalizing from experience, but that does not explain the behavior which he does seem to accept, that of seeking shelter in timbered areas during the winter storms or for shade from the intense heat of the sun. A suggestion is offered here that possibly guidance may be derived from the physicists' theory of the unpredictable behavior of individual particles, but the high degree of predictability in the sense of statistical probability as applied to behavior of large numbers.

Although Roe emphasized the fact that he was not a scientist and was deliberately excluding scientific aspects from his book, he did not escape making scientific blunders. One of these may be mentioned as illustrative of the importance of the historian's knowing something about science. In discussing the extensive deposits of buffalo bones scattered over the plains, which must "have been broken, crushed and stamped into the earth," he suggested that: "This also may have some bearing on the enrichment of the soil. A chemical analysis of Kansas virgin prairie soils might yield some interesting information" (p. 515, Note 116). He was unaware of the epoch-making monograph of E. W. Hilgard, as long ago as 1892, which demonstrated conclusively the fact of lime accumulation in soil of low rainfall climates, regardless of the parent materials from which they were derived.[3] Such additions to the lime content of the soil as buffalo bones or any other artificial additions of lime to lime-rich soil contributed nothing to soil properties or to productivity.

In reviewing the literature dealing with the buffalo, Roe has demonstrated that the most recent is not necessarily the best. Joel A. Allen's book, *The American Bisons, Living and Extinct* (1876), is among the earliest formal treatments and was the best of the lot. Roe demonstrated also that the scientist is not necessarily the best authority, both Hornaday and Seton being proved quite unreliable except upon limited aspects of the subject. George Catlin, an artist, emerged conspicuously as one of the most reliable observers of buffalo and Indian lore. Furthermore, Roe's study made embarrassingly clear that the medium through which a supposedly scholarly study is published is not necessarily an index of its authenticity. Again, Hornaday's monograph, the work of a scientist, published by the Smithsonian Institution, is the horrible example. This mid-twentieth-century culture is the victim of a naïve worship of formal training and specialization, forgetting too much the first principle, that competence in any field is grounded in the quality of the individual. Roe is not a professional historian, and he disavows explicitly any scientific pretensions, yet he has produced a major historical work that is fundamental to both historians and scientists. The only plea that may be appropriately advanced in this connection is that some formal discipline in history and science and their respective methodologies might have enabled Roe to produce a still better book. There is room for an argument in rebuttal, however, that the requirement of formal methodology and training might have killed all incentive to write the book.

As a result of Roe's study of the buffalo, both the historian and the scientist must largely rethink the whole problem of the interrelations of the buffalo and of man, and of many corollaries or inferences that are applicable to the grazing of domestic livestock on grass. One more point in emphasis from Roe may be permissible, one which is in a sense the major point of the present paper. Roe defined his idea of the relation of history to science, and with the qualifications given above, that declaration stands as the view of the present author:

> The first task is to ascertain and classify the historical evidence; and not until that has been done can biological investigation proceed with much profit.

III. Animal Exclusion Studies

Attention is now directed to another type of study—this one an experimental project conducted in the field. To facilitate objectivity, an English work is used, that of V. S. Summerhayes, "The Effect of Voles (*Microtus agrestis*) on Vegetation."[4] By exclusion of voles from the test plots over a period of seven years the conclusion was arrived at that the yield of dominant grasses was increased:

> On the removal of the vole attack the non-dominants, particularly the mosses, decreased in abundance, apparently as a result of the increased competition with the more luxuriant dominants. Voles therefore tend to preserve a relatively open vegetation, comparatively rich in species ["flowering plants and especially mosses"]. This is presumably effected by the direct eating or cutting up of the aerial parts of the dominants, and by the complicated series of burrows below the main surface of the vegetation, the formation and maintenance of these burrows preventing the development of large tussocks of grasses like *Molinia,* or thick-matted turf-like growth as in *Holcus Mollis* (p. 45).

The author let the matter rest with those conclusions, and refrained from any policy recommendations, but the customary policy conclusions drawn from such animal exclusion studies are that the predators should be exterminated to increase the grass yield available for livestock. Such policy conclusions do not necessarily follow. The duration of the experiment was seven years, but what might have been the result if it could have been continued one hundred years? The central point is that soil as an object of study had no place in the experiment, yet any assumption about indefinite maintenance of the increase in yield of dominant grasses must be posited upon a parallel assumption of an indefinite maintenance of soil productivity. Questions that require answers on such a long-term basis include the status in the investigation of legumes and of deep-rooted forbs, and of the activities of animals and of the soil population, all considered in relation to the soil as an object of study. Had soil been included within the scope of the project under consideration, the seven years of effort might have meant being seven years further along on the study of the changes that occur in soil under the conditions of the experiment. What has just been said about the particular project under consideration applies substantially to similar work in the United States.

IV. Man within the Ecosystem

The process of the expansion of European culture throughout the world, a four- to five-century drive that has about spent itself, was characterized conspicuously by a contempt for the "savage" and the "backward" peoples of the globe. No branch of that culture was more conspicuous in that respect than the Anglo-American tradition.. Belatedly, the situation is changing, during the mid-twentieth century in particular, and re-examination of old evidence and discovery of new facts are revealing fresh perspectives which impart to aboriginal culture a historical significance of outstanding importance. The conventional or traditional concept of the state of nature must be abandoned —that mythical, idealized condition, in which natural forces, biological and physical, were supposed to exist in a state of virtual equilibrium, undisturbed by man. The role of aboriginal man within the ecosystem must be recognized as a major ecological fact. The task of re-examination, largely historical in character, cannot be done in a day, and it has not been done for the North American grassland.

V. The Great American Desert: Semantic Problems, Myths, and Legends

In dealing with the North American Grassland historically, one of the first problems to be met is that of the semantics of the word "desert." Approached from the standpoint of the history of the usage of the word, many of the difficulties are revealed. The meanings varied widely in time, and otherwise. In the eighteenth century, and even during the early nineteenth century, good usage included the idea of an area that was deserted—especially deserted by man—therefore, desert, even though it was covered with forest. The word did not necessarily have reference to consequences of the relations of climate to vegetational cover.[5] Some forest men used the word in such a manner as to imply that a lack of trees and running water made a desert, even though a grass cover was present.[6] For the accurate interpretation of written documentary evidence, therefore, the necessity arises of determining what the original observer meant by desert, as well as the concept that existed in the mind of the person using the document.

A major myth developed during the nineteenth century, that a Great American Desert stretched across much of the western interior of the continent, and the label was placed upon some maps. That fact led to another legend that the myth of the Great American Desert was held universally, but of course that was not the case. At no time were either the literature or the maps in general agreement on the existence of a great desert or of its extent. In the monumental *Atlas of Historical Geography of the United States* (1932), edited by C. O. Paullin, ten maps were selected to illustrate the development of the cartography of the western United States, bearing dates from 1804 to 1867.

Only two of these used the "Desert" label. Pioneers, eager to occupy the land, were optimistic about the possibilities of the country, as were promoters of railways to the Pacific coast, unless describing the route of a rival.

The gold rushes of 1849–59 contributed substantially to public education in the geography of the West. R. T. VanHorn, editor of the *Journal of Commerce,* Kansas (City), Missouri, on November 10, 1859, commented optimistically that the desert of the myth had retreated from Illinois westward and the gold rush of 1859 had finally extinguished it, except the Senatorial Desert, which existed only in the Senatorial Mind at Washington.[7]

To put the question more broadly, there was a general tendency for those who opposed the rapid settlement and development of the trans-Mississippi West to be receptive to the desert myth, whereas those favorable to the aggressive westward expansion were sure and determined that all that was necessary to make the grassland blossom like the rose was to let in the population. In an able editorial in the St. Joseph *Gazette,* June 14, 1854, Lucian J. Eastin reviewed explicitly this conflict in outlook, and the reversal in point of view after the annexation of the Southwest, the opening of emigrant roads, the gold rushes, and the establishment of trade routes.

VI. The Problem of Origin of the Grassland and Climate Change

Closely related to the desert myth problem is that of the "origin of the prairie." The single point dealt with here is the factor of fire, whether natural, accidental, or used deliberately by the aboriginal population. The notion of the fire origin of the grassland may be dismissed, but in transitional country, so far as climatic and local factors were concerned, fire did act generally to restrict tree growth. Recognizing that fact, some important historical conclusions are in order. During the years following the Civil War, the idea became widespread that the climate was becoming more favorable as a result of settlement. That interpretation was a quite reasonable one, if viewed against the background just indicated. Settlement eliminated fires, and woody growth spread at the expense of grass.* The assumption became easy that this would continue until the whole area would support tree growth, if permitted. The fact of the surprising extent of the spread of trees was inescapable, but the interpretation of the facts in terms of climate change was erroneous. The white occupants of the grassland did not understand the role of the multiple factors in the situation that had operated under aboriginal culture; hence, the misinterpretation of causal relations.†

VII. Soil as an Object of Study

The introduction into this discussion of the subject of soil as an object of study is a sharp reminder that the literature from which the history of the soil conditions under aboriginal culture and European man's attitudes toward them is as contradictory as the buffalo literature with which Roe dealt. Any clear and reliable understanding of the soil problem in its essential historical aspects awaits comprehensive historical treatments on a comparable scale.

Some accounts emphasize the hardness and impervious character of the soil as found under aboriginal culture, a compactness so repellent to water that the rainfall ran off into the streams, producing floods or severe erosion cutting deep

* In his book, *Colorado, A Summer Trip* (New York, 1867), Bayard Taylor gave a vivid description of the landscape in transition in eastern Kansas and eastern Nebraska. It is a significant document so far as it was descriptive of what he observed.

† In Book Three of his book *Virgin Land* (Cambridge, Mass.: Harvard Univ. Press [1950]) Henry Nash Smith has performed the most complete job thus far in confusing the problems of the desert and of climate change, along with land policies.

channels. Thus in the upper Canadian River, about 103° west longitude, Lieut. J. W. Abert commented, in 1845, that the prairie, "baked in the hot sun, absorbs but little water. . . ."[8] Other accounts stressed the soft, yielding character of grassland geological structure, and the rapidity of erosion, by water and wind, of the unstable soils.[9]

Many of the early travelers and explorers were impressed by the activities of such burrowing animals as ants, pocket gophers, ground squirrels, and prairie dogs, and the activities of buffalo disruption of stabilized soil conditions. In 1846, on a military mission in the opening months of the Mexican War, Lieut. Abert commented on the activities of pocket gophers along the Santa Fe Trail between 96° and 97° west longitude. His journal entry for June 28 reads:

> Whenever we rode to the side of the road we noticed that our horses would frequently sink to the fetlock, and saw on the ground little piles of loose earth . . . formed by the sand rats, or gophers. . . . [Four days later he added]: The mounds of the gophers . . . were more abundant than heretofore, and in several places a number of these mounds had been so close together that the distinctness of each was completely lost in the mass, covering an area of five or six feet.[10]

This description applied to tall grass prairie, but Abert commented later on pocket gophers in the Arkansas Great Bend area. All the country in question was in the condition commonly defined as "virgin prairie," or "in the state of nature."

In the narrative compiled by the botanist Edwin James for the Stephen H. Long expedition of 1820, descriptions are given of extensive prairie dog towns in what is now Nebraska and Colorado.[11] Among the most detailed descriptions of prairie dog towns are those of Captain R. B. Marcy, covering exploring expeditions on the southern Great Plains in 1849, 1852, and 1854, especially the Red River report of 1852 which included the valley of the South Fork, a stream which the Comanches called Prairie Dog Town River.

The contradictory character of the literature,

both historical and scientific, on these problems seems to call for comprehensive investigations of both types, not only of the animals, but also of soil, as an object of study under the influence of these animals, and after they have been eradicated. Sites that are known to have been occupied in Nebraska and Colorado in 1820 or in Texas in 1852 might profitably be studied to determine what influence such occupation imposed upon the soil. Sites might be selected where the date of eradication can be established, to determine what has happened to such soils without the presence of prairie dogs, or of other burrowing animals that may properly be studied in the same fashion. Such studies as are suggested here require as a preliminary step the same type of comprehensive historical study that Roe gave to the buffalo. Even when it is conceded that soil is benefited by such animal activities, there is no agreement upon what degree of disturbance by animals is advantageous to long-term equilibrium.

In pursuing the ecological literature about the grassland another gap is conspicuous—the function of deep-rooted plants of the nonleguminous families. To be sure, there are many studies of roots, and noteworthy are those of J. E. Weaver and associates, but they are oriented from the standpoint of plant ecology, not of soil science. The literature of the explorers contains many references to the range and distribution of such plants, which stand as a challenge to the historically minded to investigate certain of them comprehensively in relation to soil as an object of study. An example that invites investigation is the man-root, a morning glory, *Ipomea leptophylla* (Torr.), found, according to Gray's *Manual of Botany* (1889), on the "plains of Nebraska to central Kansas, Texas and westward." It produced a root the weight of which was given as ranging from 10 to 100 pounds. Lieut. Abert described his experience with it in 1846 while waiting for high water of the Pawnee Fork to subside, in the general vicinity of Larned,

Kansas, about 99° west longitude, in the hard land north of the Arkansas River.[12] A soldier spent several hours trying to dig up a specimen under Abert's direction, but the ground was so hard they finally gave up and broke it off. The stem, about half an inch in diameter, ran down about 12 inches, then enlarged suddenly to 21 inches in circumference, or about 6 inches in diameter, and extended about 2 feet deeper. Abert's comment indicated that this specimen was relatively small compared with others supposed to grow to the size of a man.

From the standpoint of soil as an object of study, what happens to soil when a root expands to 6 inches or more in diameter, displacing the soil to a depth of three feet or more? When the plant died, the root decayed, and the cavity was refilled—but how, and how rapidly, and with what effect on the soil? What was the actual floristic range and the density of distribution of this plant, and its average and maximum life expectancy? There were many other grassland forbs, with roots of smaller diameter, that penetrated the soil 10 feet or more. All these deep-rooted plants penetrated the lime accumulation zone. Abert commented that the Cheyenne Indians dug and ate the man-root. If they did it generally, they must have possessed more patience than Abert and his soldier, because the Indian had no iron tools with which to dig. Also, such digging substantially disturbed the soil.

Soil should be investigated as an object of study under aboriginal conditions as a prerequisite of scientific investigations carried out under existing conditions, or artificially controlled conditions. Such historical investigations should recognize all possible factors: aboriginal man; large animals; the smaller animals, especially the burrowing animals; insects that bury themselves in the soil; the deep-rooted plants.

Again, the *Ipomea leptophylla* may be used as an example in order to make the discussion more concrete, although prairie dogs, or pocket gophers, or ground squirrels might serve as well. Considering the extent of vegetational distribution quantitatively, this plant was engaged in a continuous soil tillage operation. New plants replaced old ones; new growth in one spot displaced the soil, while decay of old roots at another permitted the cavity formed in the soil to collapse. But from what directions: From top down, or did the sides cave in? Or both, on occasion? The vital issue is that the tillage was continuous, but without destroying the soil cover as mechanical tools tend to do, and it was to varying depths; possibly, where subsurface conditions permitted, the prevailing depths were 2–4 feet. Lesser roots penetrated much deeper. No mechanical tool has been devised for cultivation of the soil that can perform a comparable job, that can open up the soil body to a depth of 30 inches or more, and certainly none that can open up the soil to any considerable depth without destroying the vegetational cover. To what extent did these processes interrupt or modify theoretical profile-forming tendencies and the lime-accumulation zone? What happens when these factors are removed altogether by eradication programs or clear-field cultivation?

From this historical approach to the problems of the grassland the conclusion has been reached that erosion, in the much-advertised sense, is not necessarily the most important aspect of soil conservation. In any case, the critical aspects of soil conservation vary with particular spots. They depend upon time and space. But, in many respects, more fundamental than the several facets of surface erosion is more knowledge about soil in a comprehensive sense under aboriginal culture, and what happens to soil internally as a consequence of the transition from aboriginal occupation to utilization by modern society—in the transition from natural tillage by wild animals grazing, by burrowing animals and insects, and by the influence of the native legumes and deep-rooted forbs to the twentieth-century mechanized regime.

VIII. Water Table

One contention of the present author, not yet given a full-scale demonstration, is that availability of well water for livestock and domestic purposes played a role in settlement survival during the pioneer period that may have been even more decisive than rainfall for grass and field crops. The drought decade of the 1930s focused attention upon water supplies for cities, and set geologists to work with a new vigor upon Pleistocene geology.[13] Among the results of such research was the conclusion that the water table generally was essentially stable, varying temporarily with climatic fluctuations and local circumstances. Soil science placed a new emphasis upon the water table, assigning to it a role as a soil-forming factor.[14] Thus, by a rapid succession of events, the original proposal, the study of the relation of well water to pioneer settlement in the grassland, has been given a more fundamental significance expanding into the far broader issue of the water table in relation to the whole problem of human occupance of the area. The subject is so large as to offer research opportunities for a number of students equipped for the task.

IX. Mesquite

Another aspect of the problem of man within the ecosystem may be illustrated by reference to an article "Man vs. Mesquite," in *Life* magazine, August 18, 1952. The caption under the map read: "Mesquite march during last 100 years has taken it from small riverside areas in which it grew in 1850 to the 75 million acres it now covers . . ." (p. 69). Inquiry concerning the authority for the map brought the answer from the editors that

A century and a half ago, there was hardly any [mesquite] in the U. S., but during the next fifty years it was brought into this country from Mexico by Spanish ponies and by wandering herds of wild buffalo. So that, in 1850 (as shown by the map) scattered stands of mesquite were growing along the creeks and river beds of the southwest. This first generation mesquite, however, was exceedingly sparse . . . during the great cattle drives of the second half of the nineteenth century (roughly 1860 to 1880), the cattle intensified the mesquite in the areas along the watercourses, and extended it out onto the plains away from the creeks and streams, and since then mesquite has fortified its hold on the southwest to cover the area shown in the 1952 distribution on the map. . . .[15]

Another statement on the mesquite history is that in the Clements-Shelford textbook *Bio-ecology* (1939), where mesquite in the costal and mixed prairies was attributed to a disclimax induced by overgrazing which took the form of "a savanna of mesquite and cactus" (p. 279). And in still another place, V. E. Shelford declared:

For example, the cattle business of the United States had its beginning in the gulf coast tallgrass prairie. This is an area almost universally mapped as mesquite—chaparral or savannah and regarded by many as having been that type before the white man came to the area. On the contrary, since cattle eat the mesquite beans and fail to digest them, they spread the seed widely and may be responsible for the entire savannah. It is well known that the mesquite has been spread from south central Texas into west central Oklahoma by this method.[16]

Some historical data may now be brought to bear upon the mesquite problem in order to establish some factual landmarks. The Stephen H. Long expedition of 1820 found mesquite in the Canadian River country, and Edwin James, who prepared the report of the expedition is credited with the first public notice of the mesquite tree.‡ Lieut. Abert found "an abundance of mesquite" in 1854, growing about 103° west longitude as a shrub about 5 feet in height in what is now northeastern New Mexico above the headwaters of the Canadian River.[17] The R. B. Marcy expedition up the Canadian River, in 1849, found mesquite just east of the Llano Estacado escarpment, the journal

---

‡ R. B. Marcy recognized this fact in his report on the Brazos River expedition of 1854, quoting from John Torrey, by whom the mesquite species collected by James was described and named *Prosopis glandulosa*.

entry stating that "We found a great deal of the small mesquite . . . today."

Marcy's return route from Santa Fe, in 1849, turned southward down the Rio Grande to Doña Ana, thence eastward to the Pecos River down that stream to the crossing; then, skirting the escarpment of the Llano Estacado, he struck northeastward across Texas. The mesquite was brushlike in the country west of the Pecos, but increased to small tree size at that river, and eastward as the ascent was made into the high plain and in the Big Spring area it attained large tree size. From the latter point northeastward, Marcy's map indicated mesquite timber. The second day after Big Spring, the route led "over rolling and rather broken country, of good soil, and covered on each side with large mesquite trees." Near what he miscalled the Double Mountain Fork of the Brazos River, Marcy recorded: "We have been travelling through groves of mesquite timber, with a beautiful carpet of grama grass underneath, nearly all day." On the next day, on the south side of Double Mountain Fork, he continued over

as beautiful a country for eight miles as I ever beheld. It was a perfectly level grassy glade, and covered with a growth of large mesquite trees at uniform distances, standing with great regularity, and presenting more the appearance of an immense peach orchard than a wilderness. [Heading toward the Brazos River above the mouth of the Clear Fork] The mesquite wood and grass continued very abundant. . . .

Four days later, just before crossing the divide into the watershed of the West Fork of the Trinity River and west of the 98th meridian, mesquite and oak openings were reported, with occasional prairies. In summing up the estimate of his line of march as a route for the Pacific Railroad, Marcy reported 200 miles "over a gently undulating country, with prairies and timber," springs and streams, "in many places covered with large groves of mesquite timber, which makes the very best fuel," and later he made a more positive commitment to the existence of "an inexhaustible amount of mesquite timber, which, for its durability, is admirably adapted for use as sleepers, and for fuel."[18]

In 1852, Marcy explored the headwaters of the Red River. When west of the 101st meridian on the South Fork of that stream he wrote:

> We find much more mesquite timber upon this branch of the river than upon the other. Indeed, I have never seen much of this wood above the thirty-sixth degree of north latitude; but south of this it appears to increase in quantity and size as far as the twenty-eighth degree. Upon the Canadian river I have observed a few small bushes; but the climate in that latitude appears too cold for it to flourish well.

In the same report, in his discussion of the Pacific Railroad by the southwest route of his exploration of 1849, Marcy wrote that after crossing the Brazos

> the road skirts small affluents of that stream and the Colorado for two hundred miles. . . . Here and there prairies present themselves, but this section is for the most part covered with a growth of trees called mesquite, which stand at such intervals that they present much the appearance of an immense peach orchard. They are from five to ten inches in diameter, their stocks about ten feet in length, and for their durable properties are admirably adapted for railway ties, and would furnish an inexhaustible amount of the very best fuel. . . .[19]

In 1854 Marcy explored the headwaters of the Brazos and Big Wichita rivers. From Fort Belknap, heading west of north, they passed over "rolling country, covered with groves of mesquite trees." The next day they crossed tributaries of Trinity River, "all of which were wooded with mesquite, and occasionally a grove of post oak seen, with here and there a cotton-wood or willow tree along the banks." Later, "On leaving the Wichita, we travelled south towards the Brazos for six miles through mesquite groves. . . ." From a low mountain near the Brazos, Marcy described the scene: "Towards the east from this elevation nothing could be seen but one continuous mesquite flat, dotted here and there with small patches of open prairie, . . ." and on the next day: "The country we are now passing is gently undulating and covered with mesquite trees."

By the time Marcy made this expedition he was much impressed by the mesquite and wrote a rather comprehensive summary of the subject some three to four pages in length:

> In the journeys I had made before upon the plains, I had observed the mesquite tree extending over vast tracts of country, and I had noticed some of its useful properties, such as its durability and its adaptation for fuel, but I was never so fully impressed with its many valuable qualities as during the past summer.
>
> It covered a great portion of the country over which we travelled. . . .

It was at this point that Marcy acknowledged that Edwin James, of the Long expedition of 1820, had given mesquite the first public notice. In commentary upon the range of distribution, Marcy admitted limitations of information, but east of the Rocky Mountains he defined its limits as between 97° and 103° west longitude, and between 28° and 36° north latitude; but west of the Rio Grande the mesquite flourished best in the valley of the Gila River. In the plains, however, he remarked that the size diminished north of 33° and to mere bushes at its northern range limits of 36° north latitude.§ In its tree form, it ranged in size from 4 to 15 inches, and was not more than 20 feet in height, and furthermore was "much used for building in southern Texas and Mexico," being well preserved in the ruins of old buildings. And then Marcy recorded information critical to the ecological problem, reporting that mesquite often grew "upon the most elevated arid prairies, far from watercourses," but it would grow only upon good soil, and that settlers competed for mesquite land.[20]

A second account of the Marcy Brazos exploration is available in the book of W. B. Parker, a

---

§ Marcy did not leave any account of having explored north of that limit, so he was indicating in part at least only the limits of his firsthand knowledge.

civilian. In many respects Parker's version is similar to Marcy's, but variation in presentation of the scenery affords some further enlargement of perspective. Between the Cross Timbers and the Little Wichita, on July 11, the entry read: "The country we had been passing over, since leaving the Cross Timbers, was a rolling prairie, very thin in soil and timber very scarce. At this point we began to find the mesquite trees in great abundance." Their size was given as 4–15 inches in diameter and not more than 20 feet tall. These were the same specifications as those given by Marcy. In addition to its qualities as fuel (burning like hickory wood), Parker added, "and not the least is its durability for building purposes— . . . invaluable to the future settlers."

An entry five days later recorded "ascending in a northwest course, a rolling country, covered with buffalo grass and mesquite timber. . . ."‖ Three days later, approaching the Little Wichita, a belt of timber marked its course, "and in front the wide prairie with its yellow coating of buffalo grass, studded with the pale green mesquite, a beautiful combination for a landscape painting." Again, a few days later: "Our course was Northwest, and ascending gradually, we came upon a very extensive plain, covered with buffalo grass and mesquite timber." Later on, a course south from the junction of the three prongs of the Little Wichita, they "entered an extensive plain covered with thin coarse grass and stunted mesquite timber." Upon arriving at a spur of the Llano Estacado, they ascended it "to a broad level plain . , . covered with buffalo grass and mesquite trees, and extending as far as the eye could reach in a perfect level toward the dim cloud like mountains at the head of the Brazos." Arriving at the escarpment of the Llano Estacado August 3, they climbed to the top, and looking eastward from an elevation estimated at 600 feet above the country below:

‖ Clearing up of confusion in the nomenclature of grasses would require a separate study.

SOIL, ANIMAL, AND PLANT RELATIONS

> The view was the most extensive and glowing in the sunset, the most striking that we had enjoyed during the whole trip, combining the grandeur of immense space—the plain extending to the horizon on every side from our point of view—with the beauty of the contrast between the golden carpet of buffalo grass and the pale green of the mesquite tree dotting its surface.[21]

Admittedly, the foregoing survey does not cover anything like all the literature, and certainly there is no intent to exaggerate the extent of mesquite occurrence, but it makes abundantly clear the fallacies widely held about the mesquite problem, especially those in evangelical conservation circles. In order to bring this discussion to a focus, a few tentative conclusions are outlined, derived from the limited historical data cited from the reports of the Long and the Marcy explorations. First, in a floristic sense, the geographical range of distribution of mesquite (*Prosopis* spp.) is about the same in 1952 as at the opening of the nineteenth century, or 150 years ago. Possible extensions of floristic range appear to be a minor aspect of the problem. Second, in a vegetational sense, the quantity of mesquite at the midpoint of the nineteenth century was substantial, and was not limited to the banks of streams; upon occasion mesquite occupied broad plains and rolling hills in west and north central Texas as far west as the Llano Estacado. Repeatedly the Marcy descriptions of the country indicated extensive reaches of mesquite savannah, with occasional patches of open prairie. Such language appeared so often, and so explicitly, as to be both significant and important. Third, in an ecological sense, the focus of interest is the change in the behavior, or growth form, of the mesquite during the century 1852–1952. As an ecological fact, the nature of mesquite occupance in much of the region under review changed from a savannah to a tangled jungle, in places almost if not quite impenetrable. The outstanding ecological problem, then, is to find an explanation of the how and the why of this change in growth form of mesquite and its associates. An accurate historical study of what has happened, establishing in fuller detail the facts of floristic range limits, quantity of vegetation, and form of growth, prior to the time the Indians handed the land over to the whites, may put the ecologist and the range manager in a position to attack the question.

A fourth and fundamental conclusion is the full acceptance, as of long standing, of the mesquite occupance of the floristic range just indicated. Marcy gave the size of mesquite as ranging from shrubs to 15 inches in diameter. This in itself is proof of long establishment. Further evidence of the long duration of mesquite occupance in southern Texas, the portion inhabited by European culture, was the reference to mesquite timber found in the ruins of old buildings. In 1884, V. Harvard compiled a growth-age table for mesquite: a trunk diameter of 7–8 inches, 30 years; 8–10 inches, 50 years; 10–12 inches, 75 years; and over 12 inches, more than 100 years old.[22] According to such a calculation, a diminishing rate of increase of diameter with age must admit, for a 15-inch diameter, a life span of 150–200 years or more. As of 1952, that would carry mesquite occupance of northwest central Texas back in time 250–300 years—possibly more, on the basis of the Marcy evidence. This does not take into consideration the possibility that earlier trees may have grown, died, and disappeared prior to those he was describing.

So far as the buffalo and other wild animals operated as a factor in scattering mesquite, they wandered over the whole area for centuries before 1800. The accounts of the Coronado and De Soto expeditions record buffalo in the area about 1540–41.

So far as domestic cattle drives or domestic overgrazing were factors, according to the census of 1880, neither operated generally in the country west of the 100th meridian prior to about 1879. The cattle drives northward during the 1850s followed a path just west of the Arkansas western boundary, many crossing the Missouri River below Kansas City. The cattle business in the plains proper awaited the breaking of the Comanche-Kiowa Indian Barrier.[23]

The savannah form of vegetation was found in other parts of North America when Europeans took over the land from the Indians, and studies of it elsewhere may be profitable to establish perspective. N. S. Shaler, by profession a geologist, but by avocation a historian of the Kentucky country, long ago attributed the prairie condition of much of the area east of the Mississippi River to fire, occurring naturally, by accident, or as an instrument used deliberately by the Indians. Shaler did not make the mistake of assigning to fire the whole responsibility. In some areas, especially westward as the rainfall diminished, climate was held to be decisive. But in the Kentucky country, what Shaler described as essentially a savannah stage was a preliminary step in the process of reducing a dominantly forest area into a prairie. He associated the Indian practice with the eastward migration of the buffalo sometime after the year A.D. 1000 and suggested that had European intervention been delayed another 500 years, the prairie might have been extended to the Alleghany Mountains. Although Shaler's is a rather extreme view, Roe's study of the buffalo gives support to the factual portion of Shaler's general contention.[24] Shaler's dating of the arrival of the buffalo was established by excavations he had made in 1868 around the salt springs at Big Bone Lick, Boone County, Kentucky. In succession from about glacial times toward the present, bone deposits accumulated, the modern buffalo species occupying the top position—in time, later than the Mound Builders, who were not acquainted with the buffalo.

In the state of Mississippi as of the late 1850s, E. W. Hilgard wrote of the country as received from the hands of the Indians:

The herbaceous vegetation and undergrowth of the Longleaf Pine Region is hardly less characteristic than the timber. Whenever the regular burning of the woods, such as practiced by the Indians, has not been superseded by the irregular and wasteful practice of the later settlers, the pine forest is almost destitute of shrubby undergrowth, and during the growing season appears like a park, where long grass is often very beautifully interspersed with brilliantly tinted flowers (p. 349).

The same writer, at another place, continued the theme under the head of "Pasturage in the Pine Woods":

In their natural state, as received from the hands of the Indians, the Pine Woods were one great pasture—as, in thinly settled regions, they still are. Nor is it, generally, the ranging of cattle which has destroyed the pasturage in other regions, but simply the injudicious burning of the woods, at seasons when the fire would destroy not only the dry leaves, but also parch the *heart* and the *roots* of the grasses. It would seem that in a region comparatively poor in agricultural resources, the maintenance of pasturage should be considered a matter of national importance. The Swiss, being unable to cultivate profitably their mountain slopes, have converted them into pastures, these form the basis of their national wealth. Why this should not be so with the inhabitants of the Pine Woods, I have been unable to discover, it is certain, however, that the pasturage of that region, is disappearing before the fires at a fearful rate, and that those who heretofore have relied on the range, during all but a few weeks in winter, for the support of their cattle, will soon be compelled, as many are now, to raise feed for them on their poor soil, which, at present, will but just furnish comfortably the prime necessities of life for the population itself. The beautiful park-like slopes of the Pine Hills are being converted into smoking desert of pine trunks, on whose blackened soil the cattle seek more vainly every year, the few scattered, sickly blades of grass, whose roots the fire has not killed.

The preceding paragraph was descriptive of past and present. Hilgard then discussed policy and procedures in terms of management:

It is not the province of this Report to suggest municipal regulations by which the burning of the woods at improper seasons might be prevented, or at least, rendered of less general occurrence; the evil, however, is a crying one to the mind of every candid observer, and the destruction of national wealth caused by it is so enormous as to deserve no less attention certainly, than the improvement of soils. However convenient and effectual may be the burning of the dry grass in order to render the young growth accessible to cattle, that advantage is certainly purchased very dearly at the cost of its total destruction within a few years—a policy little better, in fact, than cutting down a fruit-tree for its fruit; which appears more especially irrational when we consider how easily the advantage could be reaped without incurring the enormous waste, by a regular system of burning at times when, as after the first autumnal rains, and more especially in early spring, the ground is too wet to allow of injury to the roots, while yet the grass and weeds may be burnt off low enough to serve all practical purposes, and to destroy, at the same time, the Black Jack and Post Oak undergrowth, which is equally fatal to the range, with the fire itself. For the latter purpose, the burning in early spring, when the sap is rising, would be the most favorable time.[25]

In a study¶ entitled "The Recent Intrusion of Forests in the Ozarks," Beilmann and Brenner dealt particularly with the eastern and northern portions and concluded that "Within historic times this vast region was a prairie, or at least park-like in that the trees were widely spaced and confined to the water-courses and drainage-ways." In explanation of the change from prairie to forest in

¶ Some adverse criticism may be made of the Beilmann-Brenner study, although the major conclusions would not be changed. First, a more critical examination is in order of some documentary material used. For example, modern scholarship does not accept the view that the Coronado expedition of 1540–41 reached the Ozark country, and therefore the accounts of that expedition have no place in the evidence supporting the prairie interpretation of the Ozarks. Second, no discussion is included of whether earthquake disturbance may have been a contributing factor through local topographical and drainage changes, or ground-water levels. To be considered especially would be the disturbance of December 1811–March 1812, rated by geologists as of an intensity equal to the San Francisco earthquake.

The disappearance of salt licks or comparable accumulations of salt occurred elsewhere, so the assumption of a climatic change is not necessarily essential to the Ozark phenomena. Local changes in drainage and ground water incident to white occupation need more careful investigation, not only here, but as a general problem.

this transitional region they included among the factors "the extremely important rôle of fire in the perpetuation of the grassland at the expense of the trees."[26]

In 1939, H. C. Hanson summarized much of the modern research literature on the effects of fire, especially upon trees. In general, he indicated that in transition country the effect of fire was to discourage trees and favor grass, yet he warned that this was not necessarily the case, as fire increased the sprouting of some woody plants. White pine was badly damaged or destroyed by fire, but longleafed pine was resistant, fire contributing to the savannah form of vegetational structure under some conditions.[27] Braun-Blanquet, the leader of the Montpellier school of plant sociology, in Europe, took the position that "Fire is particularly destructive upon very thin, sterile soils and especially in the transitional region between forest and prairie, where both types of vegetation are struggling for control."[28]

The role of aboriginal man in influencing the mesquite problem has not been given an all-out investigation. Insect infestation should not be ignored.[29]** Any pretense at drawing conclusions now would be premature, but the preliminary analysis stated here should suggest possible investigations. Such experimental work as has been done indicates that the problem is complex, and no easy solutions are to be anticipated. The point of this discussion is to emphasize the fact that the historical perspective on the whole question is seriously deficient, and that there is need of a comprehensive historical re-examination of the mesquite problem on a scale comparable with Roe's *North American Buffalo*.

** Bartlett reported that "The tree seems to suffer from the attacks of insects in a similar manner with the locust." Harvard pointed out that insects laid eggs in the mesquite seeds, which destroyed germination.

X. Sagebrush and Cactus

The sagebrush problem is similarly the subject of contradictory treatment in the printed documentary material. Likewise, the tendency is to attribute sagebrush to overgrazing, in spite of the fact that many of the most significant descriptive accounts of the earliest explorers and travelers run to the contrary.[30] An interesting illustration is the following from a letter, written July 2, 1854, by the Rev. W. F. Boyakin, en route from St. Joseph, Missouri, westward, to a correspondent in St. Joseph:

> the whole country is from the South Pass to this place, one boundless sandy plain, for miles every way, stretching as far as the eye can see, with nothing but the wild sage to break the monotony; over which roll oceans of sand uplifted by the roughest winds, fairly darkening the horizon from ten to four o'clock every day, making traveling truly disagreeable.[31]

The cactus problem falls into the same category of treatment as sagebrush in most range conservation literature. But studies by C. W. Cook, and by G. T. Turner and D. F. Costello, demonstrated that the cactus infestations were not the result of overgrazing, but were related to insect-climate-plant relations.[32] Overgrazing must bear justly the responsibility of a number of evils, but it has become a convenient scapegoat for a multitude of situations where the proper answer should be "Nobody knows."

XI. Tame Grass

As Americans, derived from English and continental European stock, were primarily a forest people, when they met the grassland of the interior of the continent, they misunderstood it in many ways. Forest man's concept of grass was conditioned largely by his experience where desirable pasture and hay grasses were not generally native soil cover, and had to be cultivated like any other field crop. That outlook upon grass persisted tenaciously after the grass country was actually being occupied. At least three variant misconceptions, separately or in combination, are important to historical perspective: First, that prairie and plains grasses were inefficient in their utilization of soil and available moisture; second, the conviction that the prairie and plains grass cover always changed fundamentally in composition under domestic pasture; third, that prairie and plains grasses would not survive domestic pasturing. Forest man proposed and eventually undertook to introduce and cultivate in the west the tame grasses he had learned to depend upon in the forest country—timothy, orchard grass, bluegrass, etc., and the clovers—and to search for still better grasses.

Thus Edwin James wrote as follows:

> There can be little doubt that more valuable and productive grasses than the native species can with little trouble be introduced. This may easily be effected by burning the prairies at a proper season of the year, and sowing the seeds of any of the more hardy cultivated gramina. Some of the perennial plants common in the prairies will undoubtedly be found difficult to exterminate, their strong roots penetrating to a great depth and enveloping the rudiments of new shoots placed beyond the reach of fire on the surface. The soil of the more fertile plains is penetrated with such numbers of these as to present more resistance to the plough than the oldest cultivated pastures.[33]

In his "Notes on Nebraska" printed in 1852, Thomas Jefferson Sutherland, a Nebraska Boomer, agitating the opening to settlement of the Indian country, the present states of Kansas and Nebraska, advocated the planting of bluegrass, "On all of the lands of the eastern part of the Territory bluegrass in luxuriant growth may be produced; and there are spots of land scattered all over the plains of the west, possessing the requisite fertility of soil, for the growth of bluegrass, in any desired perfection."[34]

In 1883 Shaler published a paper on the "Improvement of the Native Pasture-Lands of the Far West." He advocated a search of other areas of the earth having similar characteristics, but also he made the following statement: "With the poorest grasses there are generally wide interspaces between the tussocks of high growing species. If these intervals could be filled with other forage-plants, the consequences would be a greater amount of food per acre. . . ."[35]

In Kansas, E. M. Shelton, professor of agriculture, 1874–90, at the Kansas State Agricultural and Mechanical College, was in most respects an unusual man, but he could not rid himself of the notion that tame grasses must be introduced to replace the native wild grasses. At the same institution, in 1887, W. A. Kellerman challenged the assumption that the composition of the native grass cover was undergoing a change, the tall grasses driving out the short grasses. He doubted whether the vegetation was changing.[36] The form of his remarks indicates that he was thinking especially of floristic range of distribution rather than quantitative density of the several species within the vegetational structure under fluctuations of climate.

Again, since the drought period of the 1930s much of the same debate, with suitable variants, has been carried on in connection with regrassing programs in the Western range country. For the most part, the decade of the 1940s has been favorable weatherwise for such operations, but prolonged drought may compel some revaluations. Once more, it might be appropriate to urge that comprehensive historical studies of the whole problem of grass introduction into the grassland are yet to be done.

The quotation already given from Edwin James' tame-grass proposal provides an excellent springboard for discussion. So far as a bare suggestion of the introduction of tame grasses was concerned, the first sentence quoted would have been sufficient, but the remainder of the paragraph provides the highly significant context, the clues to the conceptual equipment with which James viewed the problem.

The agronomist in James suggested how the

seeding operation might be accomplished with the least effort—burning. But as an experienced forest man he recognized the possible difficulty of killing certain deep-rooted plants by the use of fire. Also as a forest man, who was evidently acquainted with plowing among roots of trees and brush, he was impressed by the numbers and formidable character of woody root growths of the plains country. As a technologist, James envisioned the difficulties to be met in attempting to turn such plains sod with the iron-shod wooden plow or the cast-iron plow of the period. The numbers of the roots and the "resistance to the plough" gave him pause. Furthermore, from an agronomic point of view, James was thinking of clear-tilled fields of grass, in which a simple-stand crop was to be grown—free of "weeds," of course. But as an ecologist James was deficient, in spite of the remarkably accurate and comprehensive character of his observations, in recognizing and recording all these facts about the strange country he was visiting for the first time. The thought did not occur to him, apparently, that the presence of these plants in such numbers, and the character of their woody roots and their deep penetration constituted a veritable ecological system, and that the wide spacing of the grass plants on the surface, dividing space with the many species of forbs, was an integral part of that system. Neither did the thought occur to him, apparently, that to destroy any part of it would work a fundamental change in the whole system, soil and all, as deep, at least, as the deepest penetration of any root. This was a century before Tansley formulated the concept of the ecosystem, and for all of James' remarkable insight into so many things essential to this grassland system, so strange to his forest mind, he did not anticipate anything of the larger concept.

XII. Conclusion

As the central purpose of this paper is to point out the role of history in a research program,†† the discussion may properly return to the question of method. Casual or random excursions into historical material to find data that appear to fit a preconceived frame of reference, or to serve a particular purpose, are not only not sufficient, but such procedure is more likely than otherwise to lead to erroneous conclusions. Although the contention may appear paradoxical, to serve their purpose historical studies must have no purpose. In an immediate functional sense they are useless, and must be useless in the same sense that Max Planck said of science: "Scientific discovery and scientific knowledge have been achieved only by those who have gone in pursuit of them without any practical purpose whatsoever in view."[37] This dictum applied to science is equally valid as applied to history.

The documentary evidence for historical studies of the kind proposed here is so contradictory and so fragmentary that the strictest precautions and safeguards must be exercised in its use. It is possible to find evidence, especially when removed from context, to make a show of proof for almost any predetermined conclusion. If "quickie research" should arrive at a sound conclusion, it would be purely accidental. Sound and comprehensive studies are more likely to require a commitment to many years of systematic collection and analysis of data in full context. Furthermore, data must be tied explicitly to time and place. Each spot is unique in an absolute sense. No one can predict what time a historical project may require for

†† Since this was written a Unesco report has been issued submitting some significant recommendations relative to research programs for arid and desert regions: *Arid Zone Programme*: Report of the fourth session of the Advisory Committee on Arid Zone Research. Royal Society, London, 29 September–October 1952. Unesco/NS/103 Paris, 31 October 1952.

completion. Although, upon occasion, a few months may suffice, the minimum requirement is more likely to be measured in years, or a long lifetime. In a perfectionist sense a historical project can never be finished. But, within the realm of the possible, historical work, as intellectual enterprise, can be so comprehensive and complete as to render difficult, without danger of immediate exposure, any flagrant misuse of evidence by propagandists. History need not be written by historians, but whoever writes it must assume the obligation of doing it by the most rigorous historical methods.

Two other conclusions are pertinent. Every vegetational map must be dated historically. By that statement is meant that any description of vegetation is valid only for a particular historical time, and what exists at that particular time is the product of the whole situation, which must include man, whether he be primitive or so-called civilized. The other conclusion is closely related, because the vegetation of any specified time and space is an aspect of the ecosystem. If upon no other grounds, the recognition of man as a factor in the system precludes any possible recognition of the concept of climax for vegetation, for animals, for soil, for man, or for the ecosystem as a whole. Also, the traditional concept of succession must be revised. Change is the overriding fact, but disturbance of any rigid, orderly sequence is incessant, and, obviously individual events, whether acts of nature or of man, as they impinge upon any particular spot, are unpredictable. A temporary tendency toward establishment of a "steady state" is certain of interruption by the intervention of unpredictable events, and, for emphasis, man must be specified explicitly because of his characteristic of worrying about the future and of devising ways and means for trying to manipulate nature. What is said about vegetational mapping applies similarly to animals, to soil, and to man himself. The state of the ecosystem at any particular moment is the product of the factors of the past that have shaped it; and the state existing at that specified moment is the parent material upon which, or through which, succeeding states are formed—an indeterminate system.[38]

The present paper is focused upon the past, which is peculiarly in the jurisdiction of history. But there are no Rothamsteds in the United States —experimental areas, with carefully kept records of land use over more than a century. It is time such fundamental experimental areas were being set up in order that the same statement cannot be made in 2053.

The ideal of intellectual enterprise, whether history or science, is to attain at one and the same time, both depth and perspective. To attempt either without the other can lead only to futility. In many respects, twentieth-century specialization has reached that barrier, and in some areas more seriously than in others. The development of ecology, or the ecological point of view, is a healthy recognition of that fact and an earnest of a determination to do something about it.

Although the fashion has been set by major segments of the intellectual world to dismiss Aristotle and the ancient philosophers as outmoded, there can be no harm done, and possibly some good can be done, in reminding moderns of the great Aristotelian concept expressed in the phrase *in potentia*. Whatever line of descent from Aristotle one may choose to follow to the present, a positive philosophical outlook is still defensible, grounded in that concept. The potentiality of man to solve problems has not yet been exhausted, and the potentiality of the resources latent in the earth to be brought into the horizon of usefulness is still beyond the power of man to conceive. The key to the situation is not the earth, but the minds of men determined to realize their own potential in act.

References

1. MALIN, J. C. *Sci. Monthly*, **70**, 295 (1950); *Grassland Historical Studies*. Lawrence, Kan.: Author, Vol. 1, chap. 1 (1950); WHITTLESEY, D. *Ann. Assoc. Am. Geograph.*, **35**, 1–36 (1945).

2. ———. *Kansas Hist. Quart.*, **14**, 129, 265, 391 (1946); McCoy, I. *Ibid.*, **5**, 365, 366, 371, 372 (1936); Wedel, W. R. *Trans. Kansas Acad. Sci.*, **50**, 1 (1947).
3. Hilgard, E. W. *USDA Weather Bureau Bull.* 3, Washington, D. C.: GPO (1892).
4. Summerhayes, V. S. *J. Ecol.*, **29**, 14 (1941).
5. Murray, J. A. H., Ed. *A New English Dictionary on Historical Principles.* . . . Oxford: Univ. Press, 3, 240 (1897).
6. Marcy, R. B. *Explorations of the Red River of Louisiana in the year 1852.* . . . Sen. Ex. Doc. 54, 32nd Congress, 2nd session, Public Doc. 666. Washington, D. C.: Robert Armstrong, Public Printer, 99–100 (1853).
7. Malin, J. C. *Grassland Historical Studies*, Vol. 1, 135–8 (1950).
8. Abert, J. W. *Journal of Lt. J. W. Abert, from Bent's Fort to St. Louis.* Sen. Ex. Doc. 438, 29th Congress, 1st Session, Public Doc. 477. Washington, D. C.: Ritchie & Heiss, Senate Printers, 28 (1846).
9. Malin, J. C. *Grassland Historical Studies*, Vol. 1, chap. 2 (1950).
10. Abert, J. W. *Notes of Lt. J. W. Abert . . . [on] natural history. . . , during the Journey from Fort Leavenworth to Bent's Fort.* House Ex. Doc. 41, 30th Congress, 1st Session, Public Doc. 517. Washington, D. C.: Wendell & Van Benthuysen, 387, 392 (1848).
11. Long, S. A. *An Account of an expedition from Pittsburgh to the Rocky Mountains, performed in the years 1819 and '20.* . . . From Notes of Maj. Long, Mr. Say, and other gentlemen of the Exploring Party. Edwin James, Comp. (Philadelphia, 1822–23). In R. G. Thwaites, (Ed.), *Early Western Travels, 1748–1846*, Vol. 14–17. Consult index under "Prairie Dog."
12. Abert, J. W. *Notes of Lt. J. W. Abert.* . . , 399.
13. Frye, J. C., and Leonard, A. B. *Pleistocene Geology of Kansas, State Geological Survey of Kansas,* Bull. 99. Lawrence: Univ. Kansas Press (1952).
14. Jenny, H. *Soil Sci.*, **61**, 375 (1946); Crocker, R. L. *Quart. Rev. Biol.*, **27**, 144 (1952).
15. *Life*. Personal communication (Oct. 28, 1952).
16. Shelford, V. E. *Science*, **100**, 140 (1944).
17. Abert, J. W. *Journal of Lt. J. W. Abert.* . . , 28, 32–33.
18. Marcy, R. B. *Report of Exploration and Survey of Route from Fort Smith, Arkansas, to Santa Fe, New Mexico, made in 1849.* House Ex. Doc. 45, 31st Congress, 1st Session, Public Doc. 577. Washington, D. C.: Printer to the House, 41, 60, 64, 72, 74, 80, 82 (1850).
19. ———. 59, 144.
20. ———. *Message of the President of the United States, communicating a copy of the Report and Map of Captain Marcy of his Explorations of the Big Wichita and the Headwaters of the Brazos Rivers, 1854.* Sen. Doc. 60, 34th Congress, 1st Session, Public Doc. 821. Washington, D. C.: A. O. P. Nicholson, Senate Printer. 4, 10, 14, 25–26 (1856).

21. PARKER, W. B. *Notes taken during the Expedition through Unexplored Texas, in the summer and fall of 1854.* Philadelphia: Hayes & Zell, 104, 105, 119, 126, 143, 151, 162 (1856).
22. HARVARD, V. *Am. Naturalist,* **18**, 456 (1884).
23. RICHARDSON, R. N. *The Comanche Barrier to the South Plains.* . . . Glendale, Calif.: Arthur H. Clark Co. (1933).
24. SHALER, N. S. *Aspects of the Earth.* New York: Scribner's, 286–90, 295 (1889); *Nature and Man in America.* New York: Scribner's, 180–88 (1891).
25. HILGARD, E. W. *Report on the Geology and Agriculture of the State of Mississippi.* Jackson: Mississippi State Printer, 349, 361–2 (1860).
26. BEILMANN, A. P., and BRENNER, I. G. *Ann. Missouri Botan. Garden,* **38**, 261, 280 (1951).
27. HANSON, H. C. *Am. Midland Naturalist,* **21**, 415 (1939); SAMPSON, A. W. *Range Management: Principles and Practices.* New York: Wiley, chap. 13 (1952).
28. BRAUN-BLANQUET, J. *Plant Sociology.* New York: McGraw-Hill, 278 (1932).
29. BARTLETT, J. R. *Personal narrative of Explorations and Incidents in Texas . . . during the years 1850, '51, '52, and '53.* New York: D. Appleton, 1, 75 (1854); HARVARD, V. *Op. cit.,* 458.
30. MALIN, J. C. *Grassland of North America: Prolegomena to its History,* Lawrence, Kan.: Author, 149–51 (1947).
31. BOYAKIN, W. F. *The Gazette,* St. Joseph, Mo. (Sept. 6, 1854).
32. COOK, C. W. *Ecology,* **23**, 209 (1942); TURNER, G. T., and COSTELLO, D. F. *Ibid.,* 419.
33. LONG, S. H. *Op. cit.,* **16**, 142–3.
34. SUTHERLAND, T. J. *The Weekly Tribune,* Liberty, Mo. (May 7, 1852).
35. SHALER, N. S. *Science,* **1**, 186 (1883); MALIN, J. C. *Essays on Historiography.* Lawrence, Kan.: Author, chap. 2 (1946).
36. MALIN, J. C. *Winter Wheat in the Golden Belt of Kansas.* Lawrence: Univ. Kansas Press, 80–82, 84, 89, 179 (1944).
37. *Sci. Monthly,* **72**, 222 (1951).
38. MALIN, J. C. *The Grassland of North America. . . ,* 278–9.

INTRODUCTION TO CHAPTER TWENTY-SIX

The essay that appears for the first time in this, the fourth printing of the book as chapter twenty-six, "On the nature of the history of geographical area, with special reference to the Western United States," was prepared upon invitation for a symposium on the Southern Great Plains held under the auspices of "The Nature Conservancy" in cooperation with "The American Institute of Biological Sciences," at Stillwater, Oklahoma, 30 August 1960. It contains a description of the point of view and of some of the features that have been developed in my course taught at the University of Kansas from the late nineteen-thirties to the present, except for five years, 1954-1959, as "The history of the Transmissippi West"—the area as a whole and the subdivisions and regions into which, for convenience, it is divided for effective study.

Chapter Twenty-six

ON THE NATURE OF THE HISTORY OF GEOGRAPHICAL AREA, WITH SPECIAL REFERENCE TO THE WESTERN UNITED STATES

1. Introduction

In this essay attention is focused upon geographical area as an object of study: Upon the history of a chosen area and of the human cultures that have occupied it from the earliest primitive men to the present. Although applicable to any such area, for present purposes the emphasis is upon the Western United States, and especially upon the Southern Great Plains region, which is the subject of this symposium. This procedure affords a number of innovations. Treated in its own right, the history of a geographical area includes a consideration of all that has been present or is present within the bounds chosen. Proper subjects of study, from this point of view of geographical area, are its geological history, its ecological history, and the history of human culture since the beginning of occupance by primitive men—in the case of the Western United States, some 10,000 years since men reached the Folsom cultural level. The term culture, as used here, is that of the archaeologist and anthropologist and denotes the sum total of a way of life.

This is in contrast with the traditional frontier formula, in the Frederick Jackson Turner tradition, which is concerned with little more than the beginning of European-American, especially Anglo-American displacement of Indians and the beginning of the exploitation of their lands according to a form of master-race colonialism. One of the most frustrating aspects of the frontier thesis is the lack of any agreed definition, without which scholarly application to historical study is impossible. According to the frontier thesis, the frontier came to an end in 1890, and with its passing the history of the West ended also. On this practice consult the several textbooks written according to the Turner Frontier Thesis, and where there are more than one edition, compare the several editions: Frederick Logan Paxson, Robert E. Riegel, Dan E. Clark, Ray A. Billington, Thomas D. Clark. The fact should be noted also, that the traditional historian does not deal with cultures, and the word culture if used by him is not given any logical or consistent definition or usage. And furthermore, the point is emphasized that culture as used here does not mean "civilization", with its invidious implications about the uncivilized, and the supposedly superior and inferior peoples. The term culture involves no value judgments.

The study of the history of the Western United States as geographical area is not the study of 17, 20, or 22 separate states that lie within the area. To be sure, each state is a geographical area and has a history. Each was organized for political purposes and the boundaries were defined according to the exigencies of politics of a particular time. Rarely were these boundaries defensible on the basis of facts and reasoning from facts. They were for the most part arbitrary and in many respects would seem to defeat the best interests of the inhabitants. Yet, as political entities the people of each state accumulated a body of traditions, loyalties, and prerogatives and would not consider mergers or transfer of territory. The study of state history is a legitimate subject, both in its own right, and on occasion, in relation to other partitions of space. But in dealing with the Western United States as a geographical area, the emphasis is placed as largely as possible upon phenomena that concern the area as a whole and its major parts. Instead of dealing with individual states, therefore, when subdivision of the entire area is desirable, it can be done meaningfully on the basis of selected criteria suitable to the purpose.

Technology is among the aspects of human cultures that are involved in their competition anywhere, and by the term technology as used here is meant more than mere mechanical inventions. The term technology includes all manner of ideas and social inventions, and the relative efficiencies for particular purposes of differentiations in the ways of life of competing groups. This competition may occur along the boundaries of areas occupied by these groups, however precise or vague those boundaries may be; or competition may occur through the instrumentality of persons, institutions, or ideas, finding lodgment within the area occupied by competing groups. Peaceful penetration may occur effectively within a rival society, rather than by force applied along boundaries, or expended in control of the sea.[1] Such operations of culture competition occur unconsciously, as well as deliberately, among all societies that make contacts, directly or indirectly, with each other. Unquestionably, the effects of such competition of cultures are related to their capacity and success in occupance of geographical areas.

In this context, the discovery and occupance of the Americas by people of Modern European culture since the fifteenth century is only one small segment of the whole history of the Americas as a geographical area, and of their occupance by human culture 400 against some 10,000 years plus. Cultural technology gave to the people of the invading European culture an overwhelming power to overrun, and to displace, or to annihilate the more primitive peoples. From the standpoint of the peoples occupying the Americas prior to the Columbian discovery the following centuries were stark tragedy. The invading Europeans conquered or destroyed the Indians, fought each other for supremacy, and established their own cultures. For the victors, this was a glorious adventure. To possess any measure of validity, the so-called frontier thesis would have to be treated within the anthropologist's formula of competition of cultures, but historians do not deal with it in that manner.

The Western United States is a geographical area, but is not a region within the meaning of geographers, or a section within the meaning of political historians. Some subareas within this larger space may lend themselves in a significant manner to recognition as regions when such treatment may appear to be desirable. Regions and sections, being related to particular selected and limited criteria, are not necessarily of long duration. Usually, they are used as terms applied to areas and periods of time within the quite brief duration of European culture in America. To the whole study of geographical area and to its succession of occupying cultures over some 10,000 years, they may be mostly irrelevant, but some regions may suggest significant conclusions.

The passion of scientific method and social science for classification and uniformities should not mislead the student of areas into the assumption that political boundaries that cut across, or include widely different geographical characters or regional unities are necessarily in error or that they should enlist the missionary zeal of the reformer to effect uniformities as a panacea for all ills. The balancing of divers forces may result, but not necessarily so, in a sounder adjustment to realities than would be probable under overwhelming majorities dominated by a single body of regional interests. On occasion, minorities may be more right than the majority. But what is right? New technology may change radically the character of the area or regional culture. Is right to be equated as belonging to obsolescence or innovation, or neither?

2. Variety in Continental Landmasses Subject to Study as Geographical Areas

Every landmass of the earth of a size sufficient to be rated as a continent contains nearly the full range of possible physical variety except as to the greatest extremes of temperature, light, and moisture. This generalization, as is always the case with generalizations, must recognize exceptions. If the South Polar landmass is considered a continent, it is one exception, but nevertheless it exhibits variety within the limits of its temperature and light variables. Greenland would likewise rate as an exception.

But considering the generalization as applying to the conventionally accepted continents—and subject to the prevailing cultural technologies—Europe and Asia (or Eurasia), Africa, Australia, and the Americas, only Europe (separate from Asia) does not contain a dry desert or a rainforest. If Europe and Asia are classed as one continent, Eurasia, that exception disappears. All continents contain conifer and hardwood forests, grasslands, moist or shrub deserts, and dry deserts, together with all the transitional types, except Australia, which lacks the conifer and hardwood forests.

3. Habitability by Man of Continental Landmasses

Habitability by man of any area of the earth depends, not upon the properties of the area per se, but upon man's capacity to utilize the properties that exist there, and turn them to his advantage. No other assumption about history is tenable, unless an outright geographical determinism is imposed. Of course, in terms of geological time, paleontology reveals the succession of genera and species that have come and gone, and man may pass from the scene in the course of this succession principle. But short of this kind of liquidation of the human race, the properties of man as an inventive animal afford him science and technology for seemingly unlimited potentiality to convert the properties of the earth to his use. Natural resources cannot be exhausted until his contriving brain and skillful hand are exhausted. The first requisite of a natural resource is an idea. In consequence, as long as man is capable of new ideas he may continue to reappraise the properties of the earth, bringing new ones continuously into the horizon of his utilization. On the basis of this approach to history, no continental landmass can be denominated uninhabitable, sometime, not even the South Polar continent.

4. The Problem of Time: Geological and Calendar Time; History and the Sciences

As time is the major organizing principle of history, the historian must consider the several concepts of time. The time elapsed since the Creation, according to Judaeo-Christian Biblical calculations, was about 6,000 years. For present purposes, omitting the philosophical problem of finite and infinite time, scholars of all disciplines have been confronted with the geological time of rock records, and calendar time of writing men. The acceptance of the theory of evolution by nineteenth century men was delayed by the Time barrier necessary to account for it. The duration of time attributed to Biblical authority was not sufficient. Geology was largely responsible for breaking down this Time barrier, and for introducing Western Culture to the conception of such immensity of time as made biological evolution appear as a feasible and reasonable explanation.

The task was too well done, however, and the geologist and his academic allies became victims of their own victory over Genesis. To them, geological and historical time of written documentation became irreconcilable. The very immensity of geological time tended to isolate modern historical man from all geological time. Man's written history was of so short a duration as to seem insignificant, even meaningless to the larger frame of reference. Men must learn not to be overwhelmed either by the immensity of geological time, or by the immensity of recorded detail of recent and present time. Geological processes still continue in the geological time scale, only in such multiplicity of detail that the mind fails to comprehend it all as a continuing whole. But, on the other hand, the records of anonymous men with whom the archaeologist deals left their imprint upon a period of geological time—the Pleistocene—with sufficient clarity to lead to an inescapable conclusion that this overlapping of geological and human calendar time is not merely a case of parallel records, but that they are of the same order of magnitude and are an integrated whole. The geologist can no longer ignore man and his capacity to change the face of the earth, nor the fact that he had done so within the range that the geologist has been accustomed to claim as his own monopoly. The realization is not impossible, by both the geologist and the historian, that the transition from impersonal, anonymous man of geological history to the named, individualized man of written history can be comprehended, and that it can be done without doing violence to either discipline. Archaeology and anthropology have already gone a long way toward building the bridge which renders attainable, at least among those scholars who have been working in these overlapping borderlands of knowledge and have been integrating them. To be sure, the rank and file of all the academic disciplines related to this complex task have been little affected, and are scarcely aware of what is taking place. Although slowly, it has been gaining in acceptance and in effectiveness.

5. The Historian's Orientation among Specializations: General History (Synthesis) and Histories of Individual Sciences—Innovations in Science since about 1930

Of course, the historian could not be expected to master all the disciplines necessary to the study of history according to the comprehensive plan outlined here. Nevertheless, an orientation was possible in these disciplines, and sufficient to prepare him to cooperate effectively in larger syntheses of the whole of the history of a chosen geographical area. Within the range of the possible also was the comprehension of the full significance of this point of view, and that alone should work a revolution in thinking about history.

Something may be said in extenuation of traditional history and historians in their failure to respond to the opportunities available. The disciplines concerned with either new or the rethinking in the older ones necessary to the new point of view had been quite recent. Much of the intensive research, and the consequent revolution in thought in Pleistocene geology dated from about 1930. Even more revolutionary had been the thought about primitive man in North America, and especially in the Great Plains. Instead of a maximum of 3,000 years formerly allowed for man's occupance of North America, the discovery of Folsom level culture and subsequent findings, place man at that cultural level in the Great Plains area at least 10,000 years ago. This new point of view began to be accepted during the nineteen-thirties, but its full impact only became adequately appreciated after World War II. Paralleling these re-orientations in knowledge about geology and anthropology, ecology made a partial adjustment in theory and in substance to the fact of primitive man and his capacity to change the face of the earth, including profound influences upon both plant and animal life. Older scholars, if they made their adjustments to all this new knowledge, were obliged to reeducate themselves. Few historians, indeed, knew what was going on. Much allowance must be made, however, for these older scholars who failed to recognize their opportunities. No such excuse was available to the generation trained after World War II, when the validity of these revolutions in thinking had already been proven out by specialists in their respective fields to a point where no competent scholarship could possibly ignore them. Graduate school training programs did not reflect any real understanding of the requirements and the opportunities.

The history of the Western United States offered in perspective, and in a relatively abbreviated form, the full gamut of the human story. An unusual opportunity was available to trace a unique version of the old story of competition of widely different cultures running their full course. The relatively brief time-perspective on the history of the geographical area, in terms of human occupance, was an advantage in some respects, and one that should have been capitalized, not lamented or abused.

6. Historical Orientation: Directional

When and how did the more complex primitive cultures reach the interior grassland of North America? Not enough is known as yet to justify much in the way of generalization, but some major outlines appear to be taking shape.[2] In terms of cultural development, Neuclear America extended from Peru to Southern Mexico. From that general area and by routes not clear, cultural patterns were diffused northward. North of the Rio Grande and of the Gulf of Mexico there was intermingling in various complex relationships within the main continental mass of North America. Seemingly, the great interior grassland received cultural contributions from all directions, but predominantly from the south, southwest, and southeast, but in any case, not from the advances of a single uniform front line of migration.

In this perspective, one of the main contentions of Herbert E. Bolton took on an added interest. Although dealing only with the traditional European occupance of North America, he insisted that that process should be studied as primarily a movement of European occupance from south to north, with some invasion from the Pacific coast to the eastward, and from the Atlantic coast westward, and from some relatively isolated points in the interior outward where a foothold was established. Bolton had challenged, and rightly, the manner in which historians treated the history of the United States almost exclusively from the British point of view—English colonies in America, and their expansion westward across the continent, thus defying the simplest facts of chronology and geography.[3] True, the striking similarities between the movement of primitive cultures and modern European occupance, south to north, together with the necessary qualifications, may be merely coincidence, yet even as coincidence, the phenomena are of interest.

If the situation was described in terms of cultures instead of political jurisdiction, thus going beyond the Bolton methodology and transforming the historical study into one of a history of competition of cultures, a wide scope of the traditions of political history would undergo revision. Spain had lost political jurisdiction to Mexico, and Mexico in turn to the United States in 1848, and for more than a century United States "Americanization" has been the order of the day. But with what success? The process of culture competition continued despite changes in political jurisdiction. The Spanish-Indian culture yielded but slowly and only in limited aspects. In others it gained adherents. The way of life possessed values that attracted some people from other cultures. It even expanded the influence of certain of its culture traits beyond the geographical limits of 1848.

7. The Recent End of the Time-Scale

In revising attitudes toward the Western United States and history of geographical area, the recent end of the Time-scale was entitled to some fresh thinking. For frontier historians, the area had no history after 1890, the date of the miscalled "end of the frontier." Their interest ends with "frontier beginnings." This point of view violated the basic principle of continuity of history. On the contrary, whatever the shifts in culture occupying the given area, the history of the area continued.

Human cultures occupying the area were constantly subject to scientific and technological innovations. Men were engaged in reappraisals of the potentialities of the properties of the earth and of its culture. Indian, buffalo, and furs gave way to grass and cattle, and partially, grass and cattle gave way to field crop agriculture. During periods of adversity, some pessimists of an earlier day recommended that the Great Plains be given back to the Indians. Of more recent date, especially during the nineteen-thirties, the recommendation was modified—give the country back to the cattleman and grass. But this proposal was just as absurd as the first and fully as impossible.

A single illustration must suffice: a look at the United States Census figures for selected areas of the Southern Great Plains, with emphasis upon those for 1940-1960. According to the accompanying tables, the areas associated with Odessa and Midland, Texas, which acquired Metropolitan area status for the first time in the census of 1960, experienced the most phenomenal gains in population. For the twenty-year period 1940-1960, Odessa's increase was 500 per cent, and Midland's was 474 per cent. The third largest gain for the same period came to the Lubbock Metropolitan area, 200 per cent, and among the strictly High Plains cities the Metropolitan area of Amarillo took fourth place in growth, 141 per cent. Among cities just eastward of the High Plains, the well established Metropolitan areas of Dallas and Fort Worth continued their sensational pace, 169 per cent and 152 per cent.

Denver, Colorado, long a substantial commercial center generally looking eastward from near the continental divide now divided its allegiance, if it did not actually look westward. But Albuquerque, New Mexico, was even more dazzling. From a town of 6,238 in 1900, the skyrocketing began during the nineteen-forties into a city of nearly 200,000, the nucleus of an urbanized area of 260,318 in 1960. Conspicuously, it was a link along with Denver, between the Pacific coast and the interior, tending to turn the orientation of the Southern and Middle Great Plains toward the West Coast. By 1960 the economic continental divide lay well to the eastward of the physiographic continental divide.

The outstanding fact population-wise was urbanization and population explosion in these areas of low rainfall. Obviously the explanation was not grass and cattle. At some points significant developments other than these were first evident in retrospect at fairly early dates: at San Antonio during the first decade of the twentieth century; at Tulsa, El Paso, and Wichita Falls in the second decade; and at Abilene, Amarillo, Lubbock, and San Angelo in the third decade. Lubbock had appeared in the census as an incorporated city in the enumeration of 1910, and Odessa in that of 1930 (incorporated, 1927). Even during the Great Depression and Drouth of the nineteen-thirties only El Paso recorded a population loss. It would seem that these facts should have made clear something of what was involved, but if they did not, then the startling combination of circumstances that account for the population figures, 1940-1960, challenge all conventional interpretations of "the arid Great Plains" fit only for livestock. Technology afforded the realization of new potentials within the area, and in its relationships with other areas--oil and gas developments, irrigation grown cotton, automobile and air communications, the beginning of the atomic age, rare minerals, and new commercial exchange relations between the Pacific coast and the interior east of the continental divide. Science and technology, and many innovations induced by the mechanization of society during the mid-twentieth century require new explanations related to the new resources brought into the horizon of utilization.

These are only suggestions. Other areas of the Western United States may be analyzed in a similar manner to reveal their different and unique stories, some more than others less sensational during the last two decades.[4]

8. The Historian not a Prophet

To the historian, the future is out of bounds, but there is no reason to assume that the phenomenal developments of the twenty years, 1940-1960, will continue as straight line cumulations. The unpredictability of particular events in history is an inexorable principle. If the particular cities or the area are to continue their growth, or even to hold their own, their populations must be alert to constant reappraisals of potentials, and to rivals who may exhibit more originality, judgment, and energy. In any case, it would be suicide to expect the current social structure to survive on grass and cattle alone. That simple solution has long since been outmoded. Thought about the future of the Southern Great Plains must gamble on the unknown, not on the past that no longer exists and is beyond revival. Grass and cattle are and should continue to be indefinitely of importance, but only as one of several resources. The features which are most emphasized about the area as characterizing it to its disadvantage, are not necessarily handicaps. Under new technologies those features may be turned into advantages.

The concept of conservation itself, both in theory and application, is in constant need of reexamination, especially to free it and keep it free from the naive implication that natural resources are in imminent danger of exhaustion. If utilization of the natural resources of an area were limited to one or two or just a few possibilities, then exhaustion might be a danger. But there is no reason, even if it could be done, to impose such a restriction upon the area or upon its interrelations with other areas. The first requisite of a natural resource is an idea, and so long as men's minds are not exhausted of ideas the existent properties of the earth new to men may be brought into the horizon of utilization. In that perspective there is no threat of exhaustion except that men betray themselves. Conservation is extension of use of the properties of the earth, and the heart of any conservation program lies in the discovery of properties of the earth that can be utilized, thus expanding natural resources available to men. No valid reason exists for assuming that men will not be present in the area for at least another 10,000 years or indefinitely longer. Of course, there is no magic in that particular number, but it so happens that human culture of at least Folsom level of complexity has existed in the area, supposedly, for that length of time, so if man has learned anything during this immediate past 10,000 years, what more may he learn in the next 10,000! The answer lies in ideas in men's minds, and only incidentally in the earth.

9. Illustrative Resource Application: Interpretation

As a general proposition, a focus must be kept on the fact that throughout the history of the human race, a concentration of population always raised the question of scarcity of those resources most in demand at the particular time and place. The nature and extent of requirements depended upon the state of cultural technology. Quite understandably, among the resources most heavily drawn upon were water, food, and fuel. In England of the fifteenth century, during the reigns of Henry VIII and Elizabeth I, a wood fuel scarcity became a matter of concern, and coal as a substitute, or even as a more efficient source of heat, was the subject of legislation. Technology made available an abundance of a more efficient resource. This was prior to the opening of the New World to English exploitation, and it is important to historical perspective to appreciate fully, that present problems in this department are not new, nor peculiar to the era following this miscalled passing of the frontier, or to the low rainfall areas of the earth.

Likewise, water supply was always a problem where large populations gathered, and regardless of the amount of the annual rainfall. Babylon, ancient Rome, Paris, London, and New York, as well as the interior valleys and cities of southern California, undertook to solve their water scarcity by drawing water from wider and wider geographical areas and carrying it through conduits. Some cities drew upon ground water supplies to such an extent as to result in threatened or actual seapage of seawater into the fresh ground water. Eventually, the next step must be to purify seawater, which is technologically possible, and when the pressure is sufficient, will be made economically feasible. The southern plains are only another step removed. It would be foolish to predict, especially when knowledge of present supplies and their possibilities is so inadequate.[5] When and if the need is sufficiently urgent the way will be found.

The fact must be kept in mind also, that population concentrations are often mobile. Shifts in geographical location occur, and as already pointed out, are related to technology. Those concentrations of people that have persisted longest have been related over long periods to man's basic needs under the prevailing culture. The more specialized technologies and their resources are necessarily more ephemeral in their influences. The more fertile the mind of man in his inventive activities the more risk must be assumed. That fact is one that technological man must learn to live with and take in stride.

Some conclusions to such lines of reasoning are clear. Solutions to resource problems that are technologically possible will be made economically feasible if and when the controlling society is convinced of the necessity. This means successive reappraisals of the values contributed by the culture of the limited geographical area to the controlling society, and decisions about whether or not the contribution of the area justified the cost. The issue cannot be evaded by falling back upon geographical determinism and defeatism.

A second conclusion is that, technologically, society has become so mechanized and the division of labor and skills so specialized, that no longer does justification exist for talking about rural and urban society, and agriculture and industry, as separate, or rival, or competing systems in the nineteenth century and earlier sense. Integration of society has occurred to such a degree in fact as to render those traditional concepts obsolete. No definite boundary exists where rural-agricultural social organization leaves off and suburban-urban-industrial structure begins. For the transitional state of society now existing at mid-twentieth century, the term metropolitan society may be pressed into service until a new and more accurate terminology is available.

In the study of the histories of ancient cultures no one has proved that any culture declined because of water and/or fuel shortage, or of soil depletion. Many times such cause-effect sequences have been alleged, but never proven. So far as generalization in such matters is permissible, more plausible would be the contention that these alleged causes were results; that a rival culture appeared which exercised a more effective power over its geographical area. In consequence, the first culture was no longer able to compete successfully with its neighbors, and the relative inefficiency of its cultural technology led to the neglect and to the deterioration of its natural resources, and finally to their substantial destruction. If such a model for the study of decline of a culture is valid, then the exhaustion of its natural resources was a result, a consequence, not a cause. In such a culture, man's contriving brain and skillful hand had failed to function with sufficient originality or creativity. The peoples of the twentieth century world may well ponder these propositions. Already, for some of them, time may have run out?

Population
Southern Plains Cities

	1900	1910	1920	1930	1940	1950	1960
Oklahoma							
Oklahoma City	32,452	64,205	91,295	185,389	204,424	243,504	317,542
Tulsa	7,298	18,182	72,075	141,258	142,157	182,740	258,563
Texas							
Abilene	3,411	9,204	10,274	23,175	26,612	45,570	89,428
Amarillo	1,442	9,957	15,494	43,132	51,686	74,246	137,083
Dallas	42,638	92,104	158,976	260,475	294,734	434,462	672,029
El Paso	15,906	39,279	77,560	102,421	96,810	130,485	272,239
Fort Worth	26,688	73,312	106,482	163,447	177,662	278,778	393,388
Lubbock Co.	293	1,938	4,052	20,520	31,853	71,747	128,068
Midland	1,741	2,192	1,795	5,484	9,352	27,713	62,497
Odessa	-	-	Inc. 1927	2,407	9,573	29,495	79,123
San Angelo	-	10,321	10,050	25,308	25,802	52,093	57,811
San Antonio	53,321	96,614	161,379	231,542	253,854	408,442	584,471
Waco	20,686	26,425	38,500	52,848	55,982	84,706	96,776
Wichita Falls	2,480	8,200	40,079	43,690	45,112	68,042	103,204
Colorado							
Denver	133,850	213,381	256,401	287,861	322,412	415,786	489,217
New Mexico							
Albuquerque	6,238	11,020	15,157	26,570	35,449	96,815	198,856

Population
Southern Plains Metropolitan Areas

	1940	1950	1960	1950-1960 Increase %	1940-1960 Increase %
Oklahoma					
Oklahoma City	244,159	392,439	502,707	28.1	106
Tulsa	193,363	327,900	414,117	26.3	114
Texas					
Abilene	67,525	85,517	120,377	47.6	78
Amarillo	61,450	87,140	148,433	70.3	141
Dallas	398,564	743,501	1,073,573	44.4	169
El Paso	131,067	194,968	310,690	59.4	137
Fort Worth	225,521	392,643	568,484	44.8	152
Lubbock	51,782	101,048	155,485	53.9	200
Midland	11,721	25,785	67,332	161.8	474
Odessa	15,051	42,102	90,298	114.4	500
San Angelo	39,302	58,929	63,415	7.0	61
San Antonio	338,176	500,460	683,262	36.5	102
Waco	101,898	130,194	148,336	13.9	45
Wichita Falls	73,604	105,309	129,866	23.3	76
Colorado					
Denver	-	612,128	923,161	50.8	
New Mexico					
Albuquerque	-	145,673	260,318	78.7	

Note: The numbers for 1960 are the preliminary United States Census reports, except Abilene, Texas, which are final.

REFERENCES

1. In a totally different but stimulating context, boundaries and related matters are discussed by Stephen B. Jones. "Boundary concepts in the setting of place and time." Annals of the Association of American Geographers, 49 (September 1959) 241-255.

2. Gordon R. Willey, "New World Prehistory: The main outlines of the pre-Columbian past are only beginning to emerge." Science, 131 (8 January 1960) 73-86. Joseph R. Caldwell, "The New American Archaeology," Science, 129 (6 February 1960) 303-307.

3. Herbert E. Bolton, "The Epic of Greater America," American Historical Review, 38 (April 1933) 448-473. The Spanish Borderlands, (Chronicle of America Series, New Haven, 1921); with Thomas W. Marshall, Colonization of North America, 1492-1783, (New York, 1921).

4. Since the present paper was written a somewhat similar point of view was presented using the new population data of 1960 as applying to the southwestern United States west of the continental divide, Andres W. Wilson, "Urbanization of the arid lands," The Professional Geographer, 12 (November 1960) 4-7.

5. See R. L. Nace, (United States Geological Survey), "The water outlook for Texas," The Cattleman, (Fort Worth, Texas), 45 (April 1959) 42-44, 80, 82, 84, 86. Nace emphasized, and quite properly, how little is known certainly about what constitutes water conservation, and how many interests are involved.

POSTSCRIPT

A book such as this is written once and cannot be revised. It fitted into a particular time and situation in such a manner that whatever significance it possesses lies in preserving the original text. In the form in which it was issued, the book itself is an important historical document. Yet, during the twenty years since the planning and writing of it, the body of ecological and other scientific knowledge, and the prevailing points of view have been substantially modified in some areas. These facts of change have been recognized from time to time in periodical publications and then as collected into the Addenda have supplemented the original version. The current printing offers this opportunity to specify further innovation.

Adverse criticisms that might be directed at the original text of the book would suggest two emendations. First, in the perspective of twenty years, a full chapter on archeology and anthropology should have been included in Part One. Secondly, the role of H. A. Gleason's individualistic approach to the study of plant associations should have been made as explicit in Chapters 7 and 10 as in Chapter 11. These facts were not so clear in 1945 as in 1955 or in 1965.

In part at least, in consequence of Gleason's work, the special terminology of the Classical ecologists (the Clements and the Cowles schools) virtually disappeared from the new ecological literature. It was no longer relevant except historically as marking a stage in the development of the discipline. This trend was not prominent and certainly not decisively forecast in 1945.

The evidence of a completed transformation in outlook was conspicuous, however, in John T. Curtis, **The Vegetation of Wisconsin** (1959), especially in Chapters 1, 2, 3, 14, 23 and 24. His glossary omitted the terms "succession", and "climax". Among the terms defined, but not in the language or theory of the Classical ecologists, were "association", "community", and "dominant". The term "continuum" which was included, was relatively new to ecological literature.

Within a much wider range of environments, Hugh M. Raup, the Director of Harvard Forest, pioneered contributions from other directions to the new body of ecological theory and practice. In a paper, partly autobiographical, presented in 1956 at Edinburgh, he reviewed his experience in widely different geographical areas; two in northern North America (Lake Athabaska and the northern tundra), in the tropical forests of Northern Honduras and in the forests of New England.*

> Hugh M. Raup (1957), "Vegetational adjustment to the instability of site," **Proceedings and Papers**, 6th. Tech. Meeting, International Union for the Conservation of Nature & Natural Resources, June 1956, pp. 36-48.

Trained in the Classical ecology, when he met the field conditions of a new area, he found that the theory did not fit the facts. He had to unlearn his training and find a new orientation that did conform to reality. Gleason's individualistic concept of plant associations was one of few guides that afforded aid to his new thought about ecological theory and practice. A sequel to the Edinburgh paper was presented in 1964 to the Jubilee Symposium of the British Ecological Society: "Some problems in Ecological Theory and their Relation to Conservation."*

> **Journal of Ecology,** 52 (Supplement), 19-28, March 1964.

My accumulation of illustrative material for class use in the history of the Transmississippi West included photographs in color. The first such pictures were taken during May 1959, and without preconceived plan beyond recording examples that would show typical grassland scenes in various parts of Kansas, but especially in the Bluestem-Pasture area.* Once

> James C. Malin, "An introduction to the History of the Bluestem-Pasture Region of Kansas...," **Kansas Historical Quarterly,** 11 (February 1942) 3-28.

begun, however, many possibilities became apparent; among these the desirability of a photographic record of identical sites at successive seasons of the year, and from year to year. A route was selected through northern Lyons and southern Wabaunsee counties, Kansas, and with approximate regularity this path was followed again in 1960, 1961, and a part of 1962.

Remembering that explorers and travelers usually went west after the grass started in the spring, and if they returned, usually did so in the late summer and fall, and that they were more consistent record keepers at the beginning of their long journeys, the reported vegetation of these seasons tended to develop into a stereotype for the grass country.

THE GRASSLAND OF NORTH AMERICA 489

The Classical ecologists who looked for succession and climax (or disclimax), constructed their own peculiar stereotypes. In either case, such apparent uniformities were deceptive.

Instead of uniformities, my photographs recorded even greater diversity and variegation than that emphasized in the original text of this book (especially Chapters 8-11) and in the Addenda chapters of the Third printing (1956). However carefully written to support memory, word descriptions do not substitute for the realism of the photographs in color. During each season the landscape offers a different and continuously changing structure consisting of grasses and forbs (weeds) suited to the time of year; and each successive year it varies with the combination of factors that prevail during that particular year. Thus each site has a unique history, and the vegetation reflects that fact, not in uniformity, but in an endless diversity.

Only by comparisons of these pictures can the observer appreciate fully the magnitude and complexity of the seasonal differences within the same year, and the annual differences from year to year. A striking example may be cited. About 1 September 1961, certain landscapes exhibited all the appearances of Bluestem country. A few weeks later, however, the grass that gave character to the same spot was Indian grass (Sorghastrum nutans). Thus the observer who saw this particular landscape only once, or only at one time of the year, might name it according to the most conspicuous vegetation that characterized it at that one particular time. These facts do not fit into the traditions of the Classical ecologists, and their systems of climax plant formations as end products of long term (centuries) of succession leading to an approximation of equilibrium with the environment.

On the history of soil science, the most important book since the writing of this book is Hans Jenny, **E. W. Hilgard, and the Birth of Modern Soil Science.*** Professor Jenny's conclusions

---
Pisa, Italy, 3 Collana Della Revista 'Agrochimica' (xxx1961). Farallon Publications, Box 564, Berkeley, California.

---

fully justify the point of view expressed in the opening paragraph of my Chapter 6 on soil science: "E. W. Hilgard (1833-1916) anticipated much of the modern theories which are credited to the Russian School . . ." But the full range of Hilgard's scientific innovations in soil science, ecology, and other areas are available only in Jenny's book. Nevertheless, my book pioneered the restoration of Hilgard to something of his deserved place in the history of soil science.

In the substance of the history chapters of Part Two, no decisive revisionism, such as was noted in some aspects of science, has occurred, but for different reasons. Unfortunately, the cult of the patheological and the sensational, "triggernometry" and Indian wars, continues to make the largest appeal to the general public. It is otherwise with the historians of the Transmississippi area who are doing the fundamental work.

Historians generally have not integrated into their own thought and work the newer historical dimension of archaeology, anthropology, and related Pleistocene geology. Also, few historians have been adequately equipped in the sciences to use them to their full advantage in historical work. On the other hand, history has made substantial headway among scientists, especially in the form of biography and history of science.

Geographical determinism, the Turner frontier thesis, and neomalthusianism, as related themes, have tended to move together in ebb and flow, especially since World War II. Neomalthusianism, more popularly designated as the world-wide "population explosion", has been generally interpreted as giving support to geographical determinism, and to the expansion of the frontier thesis of American history into a world phenomenon, such as Walter P. Webb's Great Frontier. It is important to recognize, however, that the Turner thesis, the Great Frontier, Malthusianism, geographical determinism, etc., are not history. Regardless of their scope and form they are related primarily to philosophies of history.

James C. Malin

July, 1967

DATE DUE